Lecture Notes in Bioinformatics 11773

Subseries of Lecture Notes in Computer Science

More information about this series at http://www.springer.com/series/5381

Luca Bortolussi · Guido Sanguinetti (Eds.)

Computational Methods in Systems Biology

17th International Conference, CMSB 2019
Trieste, Italy, September 18–20, 2019
Proceedings

 Springer

Editors
Luca Bortolussi 🆔
University of Trieste
Trieste, Italy

Guido Sanguinetti
University of Edinburgh
Edinburgh, UK

ISSN 0302-9743   ISSN 1611-3349   (electronic)
Lecture Notes in Bioinformatics
ISBN 978-3-030-31303-6   ISBN 978-3-030-31304-3   (eBook)
https://doi.org/10.1007/978-3-030-31304-3

LNCS Sublibrary: SL8 – Bioinformatics

This Springer imprint is published by the registered company Springer Nature Switzerland AG
The registered company address is: Gewerbestrasse 11, 6330 Cham, Switzerland

Preface

This volume contains the papers presented at CMSB 2019, the 17th Conference on Computational Methods in Systems Biology, held during September 18–20, 2019, at the University of Trieste, Italy.

The CMSB annual conference series, initiated in 2003, provides a unique discussion forum for computer scientists, biologists, mathematicians, engineers, and physicists interested in a system-level understanding of biological processes. Topics covered by the CMSB proceedings include: formalisms for modeling biological processes; models and their biological applications; frameworks for model verification, validation, analysis, and simulation of biological systems; high-performance computational systems biology and parallel implementations; model inference from experimental data; model integration from biological databases; multi-scale modeling and analysis methods; computational approaches for synthetic biology; and case studies in systems and synthetic biology.

This year there were 53 submissions in total for the 4 conference tracks. Each regular submission and tool paper submission were reviewed by at least three Program Committee members. Additionally, tools were subjected to an additional review by members of the Tool Evaluation Committee, testing the usability of the software and the reproducibility of the results. For the proceedings, the Program Committee decided to accept 14 regular papers, 7 tool papers, and 11 short papers. This rich program of talks was complemented by a poster session, providing an opportunity for informal discussion of preliminary results and results in related fields.

In view of the broad scope of the CMSB conference series, we selected the following five high-profile invited speakers: Kobi Benenson (ETH Zurich, Switzerland), Trevor Graham (Barts Cancer Hospital, London, UK), Gaspar Tkacik (IST, Austria), Adelinde Uhrmacher (Rostock University, Germany), and Manuel Zimmer (University of Vienna, Austria). Their invited talks covered a broad area within the technical and applicative domains of the conference, and stimulated fruitful discussions among the conference attendees.

Further details on CMSB 2019 are available on the following website: https://cmsb2019.units.it.

Finally, as the program co-chairs, we are extremely grateful to the members of the Program Committee and the external reviewers for their peer reviews and the valuable feedback they provided to the authors. Our special thanks go to Laura Nenzi as local organization co-chair, Dimitrios Milios as chair of the Tool Evaluation Committee, and to François Fages and all the members of the CMSB Steering Committee, for their advice on organizing and running the conference. We acknowledge the support of the EasyChair conference system during the reviewing process and the production of these proceedings. We also thank Springer for publishing the CMSB proceedings in its *Lecture Notes in Computer Science* series.

Additionally, we would like to thank the Department of Mathematics and Geosciences of the University of Trieste, for sponsoring and hosting this event, and Confindustria Venezia Giulia, for supporting this event and providing administrative help. Finally, we would like to thank all the participants of the conference. It was the quality of their presentations and their contribution to the discussions that made the meeting a scientific success.

September 2019

Luca Bortolussi
Guido Sanguinetti

Organization

Program Committee

Alessandro Abate	University of Oxford, UK
Ezio Bartocci	Vienna University of Technology, Austria
Nikola Benes	Masaryk University, Czech Republic
Luca Bortolussi	University of Trieste, Italy
Giulio Caravagna	The Institute of Cancer Research, UK
Luca Cardelli	University of Oxford, UK
Milan Ceska	Brno University of Technology, Czech Republic
Claudine Chaouiya	Insituto Gulbenkian de Ciência, Portugal
Eugenio Cinquemani	Inria, France
Thao Dang	CNRS/VERIMAG, France
Hidde De Jong	Inria, France
François Fages	Inria, Université Paris-Saclay, France
Jerome Feret	Inria, France
Jasmin Fisher	University of Cambridge, UK
Christoph Flamm	University of Vienna, Austria
Elisa Franco	University of California, Los Angeles, USA
Tomas Gedeon	Montana State University, USA
Calin Guet	IST, Austria
Monika Heiner	Brandenburg Technical University Cottbus-Senftenberg, Germany
Jane Hillston	The University of Edinburgh, UK
Heinz Koeppl	TU Darmstadt, Germany
Jean Krivine	CNRS, France
Tommaso Mazza	IRCCS Casa Sollievo della Sofferenza, Italy
Laura Nenzi	University of Trieste, Italy
Marco Nobile	Universitá degli Studi di Milano-Bicocca, Italy
Diego Oyarzún	The University of Edinburgh, UK
Nicola Paoletti	Royal Holloway University of London, UK
Loïc Paulevé	CNRS/LaBRI, France
Ion Petre	University of Turku, Finland
Tatjana Petrov	University of Konstanz, Germany
Carla Piazza	University of Udine, Italy
Ovidiu Radulescu	University of Montpellier 2, France
Olivier Roux	IRCCyN, France
Jakob Ruess	Inria Saclay, France
Guido Sanguinetti	The University of Edinburgh, UK
Thomas Sauter	University of Luxembourg, Luxembourg
Abhyudai Singh	University of Delaware, USA

Carolyn Talcott	SRI International, USA
Chris Thachuk	California Institute of Technology, USA
P. S. Thiagarajan	Harvard University, USA
Adelinde Uhrmacher	University of Rostock, Germany
Verena Wolf	Saarland University, Germany
Boyan Yordanov	Microsoft, USA
Paolo Zuliani	Newcastle University, UK
David Šafránek	Masaryk University, Czech Republic

Additional Reviewers

Angaroni, Fabrizio	Clarke, Matthew
Backenköhler, Michael	de Franciscis, Sebastiano
Bellot, Eléonore	Hall, Ben
Boutillier, Pierre	Madari, Ahmad
Carcano, Arthur	Paul, Soumya
Chen, Hongkai	Shmarov, Fedor
Chodak, Jacek	Tognazzi, Stefano

Contents

Regular Papers

Sequential Reprogramming of Boolean Networks Made Practical   3
Hugues Mandon, Cui Su, Stefan Haar, Jun Pang, and Loïc Paulevé

Sequential Reprogramming of Biological Network Fate 20
Jérémie Pardo, Sergiu Ivanov, and Franck Delaplace

Control Variates for Stochastic Simulation of Chemical
Reaction Networks 42
Michael Backenköhler, Luca Bortolussi, and Verena Wolf

Effective Computational Methods for Hybrid Stochastic Gene Networks 60
*Guilherme C. P. Innocentini, Fernando Antoneli, Arran Hodgkinson,
and Ovidiu Radulescu*

On Chemical Reaction Network Design by a Nested Evolution Algorithm . . . 78
Elisabeth Degrand, Mathieu Hemery, and François Fages

Designing Distributed Cell Classifier Circuits Using a Genetic Algorithm . . . 96
Melania Nowicka and Heike Siebert

Extending a Hodgkin-Huxley Model for Larval *Drosophila* Muscle
Excitability via Particle Swarm Fitting 120
*Paul Piho, Filip Margetiny, Ezio Bartocci, Richard R. Ribchester,
and Jane Hillston*

Cell Volume Distributions in Exponentially Growing Populations 140
Pavol Bokes and Abhyudai Singh

Transient Memory in Gene Regulation. 155
*Calin Guet, Thomas A. Henzinger, Claudia Igler, Tatjana Petrov,
and Ali Sezgin*

A Logic-Based Learning Approach to Explore Diabetes Patient Behaviors . . . 188
*Josephine Lamp, Simone Silvetti, Marc Breton, Laura Nenzi,
and Lu Feng*

Reachability Design Through Approximate Bayesian Computation 207
Mahmoud Bentriou, Paolo Ballarini, and Paul-Henry Cournède

Fast Enumeration of Non-isomorphic Chemical Reaction Networks. 224
Carlo Spaccasassi, Boyan Yordanov, Andrew Phillips, and Neil Dalchau

A Large-Scale Assessment of Exact Model Reduction
in the BioModels Repository 248
   Isabel Cristina Pérez-Verona, Mirco Tribastone, and Andrea Vandin

Computing Difference Abstractions of Metabolic Networks
Under Kinetic Constraints 266
   Emilie Allart, Joachim Niehren, and Cristian Versari

Tool Papers

BRE:IN - A Backend for Reasoning About Interaction Networks
with Temporal Logic... 289
   Judah Goldfeder and Hillel Kugler

The Kappa Simulator Made Interactive 296
   Pierre Boutillier

Biochemical Reaction Networks with Fuzzy Kinetic Parameters in Snoopy 302
   George Assaf, Monika Heiner, and Fei Liu

Compartmental Modeling Software: A Fast, Discrete Stochastic Framework
for Biochemical and Epidemiological Simulation. 308
   *Christopher W. Lorton, Joshua L. Proctor, Min K. Roh,
   and Philip A. Welkhoff*

Spike – Reproducible Simulation Experiments with Configuration
File Branching 315
   Jacek Chodak and Monika Heiner

KAMIStudio: An Environment for Biocuration of Cellular
Signalling Knowledge 322
   Russ Harmer and Eugenia Oshurko

A New Version of DAISY to Test Structural Identifiability
of Biological Models... 329
   M. P. Saccomani, G. Bellu, S. Audoly, and L. d'Angió

Extended Abstracts (Posters and Highlight Talks)

Semi-quantitative Abstraction and Analysis of Chemical Reaction
Networks (Extended Abstract) 337
   Milan Češka and Jan Křetínský

Bayesian Parameter Estimation for Stochastic Reaction Networks
from Steady-State Observations................................... 342
   Ankit Gupta, Mustafa Khammash, and Guido Sanguinetti

Wasserstein Distances for Estimating Parameters in Stochastic
Reaction Networks 347
    Kaan Öcal, Ramon Grima, and Guido Sanguinetti

On Inferring Reactions from Data Time Series by a Statistical Learning
Greedy Heuristics 352
    Julien Martinelli, Jeremy Grignard, Sylvain Soliman,
    and François Fages

Barbaric Robustness Monitoring Revisited for STL* in Parasim 356
    David Šafránek, Matej Troják, Vojtěch Brůža, Tomáš Vejpustek,
    Jan Papoušek, Martin Demko, Samuel Pastva, Aleš Pejznoch,
    and Luboš Brim

Symmetry Breaking for GATA-1/PU.1 Model 360
    Lenka Přibylová and Barbora Losová

Scalable Control of Asynchronous Boolean Networks 364
    Cui Su, Soumya Paul, and Jun Pang

Transcriptional Response of SK-N-AS Cells to Methamidophos
(Extended Abstract). 368
    Akos Vertes, Albert-Baskar Arul, Peter Avar, Andrew R. Korte,
    Lida Parvin, Ziad J. Sahab, Deborah I. Bunin, Merrill Knapp,
    Denise Nishita, Andrew Poggio, Mark-Oliver Stehr, Carolyn L. Talcott,
    Brian M. Davis, Christine A. Morton, Christopher J. Sevinsky,
    and Maria I. Zavodszky

Separators for Polynomial Dynamic Systems with Linear Complexity 373
    Ines Abdeljaoued-Tej, Alia Benkahla, Ghassen Haddad,
    and Annick Valibouze

Bounding First Passage Times in Chemical Reaction Networks:
Poster Abstract 379
    Michael Backenköhler, Luca Bortolussi, and Verena Wolf

Data-Informed Parameter Synthesis for Population Markov Chains 383
    Matej Hajnal, Morgane Nouvian, Tatjana Petrov, and David Šafránek

Author Index 387

Regular Papers

Sequential Reprogramming of Boolean Networks Made Practical

Hugues Mandon[1], Cui Su[2], Stefan Haar[1], Jun Pang[2,3], and Loïc Paulevé[4($\boxtimes$)]

[1] LSV, ENS Paris-Saclay, Inria, CNRS, Université Paris-Saclay, Cachan, France
[2] SnT, University of Luxembourg, Luxembourg, Luxembourg
[3] FSTC, University of Luxembourg, Esch-sur-Alzette, Luxembourg
[4] Univ. Bordeaux, Bordeaux INP, CNRS, LaBRI, UMR5800, 33400 Talence, France
loic.pauleve@labri.fr

Abstract. We address the sequential reprogramming of gene regulatory networks modelled as Boolean networks. We develop an attractor-based sequential reprogramming method to compute all sequential reprogramming paths from a source attractor to a target attractor, where only attractors of the network are used as intermediates. Our method is more practical than existing reprogramming methods as it incorporates several practical constraints: (1) only biologically observable states, viz. attractors, can act as intermediates; (2) certain attractors, such as apoptosis, can be avoided as intermediates; (3) certain nodes can be avoided to perturb as they may be essential for cell survival or difficult to perturb with biomolecular techniques; and (4) given a threshold k, all sequential reprogramming paths with no more than k perturbations are computed. We compare our method with the minimal one-step reprogramming and the minimal sequential reprogramming on a variety of biological networks. The results show that our method can greatly reduce the number of perturbations compared to the one-step reprogramming, while having comparable results with the minimal sequential reprogramming. Moreover, our implementation is scalable for networks of more than 60 nodes.

Keywords: Cell reprogramming · Boolean networks · Attractors

1 Introduction

Cell reprogramming is one of the big discoveries of regenerative medicine. Takahashi and Yamanaka in [23] demonstrated that cell fate decisions could be reversed: a mature cell can be reprogrammed into an induced pluripotent stem cell. Even though different cocktails of transcription factors have been found to switch cell phenotypes [8,22], the identification of specific transcription factors for a particular task remains a big obstacle. Blindly testing combinations of transcription factors is unfeasible due to the high cost of biological experiments.

H. Mandon and C. Su—Co-first authors.

© Springer Nature Switzerland AG 2019
L. Bortolussi and G. Sanguinetti (Eds.): CMSB 2019, LNBI 11773, pp. 3–19, 2019.
https://doi.org/10.1007/978-3-030-31304-3_1

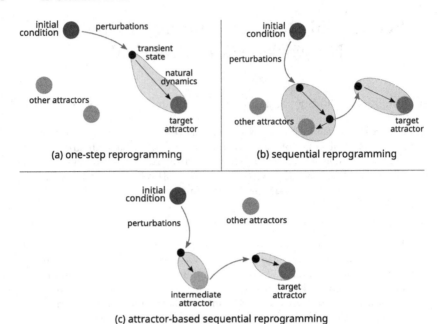

(a) one-step reprogramming

(b) sequential reprogramming

(c) attractor-based sequential reprogramming

Fig. 1. Different flavors of Boolean networks reprogramming.

Computational models of cell dynamics enable the *in silico* prediction of reprogramming targets. Qualitative models, notably Boolean networks, allow accounting the influences between numerous genes by requiring few modelling parameters. Thus, they turn out to be well suited for modelling cellular differentiation processes and thereby predict perturbations for their control [1,5–7,18,24]. In Boolean networks, each gene or protein is modelled as a binary variable, which can only take 0 or 1 as its value: a value of 0 means that the gene or protein is inactive, whereas a value of 1 means that the gene or protein is active. Each variable is assigned with a Boolean function, which determines the next value of the variable given the current values of other variables of the network. The computation of the next states depends on the chosen update mode for the variables. In this paper, we focus on the asynchronous updating mode where a single variable is updated at a time, selected non-deterministically. The long term dynamics of a Boolean network is described as attractors, which can be either single-state attractors (fixed points), or cyclic attractors.

Cell reprogramming consists of triggering a change of cellular phenotype. In the context of Boolean networks, phenotypes are modelled by the attractors. Cellular reprogramming becomes then a control problem: driving the dynamics of the network from a source attractor to a target attractor. In order to control a network, the system is perturbed out of its actual state. These perturbations can be applied instantaneously (for an instant), temporarily (for limited time), or permanently (mutations). In this paper, we focus on instantaneous perturbations. Moreover, the perturbations can take place at different "times", and as such, multiple kinds of reprogramming strategies can be found in the literature.

Existing works focus on one-step reprogramming [5,7,10,16,18], or in rare instances, on sequential reprogramming, e.g., [12]. One-step reprogramming allows applying perturbations only once as shown in Fig. 1(a). On the other hand, sequential reprogramming identifies a sequence of perturbations to be applied at different intermediate states. The intermediate states can be either a transient state or a state in an attractor. As illustrated in Fig. 1(b), a set of perturbations are applied to the initial state, which stirs the network to a transient state. After one-step spontaneous evolution, we apply another set of perturbations to the new transient state. This leads the network dynamics to a state in the strong basin of the target attractor, from which the network always eventually reaches the target attractor. By taking advantage of the natural dynamics of the network, sequential reprogramming can provide alternative predictions to one-step reprogramming, notably requiring considerably less perturbations [12]. However, in order to apply the perturbations at the correct time, sequential reprogramming requires *complete observability* of the network (i.e., the state of the network is known at any discrete time), which is rarely feasible in practice. This motivates us to develop an attractor-based sequential reprogramming as illustrated by Fig. 1(c), where perturbations should be applied only at attractors. Since the attractors can be observed experimentally, the attractor-based sequential reprogramming only requires *partial observability* of the network. Moreover, in experiments, perturbations need to take time before effectively changing the values of the variables. Attractor-based sequential reprogramming captures this requirement well, as the network dynamics remains in the attractor when perturbations are applied.

In this paper, we describe in detail our attractor-based sequential reprogramming to compute sequential reprogramming paths through other attractors of the network. We compare the performance of this new method with the minimal one-step reprogramming and the minimal sequential reprogramming. The results show that all the three methods are efficient in terms of computation time. Both sequential reprogramming methods can greatly reduce the number of perturbations compared to the minimal one-step reprogramming. Even though our attractor-based sequential reprogramming may need a few more perturbations than the minimal sequential reprogramming for some cases, the paths identified by our method are more easily transferable to biological experiment protocols.

Outline. Section 2 gives preliminary notions on Boolean networks. Section 3 addresses the attractor-based sequential reprogramming, with definitions and an algorithm to compute the solutions. Section 4 evaluates it by comparing its performance with the minimal one-step reprogramming and the minimal sequential reprogramming on several biological networks. Lastly, Sect. 5 discusses the results and reviews further the state of the art.

2 Background

2.1 Boolean Networks

A Boolean network (BN) describes elements of a dynamical system with binary-valued nodes and interactions between elements with Boolean functions. It is formally defined as follows.

Definition 1 (Boolean networks). *A Boolean network is a tuple* $\mathsf{BN} = (\mathbf{x}, \mathbf{f})$ *where* $\mathbf{x} = (x_1, x_2, \ldots, x_n)$ *such that each* $x_i, 1 \leq i \leq n$ *is a Boolean variable and* $\mathbf{f} = (f_1, f_2, \ldots, f_n)$ *is a tuple of Boolean functions over* $\mathbf{x}$. $|\mathbf{x}| = n$ *denotes the number of variables.*

In what follows, i will always range between 1 and n, unless stated otherwise. A Boolean network $\mathsf{BN} = (\mathbf{x}, \mathbf{f})$ may be viewed as a directed graph $\mathcal{G}_{\mathsf{BN}} = (V, E)$ where $V = \{v_1, v_2 \ldots, v_n\}$ is the set of vertices or nodes and for every $1 \leq i, j \leq n$, there is a directed edge from v_j to v_i if and only if f_i depends on x_j. An edge from v_j to v_i will be often denoted as $v_j \rightarrow v_i$. A path from a vertex v to a vertex v' is a (possibly empty) sequence of edges from v to v' in $\mathcal{G}_{\mathsf{BN}}$. For the rest of the exposition, we assume that an arbitrary but fixed network BN of n variables is given to us and $\mathcal{G}_{\mathsf{BN}} = (V, E)$ is its associated directed graph.

A state $\mathbf{s}$ of BN is an element in $\{0, 1\}^n$. Let $\mathbf{S}$ be the set of states of BN. For any state $\mathbf{s} = (s_1, s_2, \ldots, s_n)$, and for every i, the value of s_i, often denoted as $\mathbf{s}[i]$, represents the value that the variable x_i takes when the BN 'is in state $\mathbf{s}$'. For some i, suppose f_i depends on $x_{i_1}, x_{i_2}, \ldots, x_{i_k}$. Then $f_i(\mathbf{s})$ will denote the value $f_i(\mathbf{s}[i_1], \mathbf{s}[i_2], \ldots, \mathbf{s}[i_k])$. For two states $\mathbf{s}, \mathbf{s}' \in \mathbf{S}$, the Hamming distance between $\mathbf{s}$ and $\mathbf{s}'$ will be denoted as $\mathsf{hd}(\mathbf{s}, \mathbf{s}')$ and $\arg(\mathsf{hd}(\mathbf{s}, \mathbf{s}')) \subseteq \{1, 2, \ldots, n\}$ will denote the set of indices in which $\mathbf{s}$ and $\mathbf{s}'$ differ. For a state $\mathbf{s}$ and a subset $\mathbf{S}' \subseteq \mathbf{S}$, the Hamming distance between $\mathbf{s}$ and $\mathbf{S}'$ is defined as the minimum of the Hamming distances between $\mathbf{s}$ and all the states in $\mathbf{S}'$. That is, $\mathsf{hd}(\mathbf{s}, \mathbf{S}') = \min_{\mathbf{s}' \in \mathbf{S}'} \mathsf{hd}(\mathbf{s}, \mathbf{s}')$. We let $\arg(\mathsf{hd}(\mathbf{s}, \mathbf{S}'))$ denote the set of subsets of $\{1, 2, \ldots, n\}$ such that $I = \arg(\mathsf{hd}(\mathbf{s}, \mathbf{S}'))$ if and only if I is a set of indices of the variables that realise this Hamming distance.

2.2 Dynamics of Boolean Networks

We assume that the Boolean network evolves in discrete time steps. It starts initially in a state $\mathbf{s}_0$ and its state changes in every time step according to the update functions $\mathbf{f}$. The updating may happen in various ways. Every such way of updating gives rise to a different dynamics for the network. In this article, we focus on the fully asynchronous update mode, but the method is actually generic to any update mode, as it computes on the resulting global transition system.

Definition 2 (Asynchronous dynamics of Boolean networks). *Suppose* $\mathbf{s}_0 \in \mathbf{S}$ *is an initial state of* BN. *The asynchronous evolution of* BN *is a function* $\xi : \mathbb{N} \rightarrow \wp(\mathbf{S})$ *such that* $\xi(0) = \mathbf{s}_0$ *and for every* $j \geq 0$, *if* $\mathbf{s} \in \xi(j)$ *then* $\mathbf{s}' \in \xi(j + 1)$ *is a possible next state if and only if either* $\mathsf{hd}(\mathbf{s}, \mathbf{s}') = 1$ *and* $\mathbf{s}'[i] = f_i(\mathbf{s})$ *where* $\{i\} = \arg(\mathsf{hd}(\mathbf{s}, \mathbf{s}'))$ *or* $\mathsf{hd}(\mathbf{s}, \mathbf{s}') = 0$ *and there exists* i *such that* $\mathbf{s}'[i] = f_i(\mathbf{s})$.

Note that the asynchronous dynamics is non-deterministic – the value of exactly one variable is updated in a single time-step. The index of the variable that is updated is not known in advance. Henceforth, when we talk about the dynamics of BN, we shall mean the asynchronous dynamics as defined above.

The dynamics of a Boolean network can be represented as a *state transition graph* or a *transition system (TS)*.

Definition 3 (Transition system of BN**).** *The transition system of* BN, *denoted by the generic notation* TS *is a tuple* $(\mathbf{S}, \rightarrow)$ *where the vertices are the set of states* $\mathbf{S}$ *and for any two states* $\mathbf{s}$ *and* $\mathbf{s}'$ *there is a directed edge from* $\mathbf{s}$ *to* $\mathbf{s}'$, *denoted* $\mathbf{s} \rightarrow \mathbf{s}'$ *iff* $\mathbf{s}'$ *is a possible next state according to the asynchronous evolution function* ξ *of* BN.

2.3 Attractors and Basins of Attraction

A path from a state $\mathbf{s}$ to a state $\mathbf{s}'$ is a (possibly empty) sequence of transitions from $\mathbf{s}$ to $\mathbf{s}'$ in TS. A path from a state $\mathbf{s}$ to a subset $\mathbf{S}'$ of $\mathbf{S}$ is a path from $\mathbf{s}$ to any state $\mathbf{s}' \in \mathbf{S}'$. For any state $\mathbf{s} \in \mathbf{S}$, let $\mathsf{pre}_{\mathsf{TS}}(\mathbf{s}) = \{\mathbf{s}' \in \mathbf{S} \mid \mathbf{s}' \rightarrow \mathbf{s}\}$ and let $\mathsf{post}_{\mathsf{TS}}(\mathbf{s}) = \{\mathbf{s}' \in \mathbf{S} \mid \mathbf{s} \rightarrow \mathbf{s}'\}$. $\mathsf{pre}_{\mathsf{TS}}(\mathbf{s})$ contains all the states that can reach $\mathbf{s}$ by performing a single transition in TS and $\mathsf{post}_{\mathsf{TS}}(s)$ contains all the states that can be reached from $\mathbf{s}$ by a single transition in TS. $\mathsf{pre}_{\mathsf{TS}}(\mathbf{s})$ and $\mathsf{post}_{\mathsf{TS}}(\mathbf{s})$ are often called the set of *predecessors* and *successors* of $\mathbf{s}$. Note that, by definition, $\mathsf{hd}(\mathbf{s}, \mathsf{pre}_{\mathsf{TS}}(\mathbf{s})) \leq 1$ and $\mathsf{hd}(\mathbf{s}, \mathsf{post}_{\mathsf{TS}}(\mathbf{s})) \leq 1$. $\mathsf{pre}_{\mathsf{TS}}$ and $\mathsf{post}_{\mathsf{TS}}$ can be lifted to a subset $\mathbf{S}'$ of $\mathbf{S}$ as: $\mathsf{pre}_{\mathsf{TS}}(\mathbf{S}') = \bigcup_{\mathbf{s} \in \mathbf{S}'} \mathsf{pre}_{\mathsf{TS}}(\mathbf{s})$ and $\mathsf{post}_{\mathsf{TS}}(\mathbf{S}') = \bigcup_{\mathbf{s} \in \mathbf{S}'} \mathsf{post}_{\mathsf{TS}}(\mathbf{s})$. We define $\mathsf{pre}^{i+1}_{\mathsf{TS}}(\mathbf{S}') = \mathsf{pre}_{\mathsf{TS}}(\mathsf{pre}^{i}_{\mathsf{TS}}(\mathbf{S}'))$ and $\mathsf{post}^{i+1}_{\mathsf{TS}}(\mathbf{S}') = \mathsf{post}_{\mathsf{TS}}(\mathsf{post}^{i}_{\mathsf{TS}}(\mathbf{S}'))$ where $\mathsf{pre}^{0}_{\mathsf{TS}}(\mathbf{S}') = \mathsf{post}^{0}_{\mathsf{TS}}(\mathbf{S}') = \mathbf{S}'$. For a state $\mathbf{s} \in \mathbf{S}$, $\mathsf{reach}_{\mathsf{TS}}(\mathbf{s})$ denotes the set of states $\mathbf{s}'$ such that there is a path from $\mathbf{s}$ to $\mathbf{s}'$ in TS and can be defined as the fixpoint of the successor operation which is often denoted as $\mathsf{post}^{*}_{\mathsf{TS}}$. Thus, $\mathsf{reach}_{\mathsf{TS}}(\mathbf{s}) = \mathsf{post}^{*}_{\mathsf{TS}}(\mathbf{s})$.

Definition 4 (Attractor). *An attractor A of* TS *(or of* BN*) is a minimal subset of states of* $\mathbf{S}$ *such that for every* $\mathbf{s} \in A, \mathsf{reach}_{\mathsf{TS}}(\mathbf{s}) = A$.

Remark that attractors are the bottom strongly connected component of TS.

Any state which is not part of an attractor is a transient state. An attractor A of TS is said to be reachable from a state $\mathbf{s}$ if $\mathsf{reach}_{\mathsf{TS}}(\mathbf{s}) \cap A \neq \emptyset$. Attractors represent the stable behaviour of the BN according to the dynamics. Assuming strong fairness, the network starting at any initial state $\mathbf{s}_0 \in \mathbf{S}$ will eventually end up in one of the attractors of TS and remain there forever unless perturbed.

For an attractor A of TS, we define a subset of states of $\mathbf{S}$ called the strong basins of A, denoted as $\mathsf{bas}^{S}_{\mathsf{TS}}(A)$, as follows.

Definition 5 (Strong basin). *Let A be an attractor of* TS. *The* **strong basin** *of attraction of A with respect to* TS, *is defined as* $\mathsf{bas}_{\mathsf{TS}}(A) = \{\mathbf{s} \in \mathbf{S} \mid \mathsf{reach}_{\mathsf{TS}}(\mathbf{s}) \cap \bigcup A' = \emptyset\}$ *where the union is over all attractors A' of* TS *such that $A' \neq A$.*

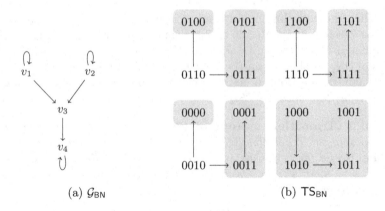

(a) $\mathcal{G}_{BN}$ (b) TS_{BN}

Fig. 2. The graph of BN and its transition system, with the attractors in red. (Color figure online)

The definition of strong basin guarantees that any state $\mathbf{s}$ in $bas_{TS}(A)$ can only reach the attractor A and cannot reach any other attractor A', $A' \neq A$ of BN.[1]

Example 1. Consider the following four-node network $BN = (\mathbf{x}, \mathbf{f})$ where $\mathbf{x} = (x_1, x_2, x_3, x_4)$, and $\mathbf{f} = (f_1, f_2, f_3, f_4)$ where $f_1 = x_1$, $f_2 = x_2$, $f_3 = x_1 \wedge \neg x_2$ and $f_4 = x_3 \vee x_4$. The graph of the network $\mathcal{G}_{BN}$ and its associated transition system TS are given in Fig. 2. TS has seven attractors marked in red. Their corresponding strong basins of attractions are shown by enclosing grey regions of a lighter shade.

3   Attractor-Based Sequential Reprogramming

3.1   Motivation

In most methods on cellular reprogramming using Boolean networks [7,16,18], all perturbations are done at once, and the system is left to stabilize itself towards the desired target attractor. However, allowing perturbations to be performed at different points in time opens alternative reprogramming paths, possibly less costly. In general, sequential reprogramming allows the network to be perturbed in any state (transient states or states in an attractor) [12,19]. This requires complete observability of the system, which is very hard to obtain experimentally.

To make the sequential reprogramming practical, we design an attractor-based sequential reprogramming, which only requires partial observability of the network. The principle of this method is to use other attractors as intermediate states for the reprogramming. At each step, we apply a set of perturbations to stir the dynamics towards a state in the strong basin of an intermediate attractor (or a target attractor). We then let the network evolve spontaneously to the

---

[1] Henceforth, we drop the subscript TS for the sake of simplicity.

intermediate attractor (or the target attractor). We repeat the above procedure until the network reaches the target attractor. In this paper, we focus on instantaneous perturbations, while applying the perturbations longer will not affect the reachability of the target attractor. In practice, based on empirical experience, biologists may be able to determine how long it takes for the network to stabilize in an intermediate attractor, i.e., the timing to apply the next perturbations. In that case, if the intermediate attractors are single-state attractors, partial observability is not required. However, if the intermediate attractors are cyclic attractors, an observation of the state might still be required.

A feasible reprogramming method has to encode practical considerations. In most cases, some variables cannot be perturbed, either because they represent an external cause the experimenter cannot change, or a set of multiple genes and proteins that would require a lot more work to perturb, or a transcription factor impacting only the gene or protein hasn't been found. Moreover, some attractors might not be suitable as intermediate states, because they lead to the death or disease of the cell. Thus, the algorithm we will describe in Sect. 3.3 provides options to avoid perturbing user-specified variables and/or avoid passing user-specified attractors.

The general principle of this method can be applied to other means to compute the required perturbations for the system to reach a target attractor, given an initial state in an attractor.

3.2   The Reprogramming Problem

In this work, we are interested in instantaneous perturbations, thus we define reprogramming of a BN as follows.

Definition 6 (Reprogramming). *A reprogramming set* C *of a* BN *is a (possibly empty) subset of* $\{1, 2, \ldots, n\}$. *For a state* $\mathbf{s} \in \mathbf{S}$, *the application of* C *to* $\mathbf{s}$ *reprograms the state of* BN *from* $\mathbf{s}$ *to* $\mathbf{s}' \in \mathbf{S}$, *such that* $\mathbf{s}'[i] = 1 - \mathbf{s}[i]$ *if* $i \in$ C *and* $\mathbf{s}'[i] = \mathbf{s}[i]$ *otherwise.*

Since the perturbations are applied instantaneously, only the state of BN is changed while the Boolean functions remain the same. Based on the above definition, we define one-step reprogramming of a BN as follows.

Definition 7 (One-step reprogramming). *Given a source attractor* A_s *and a target attractor* A_t *of* BN, *find a state* $\mathbf{s} \in A_s$ *and a reprogramming set* C, *such that the dynamics of* BN *always eventually reaches* A_t *after the application of* C *to* $\mathbf{s}$.

According to [16], we can easily obtain the following proposition.

Proposition 1. *A one-step reprogramming* $C^{A_s \to A_t}(\mathbf{s})$ *($\mathbf{s} \in A_s$) from* A_s *to* A_t *is minimal if and only if*

1. $C(\mathbf{s}) \in \mathsf{bas}(A_t)$ *and* $C = \arg(\mathsf{hd}(\mathbf{s}, \mathsf{bas}(A_t)))$.
2. $\forall \mathbf{s}' \in A_s$, $\mathsf{hd}(\mathbf{s}', \mathsf{bas}(A_t)) \geq \mathsf{hd}(\mathbf{s}, \mathsf{bas}(A_t))$.

We denote a minimal one-step reprogramming from A_s to A_t as $\mathsf{C}_{min}^{A_s \to A_t}(\mathbf{s})$. A minimal one-step reprogramming drives the dynamics of BN from A_s to a state in the strong basin of A_t, from which spontaneous evolution will eventually guide the network to A_t.

As explained in Sect. 3.1, attractor-based sequential reprogramming can provide new solutions apart from the one-step reprogramming paths. Let $|\mathcal{A}_{\mathsf{BN}}|$ denote the total number of attractors of BN. We define attractor-based sequential reprogramming as follows.

Definition 8 (Attractor-based sequential reprogramming). *Given A_s (a source attractor) and A_t (a target attractor) of BN, find a sequence of attractors $\{A_1, A_2, \ldots, A_m\}$ of BN, where $A_1 = A_s, A_m = A_t, A_i \neq A_j$ for any $i, j \in [1, m]$ and $2 \leq m \leq |\mathcal{A}_{\mathsf{BN}}|$, such that a sequence of minimal one-step reprogramming $\{\mathsf{C}_{min}^{A_1 \to A_2}, \mathsf{C}_{min}^{A_2 \to A_3}, \ldots, \mathsf{C}_{min}^{A_{m-1} \to A_m}\}$ always eventually reaches A_t (A_m). We call it an attractor-based sequential path, denoted as*

$$\rho : A_1 \xrightarrow{\mathsf{C}_{min}^{A_1 \to A_2}} A_2 \xrightarrow{\mathsf{C}_{min}^{A_2 \to A_3}} A_3 \xrightarrow{\cdots} \ldots \xrightarrow{\mathsf{C}_{min}^{A_{m-1} \to A_m}} A_m$$

$(|\mathsf{C}_{min}^{A_1 \to A_2}| + |\mathsf{C}_{min}^{A_2 \to A_3}| + \ldots + |\mathsf{C}_{min}^{A_{m-1} \to A_m}|)$ *is the total number of perturbations.*

Due to the diversity of biological networks, there does not exist one universal reprogramming strategy that suits all different networks. Hence, we develop an algorithm to compute all attractor-based sequential reprogramming paths satisfying the following constraints:

1. the total number of perturbations is less than a threshold;
2. certain attractors can be avoided as intermediates;
3. certain nodes of the network can be avoided to be perturbed.

These constraints encode practical considerations described in Sect. 3.1 and thus lead to biologically feasible reprogramming paths. We describe such an algorithm in the next section.

3.3   Algorithm

Let $\mathsf{BN} = (\mathbf{x}, \mathbf{f})$ be a Boolean Network of size $n = |\mathbf{x}|$. Let U be the set of variables that cannot be perturbed, A_s be an attractor of the network, which is the initial state of the system, and A_t be another attractor of the network, which is the target to reprogram to.

Algorithm 1 describes the algorithm to compute sequential paths from A_s to A_t, using other attractors as intermediate steps. The inputs are: the Boolean Network BN, the initial attractor A_s, the target attractor A_t, the set of attractors $\mathcal{A}$ that can act as intermediate states, and the set of variables U that can not be

---

Algorithm 1. Inevitable reprogramming of BNs from A_s to A_t

1: **procedure** COMPUTATION_OF_INEVITABLE_PATHS($BN, A_s, A_t, \mathcal{A}, U$)
2:     max_dist = $HB_m(BN, U, A_s, A_t))$
3:     $\mathcal{L}_{A_s}$ = new empty dictionary
4:     **if** max_dist $< \infty$ **then**
5:         $a = $ arg_HB(BN, U, A_s, A_t)
6:         ▷ Associate (distance, [perturbations list]) to the [path]
7:         $\mathcal{L}_{A_s}$.add($[A_t] : ($max_dist$, [a]))$
8:         ▷ Associate the minimal length of all paths from A_s to A_t
9:         $\mathcal{L}_{A_s}$.add("min" : max_dist)
10:     list := $\emptyset$
11:     **for** $A \in \mathcal{A}$ **do**
12:         $d = HB_m(BN, U, A, A_t)$
13:         **if** $d < $ max_dist **then**
14:             list.add(A)
15:             $\mathcal{L}_A = $ map()
16:             $a = $ arg_HB(BN, U, A, A_t)
17:             ▷ Associate (distance, [perturbations list]) to the [path]
18:             $\mathcal{L}_A$.add($[A_t] : (d, [a])$)
19:             ▷ Associate minimal length of all paths from A to A_t
20:             $\mathcal{L}_A$.add("min" : d)
21:     ▷ Recursively computes the paths with attractors as intermediate steps
22:     **while** list $\neq \emptyset$ **do**
23:         $l := \emptyset$
24:         **for** $A_1 \in \mathcal{A}$ **do**
25:             **for** $A_2 \in $ list **do**
26:                 $d = HB_m(BN, U, A_1, A_2)$
27:                 **if** $d \neq \infty$ and $d + \mathcal{L}_{A_2}[$"min"$] \leq$ max_dist **then**
28:                     l.add(A_1)
29:                     **for** path $\in \mathcal{L}_{A_2} \setminus \{$"min"$\}$ **do**
30:                         $td = d + \mathcal{L}_{A_2}[$path$][0]$ ▷ total length of the new path to A_t
31:                         **if** $td \leq$ max_dist and $A_2 \notin$ path **then**
32:                             **if** $\mathcal{L}_{A_1}$ does not exists **then**
33:                                 $\mathcal{L}_{A_1} = $ map()
34:                                 ▷ Associate minimal length of all paths from A to A_t
35:                                 $\mathcal{L}_{A_1}$.add("min" : td)
36:                             $a = $ arg_HB(BN, U, A_1, A_2)
37:                             ▷ Associate (distance, [perturbations]) to the [path]
38:                             $\mathcal{L}_{A_1}$.add($[A_2] + $ path : $(td, [a] + \mathcal{L}_{A_2}[$path$][1]))$
39:                             **if** $td < \mathcal{L}_{A_1}[$"min"$]$ **then**
40:                                 $\mathcal{L}_{A_1}[$"min"$] = td$
41:         list $= l$
42:     **return** $\mathcal{L}_{A_s}$

---

perturbed. The set $\mathcal{A}$ excludes the attractors that cannot act as intermediates, such as the source attractor and the undesired attractors.[2]

---

[2] We refer details on attractor detection to [13].

Algorithm 2. Distance functions

1: **function** $\mathsf{HB}_m(\mathsf{BN}, \mathsf{U}, S, T)$
2:     $B = \mathsf{bas}(T)$
3:     ▷ Details on the computation of basinS can be found in [16,14]
4:     return $min_{s \in S, t \in B}(\mathsf{hd}_m(\mathsf{U}, s, t))$
5: **function** $\mathsf{hd}_m(\mathsf{BN}, \mathsf{U}, s, t)$
6:     $sum = 0$
7:     **for** $i = 1, i \leq n, i + +$ **do**
8:         **if** $s[i] \neq t[i]$ **then**
9:             **if** $i \in \mathsf{U}$ **then**
10:                 return ∞
11:             $sum = sum + |s[i] - t[i]|$
12:     return sum
13: **function** $\mathsf{arg\_HB}(\mathsf{BN}, \mathsf{U}, S, T)$
14:     $min = \mathsf{HB}_m(\mathsf{BN}, \mathsf{U}, S, T)$
15:     **if** $min = \infty$ **then**
16:         Fail("infinite distance")
17:     $D = \mathrm{map}()$
18:     **for** $s \in S$ **do**
19:         **for** $t \in T$ **do**
20:             **if** $\mathsf{hd}_m(\mathsf{U}, s, t) = min$ **then**
21:                 **for** $i = 1, i \leq n, i + +$ **do**
22:                     **if** $s[i] \neq t[i]$ **then**
23:                         ▷ Associate the desired value of the variable i to t_i
24:                         $D.\mathrm{add}(i : t_i)$
25:     return D

The algorithm uses a modified Hamming distance hd_m between the states of the transition system. Between a state s and a state t, this modified Hamming distance $\mathsf{hd}_m(s, t)$ is defined as:

$$\mathsf{hd}_m(s, t) = \begin{cases} \infty & \text{if } \exists v \in \mathsf{U}, s[v] \neq t[v] \\ \mathsf{hd}(s, t) & \text{otherwise} \end{cases}$$

The modified Hamming distance between two sets of states S and T is defined as: $\mathsf{hd}_m(S, T) = min_{s \in S, t \in T}(\mathsf{hd}_m(s, t))$.

To compute the sequential paths from $\mathsf{A_s}$ to $\mathsf{A_t}$ using other attractors as intermediate states, we have to compute the strong basin of $\mathsf{A_t}$, which is $\mathsf{bas}(\mathsf{A_t})$. Since we only use the distance between a state and a basin, let HB_m, the distance between a set of states S and the basin of a set of states T, be defined as $\mathsf{HB}_m(S, T) = \mathsf{hd}_m(S, \mathsf{bas}(T))$. Algorithm 2 describes how to compute both of these distances, as well as how to compute the argument of HB_m, including the set of variables that realize the minimum Hamming distance and the desired value of these variables. The distance between $\mathsf{A_s}$ and the basin of $\mathsf{A_t}$, $\mathsf{max\_dist} = \mathsf{HB}_m(\mathsf{A_s}, \mathsf{A_t})$, will be used as a benchmark for the next computations: this is the maximum number of perturbations allowed to reach $\mathsf{A_t}$.

An empty dictionary $\mathcal{L}_{A_s}$ is created, to store the possible paths. If max_dist $<$ ∞, the perturbed variables, $a = $ arg_HB(BN, U, A_s, A_t) are computed. The path, represented by a list of targets to reach in order to reach the next one, $[A_t]$ is added as an entry of the dictionary, with the value (max_dist, $[a]$). This dictionary regroups all paths from A_s to A_t, the first value is the length of the path, and the second is how to get from one attractor to the next one in the list. A special value is added to the dictionary, "min", which is the minimal length of all the paths from A_s to A_t, and it is given the value max_dist.

Then, for all attractor A in $\mathcal{A}$, the distance $d = $ HB$_m$(A, A_t) is computed. If this distance d is strictly lower than max_dist[3], then A is added to a list of attractors list and a dictionary $\mathcal{L}_A$ is created. We add to $\mathcal{L}_A$ the entry $[A_t]$ to which we associate the length of the path, d, and the perturbations made in a list, [arg_HB(BN, U, A, A_t)]. The path is a list of the attractors to reach in the right order. The perturbations made are a set, a dictionary in our case, containing the variables to perturb and the desired values. This set is put in a list: each set of the list is a set of perturbations to go from the current attractor to the next one in the path defined above. A special value "min" is added to the dictionary, in the same way as for $\mathcal{L}_{A_s}$, to store the minimal length of paths from A to A_t.

The list list is used to recursively compute the shortest paths. As long as list is not empty, the following steps are done:

1. First, create an empty list l.
2. Then, from all attractor A_1 in $\mathcal{A}$, for all attractor A_2 in list, the distance $d = $ HB$_m$(A_1, A_2) is computed. If this distance plus $\mathcal{L}_{A_2}$["min"][4] is lower than max_dist, then for every path path in $\mathcal{L}_{A_2}$, the total distance $d + \mathcal{L}_{A_2}$[path][0][5] is computed. If this distance is lower or equal to max_dist and if $A_1 \notin$ path, a new entry $[A_2] + $ path[6] is added to $\mathcal{L}_{A_1}$, with the value $(d + \mathcal{L}_{A_2}$[path][0], [arg_HB(A_1, A_2)] + $\mathcal{L}_{A_2}$[path][1]). The first value, the distance, is the one to go from A_1 to A_t using A_2 as an intermediate step, and the paths from A_2 to A_t already computed. The second value is the set of variables to perturb, using the same principle. If the dictionary does not exist, it is created, and "min" is updated or created. Moreover, A_1 is added to l.
3. Lastly, the value of list is changed to match l, list $= l$.

When this loop is over, all paths are in $\mathcal{L}_{A_s}$, with the associated length and steps of variables to perturb, and $\mathcal{L}_{A_s}$ is returned.

4   Evaluation

To demonstrate the efficiency and the efficacy of our attractor-based sequential reprogramming described in Algorithm 1, we compare its performance with the minimal one-step reprogramming [16] and the minimal sequential reprogramming [12] on a variety of biological networks.

---

[3] In this case, if max_dist $= \infty$, any non infinite distance is considered strictly lower.
[4] This value is the minimal length path from A_1 to A_t.
[5] As A_2 is in list, $\mathcal{L}_{A_2}$ exists.
[6] Here, the $+$ is the usual concatenation for lists.

4.1   Reprogramming Strategies

To drive a network from a source state to a target attractor, the minimal one-step reprogramming [16] computes a minimal set of perturbations to be conducted simultaneously, and the minimal sequential reprogramming [12] computes shortest sequential paths, where any state may act as an intermediate state. Different from [12], the attractor-based sequential reprogramming (this work) identifies all the sequential paths with at most k perturbations, where only attractors (biologically observable states) can play the role of intermediate states. We compute the reprogramming paths for all combinations of source and target attractors of the studied networks with the three methods. For the attractor-based sequential reprogramming, the maximal number of perturbations allowed is set as the number of perturbations required by the minimal one-step reprogramming; and we assume all the nodes can be perturbed, thus $U = \emptyset$ due to the lack of relevant biological knowledge. The three methods are implemented as part of the software tool ASSA-PBN [14]. All the experiments are performed on a computer with a CPU of Intel Core i7 @3.1 GHz and 8 GB of DDR3 RAM[7].

4.2   Benchmark Biological Networks

We give a short description of the biological networks on which we test the three different reprogramming methods of Boolean networks. Table 1 gives an overview of the sizes and number of attractors for these networks. All the attractors of the networks are single-state attractors.

- The myeloid differentiation network is designed to model myeloid differentiation from common myeloid progenitors to megakaryocytes, erythrocytes, granulocytes and monocytes [11].
- The cardiac gene regulatory network is constructed for the early cardiac gene regulatory network of the mouse, including the core genes required for the cardiac development and the FHF/SHF determination [9].
- The ERBB receptor regulated G1/S transition network enables us to identify potential targets in the treatment of trastuzumab resistant breast cancer [20].
- The tumour network is constructed to study the role of individual mutations or their combinations in the metastatic process [5].
- The PC12 cell network models the temporal sequence of protein signalling, transcriptional response and subsequent autocrine feedback [15].
- The model of hematopoietic cell specification recaps cytokine induced differentiation, several reported gene knockdowns and the reprogramming of pre-B cells [6].
- The model of bortezomib responses can predict responses to the lower bortezomib exposure and the dose-response curve for bortezomib [4].

---

[7] Executable and data are available at the following link: https://github.com/cuisu/attractor_based_sequential_reprogramming.

Table 1. An overview of the networks and the evaluation results. O, A and S stand for the minimal one-step reprogramming, the attractor-based sequential reprogramming and the minimal sequential reprogramming, respectively.

Network	# nodes	# edges	#A	Range of $|C|$			Time (seconds)		
				$|C_O|$	$|C_A^{min}|$	$|C_S|$	T_O	T_A	T_S
Myeloid	11	30	6	$1-5$	$1-4$	$1-4$	0.02	0.04	0.21
Cardiac	15	39	6	$1-9$	$1-8$	$1-4$	0.23	0.63	2.28
ERBB	20	52	3	$1-9$	$1-8$	$1-5$	0.05	0.07	0.49
Tumour	32	158	9	$1-10$	$1-9$	$1-6$	1.54	5.99	387.04
PC12	33	62	7	$1-11$	$1-10$	$1-10$	0.39	3.21	95.10
Hematopoietic	33	88	5	$1-13$	$1-12$	$1-12$	1.89	4.87	8067.73
Bortezomib	67	135	5	$1-21$	$1-15$	$*$	50.24	106.91	$*$

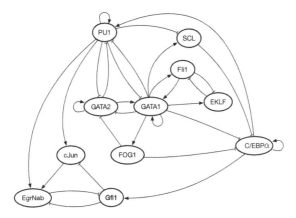

Fig. 3. Structure of the myeloid network. Rightarrow and bar arrow represent activation and inhibition, respectively.

4.3   Results on the Myeloid Differentiation Network

Let us analyse in more depth the predictions obtained on the myeloid differentiation network. Figure 3 depicts its influence graph, and Table 2 lists its six attractors, all being fixed points, four of which correspond to megakaryocyte (A_2), erythrocyte (A_3), granulocyte (A_5) and monocyte (A_6) [11]. Table 3 describes the number of perturbations required by the three methods ($|C_O|$, $|C_A|$, and $|C_S|$) for this network. The first column and the first row stand for the source and the target attractors, respectively. The minimal one-step reprogramming needs more perturbations since it only allows to apply perturbations once. By choosing appropriate states as intermediates, the sequential reprogramming can reduce the number of perturbations for a few cases (e.g. from A_2 (A_3, A_4 or A_5) to A_6). The minimal number of perturbations required by the two sequential reprogramming methods are identical for this network. Besides the shortest paths, the attractor-based sequential reprogramming also identifies paths with

Table 2. Attractors of the myeloid network.

	GATA2	GATA1	FOG1	EKLF	Fli1	SCL	C/EBPα	PU1	cJun	EgrNab	Gfi1
A_1	0	0	0	0	0	0	0	1	1	1	0
A_2	0	1	1	0	1	1	0	0	0	0	0
A_3	0	1	1	1	0	1	0	0	0	0	0
A_4	0	0	0	0	0	0	0	0	0	0	0
A_5	0	0	0	0	0	0	1	1	0	0	1
A_6	0	0	0	0	0	0	1	1	1	1	0

Table 3. The number of perturbations computed by the three reprogramming methods for the myeloid differentiation network. The first column and the first row stand for the source and the target attractors, respectively.

	A_1			A_2			A_3			A_4			A_5			A_6		
	$|C_O|$	$|C_A|$	$|C_S|$	$|C_O|$	$|C_A|$	$|C_S|$	$|C_O|$	$|C_A|$	$|C_S|$	$|C_O|$	$|C_A|$	$|C_S|$	$|C_O|$	$|C_A|$	$|C_S|$	$|C_O|$	$|C_A|$	$|C_S|$
A_1	0	0	0	3	3	3	3	3	3	1	1	1	3	3	3	1	1	1
A_2	2	2	2	0	0	0	2	2	2	2	2	2	4	4	4	5	3, 4, 5	3
A_3	2	2	2	2	2	2	0	0	0	1	1	1	3	3	3	5	3, 5	3
A_4	1	1	1	2	2	2	2	2	2	0	0	0	2	2	2	4	2, 4	2
A_5	1	1	1	3	3	3	3	3	3	2	2	2	0	0	0	3	2, 3	2
A_6	1	1	1	3	3	3	3	3	3	2	2	2	2	2	2	0	0	0

at most $|C_O|$ perturbations. For instance, there are in total three attractor-based sequential paths from A_2 to A_6:

- $\rho_1 : A_2 \xrightarrow{\text{GATA1,EgrNab,PU1,cJun, C/EBPα}} A_6,;$
- $\rho_2 : A_2 \xrightarrow{\text{GATA1, Fli1}} A_4 \xrightarrow{\text{PU1}} A_1 \xrightarrow{\text{C/EBPα}} A_6;$
- $\rho_3 : A_2 \xrightarrow{\text{GATA1,PU1}} A_1 \xrightarrow{\text{C/EBPα}} A_6.$

Path ρ_1 is also a shortest one-step path, which requires 5 perturbations. Paths ρ_2 and ρ_3 only require 4 and 3 perturbations, respectively. An interesting observation is that the sequential paths may require the perturbation of the same gene multiple times, which will not happen for the one-step reprogramming method. For instance, a sequential path from A_5 to A_6 is $\rho : A_5 \xrightarrow{\text{C/EBPα}} A_1 \xrightarrow{\text{C/EBPα}} A_6$. By perturbing 'C/EBPα' twice, we can achieve the sequential reprogramming from A_5 to A_6.

4.4   Results on the Benchmark Biological Networks

An overview of the evaluation results on the seven networks is given in Table 1. It is worth noting that Table 3 gives the number of perturbations for every pair of source and target attractors of the myeloid differentiation network, while in Table 1, columns $|C_O|$, $|C_A^{min}|$, and $|C_S|$ summarise the minimal number of perturbations required by the three methods for all pairs of source and target attractors of the seven biological networks.[8] In Table 1, $|C_A^{min}|$ only considers the shortest attractor-based sequential paths, instead of all the identified paths (see $|C_A|$ in Table 3) with less than $|C_O|$ perturbations.

In general, the sequential strategy results in less perturbations. Even though the attractor-based sequential method requires a few more perturbations than the minimal sequential reprogramming, it uses biological observable states as intermediates and thus is considered more realistic and applicable. In particular, the attractor-based sequential control can reduce up to 9 perturbations compared to the minimal one-step control. Columns T_O, T_A and T_S of Table 1 give the total computation time. We can see that all three methods are efficient and scale well for large networks. Even though the attractor-based sequential reprogramming takes a bit longer than the one-step reprogramming, it identifies a number of potential applicable solutions.

5   Discussion

Combining the available techniques from computer science with the constraints of experimental protocols in biology, in this paper, we have designed attractor-based sequential reprogramming of Boolean networks. Compared to one-step reprogramming [16], where all perturbations are applied only once, our method identifies a sequence of perturbations to be applied sequentially. Taking full advantage of spontaneous evolutions, our method requires less perturbations and thus results in lower experimental costs. Different from the sequential reprogramming [12], our method only uses other attractors as intermediates. Therefore, it does not require complete observability, except within cyclic attractors, which makes its application more feasible in biological experiments.

Moreover, our method allows avoiding some variables to be perturbed and some attractors to be used as intermediate steps, which differs from a previously developed sequential reprogramming method [12]. These constrains key observations in practice, as some biological networks have genes that cannot yet be influenced by transcription factors (or they can be influenced at a very high cost), and some attractors such as apoptosis of the cell shouldn't be viable intermediate steps. Our method sits in a middle ground between one-step reprogramming [16] and sequential reprogramming [12].

Existing works mainly focus on one-step reprogramming [2,3,5–7,21,24], considering various kinds of perturbations and targeted dynamical properties. Predictions are obtained following different techniques, with probabilistic modelling in [3,5,6], or qualitative modelling in [2,7,21,24]. Sequential reprogramming is also studied in the literature [1,12,19] using quite different approaches: Abou-Jaoudé et al. [1] applied model checking to verify that a set of perturbations can reprogram the cell correctly, using other attractors as intermediate steps if needed, Ronquist et al. [19] used a quantitative model that returns a specific time for the perturbations to be made; lastly in the work of Mandon et al. [12],

---

[8] '*' means the algorithm fails to return any result within five hours. We excluded the 'apoptosis' attractor for the tumour network for the evaluation.

the perturbations can be done at any time, but require precise knowledge of the state of the system (i.e., complete observability).

In future work, besides relaxing the observability within cyclic attractors, we plan to address attractor-based sequential reprogramming with temporary perturbations (i.e., sustained for a limited time). This corresponds to another classical experimental setting in cellular reprogramming, and should provide alternative and potentially shorter sequences of perturbations.

Acknowledgement. This research was supported by the ANR-FNR project Algo-ReCell (ANR-16-CE12-0034; FNR INTER/ANR/15/11191283); Labex DigiCosme (project ANR-11-LABEX-0045-DIGICOSME) operated by ANR as part of the program "Investissement d'Avenir" Idex Paris-Saclay (ANR-11-IDEX-0003-02); and by the project SEC-PBN funded by University of Luxembourg. Cui Su was also partially supported by the COST Action IC1405.

References

1. Abou-Jaoudé, W., et al.: Model checking to assess T-helper cell plasticity. Front. Bioeng. Biotechnol. **2**, 86 (2015)
2. Biane, C., Delaplace, F.: Abduction based drug target discovery using Boolean control network. In: Feret, J., Koeppl, H. (eds.) CMSB 2017. LNCS, vol. 10545, pp. 57–73. Springer, Cham (2017). https://doi.org/10.1007/978-3-319-67471-1_4
3. Chang, R., Shoemaker, R., Wang, W.: Systematic search for recipes to generate induced pluripotent stem cells. PLoS Comput. Biol. **7**(12), e1002300 (2011)
4. Chudasama, V., Ovacik, M., Abernethy, D., Mager, D.: Logic-based and cellular pharmacodynamic modeling of Bortezomib responses in U266 human myeloma cells. J. Pharmacol. Exp. Ther. **354**(3), 448–458 (2015)
5. Cohen, D.P.A., Martignetti, L., Robine, S., Barillot, E., Zinovyev, A., Calzone, L.: Mathematical modelling of molecular pathways enabling tumour cell invasion and migration. PLoS Comput. Biol. **11**(11), e1004571 (2015)
6. Collombet, S., et al.: Logical modeling of lymphoid and myeloid cell specification and transdifferentiation. Proc. Nat. Acad. Sci. **114**(23), 5792–5799 (2017)
7. Crespo, I., Perumal, T.M., Jurkowski, W., del Sol, A.: Detecting cellular reprogramming determinants by differential stability analysis of gene regulatory networks. BMC Syst. Biol. **7**(1), 140 (2013)
8. Graf, T., Enver, T.: Forcing cells to change lineages. Nature **462**(7273), 587–594 (2009)
9. Herrmann, F., Groß, A., Zhou, D., Kestler, H.A., Kühl, M.: A Boolean model of the cardiac gene regulatory network determining first and second heart field identity. PLoS ONE **7**, 1–10 (2012)
10. Jo, J., et al.: An integrated systems biology approach identifies positive cofactor 4 as a factor that increases reprogramming efficiency. Nucleic Acids Res. **44**(3), 1203–1215 (2016)
11. Krumsiek, J., Marr, C., Schroeder, T., Theis, F.J.: Hierarchical differentiation of myeloid progenitors is encoded in the transcription factor network. PLoS ONE **6**(8), e22649 (2011)
12. Mandon, H., Haar, S., Paulevé, L.: Temporal reprogramming of Boolean networks. In: Feret, J., Koeppl, H. (eds.) CMSB 2017. LNCS, vol. 10545, pp. 179–195. Springer, Cham (2017). https://doi.org/10.1007/978-3-319-67471-1_11

13. Mizera, A., Pang, J., Qu, H., Yuan, Q.: Taming asynchrony for attractor detection in large Boolean networks. IEEE/ACM Trans. Comput. Biol. Bioinf. **16**(1), 31–42 (2018)
14. Mizera, A., Pang, J., Su, C., Yuan, Q.: ASSA-PBN: a toolbox for probabilistic Boolean networks. IEEE/ACM Trans. Comput. Biol. Bioinf. **15**(4), 1203–1216 (2018)
15. Offermann, B., et al.: Boolean modeling reveals the necessity of transcriptional regulation for bistability in PC12 cell differentiation. Front. Genet. **7**, 44 (2016)
16. Paul, S., Su, C., Pang, J., Mizera, A.: A decomposition-based approach towards the control of Boolean networks. In: Proceedings 9th ACM Conference on Bioinformatics, Computational Biology, and Health Informatics, pp. 11–20. ACM Press (2018)
17. Paul, S., Su, C., Pang, J., Mizera, A.: An efficient approach towards the source-target control of Boolean networks. IEEE/ACM Trans. Comput. Biol. Bioinf. (2019, accepted)
18. Remy, E., Rebouissou, S., Chaouiya, C., Zinovyev, A., Radvanyi, F., Calzone, L.: A modelling approach to explain mutually exclusive and co-occurring genetic alterations in bladder tumorigenesis. Cancer Res. **75**, 4042–4052 (2015). https://doi.org/10.1158/0008-5472.CAN-15-0602
19. Ronquist, S., et al.: Algorithm for cellular reprogramming. Proc. Nat. Acad. Sci. **114**(45), 11832–11837 (2017)
20. Sahin, Ö., et al.: Modeling ERBB receptor-regulated G1/S transition to find novel targets for de novo trastuzumab resistance. BMC Syst. Biol. **3**(1), 1 (2009)
21. Samaga, R., Von Kamp, A., Klamt, S.: Computing combinatorial intervention strategies and failure modes in signaling networks. J. Comput. Biol. **17**(1), 39–53 (2010)
22. del Sol, A., Buckley, N.J.: Concise review: a population shift view of cellular reprogramming. Stem Cells **32**(6), 1367–1372 (2014)
23. Takahashi, K., Yamanaka, S.: A decade of transcription factor-mediated reprogramming to pluripotency. Nat. Rev. Mol. Cell Biol. **17**(3), 183–193 (2016)
24. Zañudo, J.G.T., Albert, R.: Cell fate reprogramming by control of intracellular network dynamics. PLoS Comput. Biol. **11**, 1–24 (2015)

Sequential Reprogramming of Biological Network Fate

Jérémie Pardo, Sergiu Ivanov(✉), and Franck Delaplace

IBISC, Univ Évry, Paris-Saclay University, 91025 Évry, France
{jeremie.pardo,sergiu.ivanov,franck.delaplace}@ibisc.univ-evry.fr
http://www.ibisc.univ-evry.fr

Abstract. A major challenge in precision medicine consists in finding the appropriate network rewiring to induce a particular reprogramming of the cell phenotype. The rewiring is caused by specific network action either inhibiting or over-expressing targeted molecules. In some cases, a therapy abides by a time-scheduled drug administration protocol. Furthermore, some diseases are induced by a sequence of mutations leading to a sequence of actions on molecules. In this paper, we extend previous works on abductive-based inference of network reprogramming [3] by investigating the sequential control of Boolean networks. We present a novel theoretical framework and give an upper bound on the size of control sequences as a function of the number of observed variables. We also define an algorithm for inferring minimal parsimonious control sequences allowing to reach a final state satisfying a particular phenotypic property.

Keywords: Dynamical systems reprogramming ·
Boolean Control Network · Control sequence · Abductive reasoning ·
Drug target prediction · Sequential therapy

1 Introduction

Cell reprogramming consists in modifying gene expression to induce a particular cell behavior naturally or artificially. The potential outcomes of reprogramming will have valuable benefits in essential challenges of health: cancerous targeted therapy, complex disease etiology, regenerative medicine, stem cells monitoring, etc. Despite the impressive progress in cell reprogramming during the last decade, more breakthroughs are required before cellular reprogramming yields routine clinical use [18]. The main issues lie on the discovery of reliable ways to trigger reprogramming process and to understand exactly how the process works. In this endeavour, the definition of suitable theoretical frameworks and computational methods are crucial for enabling the analysis and the design of *reprogramming patterns* responsible for the phenotypic switch.

Cell reprogramming mechanisms are based on the control of molecular processes to monitor the dynamics of the network fate. In [20], the authors relate mutations to their network effect: nonsense mutation, out-of-frame insertion or

L. Bortolussi and G. Sanguinetti (Eds.): CMSB 2019, LNBI 11773, pp. 20–41, 2019.
https://doi.org/10.1007/978-3-030-31304-3_2

deletion and defective splicing are interpreted as node or arc deletions whereas missense mutation and in-frame insertion or deletion can be modelled as node or arc additions. Moreover, in [7], the authors classify mutations according to the way they affect signalling networks and distinguish mutations that constitutively activate or inhibit enzymes (nodes) and mutations that rewire the interactions (arcs). In the same vein, the action of targeted therapies is interpreted as network rewiring [8]. The effect of mutations and drugs can thus be described as elementary topological actions on the network: deletion or insertion of nodes and arcs. Cell reprogramming is then viewed as network alteration based on these topological actions. The impact of the network actions should be evaluated from a model of dynamics translating the topological actions into dynamical alteration of the trajectories. Accordingly, the phenotypic changes are assessed at molecular level by the measurement of the state of peculiar molecules called *biomarkers*—observable indicators of biological processes whose molecular signature variation discriminates the phenotypes [8,17]. The signatures must be observed in a significant period of time for testifying their relevance, and thus assumed to be met at stability condition of the biological system. This approach is part of Network Medicine [1] that aims to address drug target discovery and the elucidation of disease mechanism based on network analysis by renewing the phenotype-genotype relationship into the association of a phenotype to some network perturbations [16].

Recent research in computational biology provides novel inference methods for reprogramming a system to make the dynamics converge towards an expected fate. These works use the *Boolean Control Network* (BCN) and model specifying the actions as controls on Boolean network, detailed in Sect. 2. In [12] the authors apply a stuck-at fault model for simulating drug and disease processes. A Max-SAT based algorithm is then used for inferring node actions. In [19], the authors propose a heuristic method focused on the control of key-nodes stabilizing the state of "motifs" that correspond to specific sub-networks. In [15], the authors propose a method based on Gröbner basis computation to find the node actions for generating or avoiding particular stable states. In [3], the authors use an abductive method based on prime implicants for the inference, and cover actions on nodes and arcs. These works were validated on real biological cases showing their adequacy for drug therapy prediction. The first method is restricted to acyclic network whereas the others admit any network. In summary, the state of the art related to Boolean control network shows that the majority of the works use a similar methodology that consists in computing a single network action modelled as control input for monitoring the dynamics in order to reach stable states meeting some expected properties assessed at molecular level.

However in some biological cases, a sequence of mutations is observed or a therapy involves a scheduled protocol for administering drugs. Tumorigenesis is the result of a multi-step process governed by sequential genetic alterations. Colorectal tumor offers a paradigmatic system illustrating this sequential progression shifting from a benign tumor (adenoma) to a malignant one (carcinoma) following a sequence of four gene mutations ending with the appearance of metastases [9]. Furthermore, in [11] the authors describe a systematic approach to identifying efficient drug combinations in killing cancer cells depending on changes in the order and duration of drug exposure. They found that some drug combinations (EGFR inhibitor) can synergise the apoptotic response to DNA damaging chemotherapy for a subset of triple negative breast cancers if the drugs are given sequentially but not simultaneously, leading to an appropriate dynamic rewiring of oncogenic signalling networks.

Therefore, to widen the scope of potential applications in precision medicine, we propose to extend the previous works by investigating control sequences to provide the possibility to explain the causes of diseases by sequences of perturbations and to discover therapeutic regimen as a long term perspective. A control sequence is composed of a list of control/topological network actions whose sequential application routes the dynamics of the network by steps to the expected fate. Little research has been developed on this extension. In [13], the authors study temporal reprogramming of Boolean networks based on Petri net analysis. Given a trajectory, they identify the appropriate states at which a control should be applied, and deduce the corresponding controls (perturbations) to reach an expected state. The algorithm explores the extended state graph encompassing perturbation representation, implying that the number of possible controls has to remain low to be tractable. An improved and generalised version of this algorithm is given in [14].

In this article, we define a computational method for inferring the minimal sequence of controls to reach some expected properties met at stable states. More specifically, we propose a theoretical framework describing a controlled dynamics enabling us to characterize a bound on the length of control sequences of minimal size.

The article is organised as follows. Section 2 recalls the principles of Boolean networks and introduces the extension to Boolean Control Networks by defining the main notions of controlled dynamics. In Sect. 3, we examine the properties of control sequences required to control network fate. In the section, we also detail the algorithm inferring minimal control sequences.

2   Boolean Control Network

Boolean Control Networks (BCN) extend Boolean networks by adding Boolean controls which can alter the dynamics. In this section, we briefly recall the main definitions of Boolean networks (Sect. 2.1), and then we define the extension to BCN (Sect. 2.2). We more specifically focus on a particular class of control called the *freezing control* where a control input freezes a variable state to a specific value definitively. This category truthfully models the aftermaths of the perturbations on genetic and signalling networks blocking the gene expression in a particular regulation state that are notably the consequences of mutations or drug effects.

Notations. We use the following notations. Let $E' \subseteq E$. We denote: $-E' = E \backslash E'$. Let f be a function with E as domain, $f_{\downarrow E'}$ defines the restriction/projection of the function to E' such that f is only defined for the elements of E'.

2.1 Boolean Network

A Boolean network is a discrete dynamical system defined on Boolean variables X. A state s belonging to the set of states S_X is an interpretation assigning a Boolean value to the variables (*i.e.*, $s : X \to \mathbb{B}$). A Boolean network is defined by a collection of Boolean functions,

$$F = \{x_i = f_i(x_1, \ldots, x_n) \mid 1 \le i \le n\},$$

where each f_i is a propositional formula computing the state of x_i.

The *model of dynamics* describes the evolution of states for all variables by a labeled transition system $\langle \longrightarrow, M, S_X \rangle$, where the states are updated according to an updating policy $M \subseteq 2^X$, called the *mode*, which is a cover of X ($\bigcup_{m \in M} m = X$). Each transition relation ($\longrightarrow \subseteq S_X \times M \times S_X$) is labeled by the set of updated variables m stipulating the modified variables during the transition:

$$s \xrightarrow{m} s' \stackrel{\text{def}}{=\!=} s' = (F_{\downarrow m}(s) \cup s_{\downarrow -m}). \tag{1}$$

The complement $-m$ is taken with respect to X. The global transition relation is defined as: $\longrightarrow = \bigcup_{m \in M} \xrightarrow{m}$. A path[1] $s \longrightarrow^* s'$ characterizes a trajectory from s to s'. In biological modelling two modes are preferentially used: the *synchronous* mode where all the variables are updated during a transition ($M = \{X\}$) or the *asynchronous* mode where one variable only is updated per transition ($M = \{\{x_i\}\}_{x_i \in X}$).

An *equilibrium* s is a particular state which is indefinitely reached once met *i.e.*, $\forall s' \in S_X : s \longrightarrow^* s' \implies s' \longrightarrow^* s$. A *stable state* s is a particular equilibrium satisfying the stability condition: $\text{STBL}_F(s) \stackrel{\text{def}}{=\!=} F(s) = s$. The picture on the left of Fig. 1 describes a Boolean dynamics under the synchronous mode.

2.2 Boolean Control Network

A BCN F_μ is a function generating a Boolean network from an interpretation $\mu \in S_U$ of control parameters $u_i \in U$, called a *control input*. It is defined as follows:

$$F_\mu = \{x_i = f_i(x_1, \ldots, x_n, u_1, \ldots, u_m\} \mid 1 \le i \le n\}.$$

Each application of a control input F_μ leads to a Boolean network with a particular dynamics.

The freezing control assigns a definite value to each variable. The two possible freezing outcomes, 0 or 1, are supported by two parameters with two distinct regimes: either they freeze the variable or remain idle. By convention, inspired by

---

[1] $\longrightarrow^*$ is the reflexive and transitive closure of the transition relation.

the freezing temperature of water $0\,^{\circ}\mathrm{C}$, the freezing action is triggered when the control parameter is set to 0 whereas 1 stands for the idle situation. The implementation of the freezing control on a Boolean network augments the formulas of a Boolean network by adding the control parameter to obtain the expected control behavior. For a formula f_i, the addition of the control parameters $u_i^0 \in U^0$ and $u_i^1 \in U^1$ for respectively freezing the variable x_i to 0 or 1 leads to the following specification:

$$x_i = f_i(x_1, \ldots, x_n) \wedge u_i^0 \qquad \text{for freezing to 0,} \qquad (2)$$

$$x_i = f_i(x_1, \ldots, x_n) \vee \neg u_i^1 \qquad \text{for freezing to 1.} \qquad (3)$$

U^0 and U^1 control parameters can be combined to trigger the freeze to different values (i.e., $x_i = f_i(x_1, \ldots, x_n) \wedge u_i^0 \vee \neg u_i^1$). In the sequel $U = U^0 \cup U^1$ will represent the set of indiscriminate freezing control parameters, and $u_i \in U$ a generic freezing control parameter (u_i^0 or u_i^1). The model can be extended to arc freezing [2,3]. Figure 1 depicts the application of a control to variable x_1. Three different dynamics respectively corresponding to the absence of control, the freeze of variable x_1 to 1 ($d_1^1 = 0$), and the freeze to 0 ($d_1^0 = 0$) are shown. The dynamics changes by the application of a control and leads to different equilibria.

The *active control* set of a control input, $\dot{\mu}$, represents the set collecting all the activated controls: $\dot{\mu} = \{u \mid \mu(u) = 0\}$. Notice, that μ and $\dot{\mu}$ are equivalent descriptions of the control since we can define one from the other one.

It is worth noticing that some variables are purposely uncontrolled to play the role of observers used for freely reporting the state evolution of a system. In biology, these observers play the role of biomarkers. Therefore, the uncontrolled variables are important for assessing the fate of the dynamical system. The set of controlled variables is denoted C_X and the set of uncontrolled variables is $\bar{C}_X = X \setminus C_X$. Throughout the article, the profile of uncontrolled variables is termed "$\bar{C}_X$−profile", and "C_X−profile" for controlled variables.

2.3 Control Sequence Dynamics

The *controlled dynamics* extends the Boolean network dynamics by showing how the system evolves through a sequence of control inputs. A sequence of control is formally defined by a function $\mu : \mathbb{N}^+ \to (U \to \mathbb{B})$ indexing control inputs, where $\mu_i, i \geq 1$, is the i-th control input in the sequence and $\mu_{[k]}$ stands for the sequence of size k starting from μ_1 and ending in μ_k.

Controlled Dynamics Definition. Given a Boolean control network F_μ, the model of controlled dynamics is defined as a labeled transition system including the control inputs as labels $\langle S_X, S_U \times M, \longrightarrow \rangle$ such that a transition is defined by:

$$s \xrightarrow{\mu_i, m} s' \overset{\text{def}}{=\joinrel=} s' = (F_{\mu_i})_{\downarrow m}(s) \cup s'_{\downarrow -m}, \qquad (4)$$

leading to the following trajectory (path) from a control sequence $\mu_{[k]}$:

$$s \xrightarrow{\mu_1, m_1} \ldots s_i \xrightarrow{\mu_i, m_i} s_{i+1} \ldots s_k \xrightarrow{\mu_k, m_k} s_{k+1}.$$

$$F = \begin{cases} x_1 = (x_1 \wedge \neg x_2) \vee (x_1 \wedge \neg x_3) \vee (\neg x_1 \wedge x_2 \wedge x_3) \\ x_2 = (x_1 \wedge x_3) \vee (\neg x_1 \wedge x_2) \\ x_3 = (x_1 \wedge x_3) \vee (\neg x_1 \wedge \neg x_2) \end{cases}$$

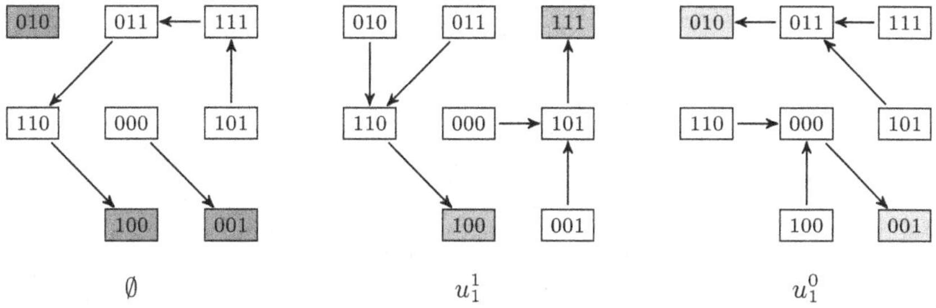

Legend : The Boolean network F is completed by the formulas of the freezing controls to produce the equivalent BCN. From left to right the respective controls are: no freeze, x_1 is frozen to 1, x_1 is frozen to 0. The active control parameter are mentioned below each dynamics. The dynamics is synchronous and the self loops on states are removed. The stable states of each dynamics are coloured in 3 shades of gray where each one is associated to a different control input.

Fig. 1. The synchronous dynamics of a Boolean control network.

For the sake of clarity we omit the mode if it is not needed for the explanation[2]. For example, the sequential application of the different controls described in Fig. 1 leads to the following trajectory in the controlled dynamics by starting at state 000. The control inputs are represented by their active control set and the stable states traversed by the trajectory are in bold face.

$$000 \xrightarrow{\emptyset} \mathbf{001} \xrightarrow{\{u_1^1\}} 101 \xrightarrow{\{u_1^1\}} \mathbf{111} \xrightarrow{\{u_1^0\}} 011 \xrightarrow{\{u_1^0\}} \mathbf{010}. \tag{5}$$

Although no paths connect 000 to 010 initially, the controlled trajectory enables the creation of a path between these two states by the successive application of the controls u_1^1 and u_1^0.

State Trace. The trace defines the sequence of visited states $(s_i)_{1 \leq i \leq k+1}$. For the example of Fig. 1, the state trace of Trajectory (5) is:

$$(000, \mathbf{001}, 101, \mathbf{111}, 011, \mathbf{010})$$

Control Evolving Based on Stable-State Dynamics. The model of controlled dynamics is said to be *Control Evolving based on Stable-State dynamics* (ConEvs) if the modification of the control only occurs at a stable state of the previous Boolean network dynamics. Hence the ConEvs dynamics fulfills the following property:

---

[2] Formally, we consider the relation with the same control input: $\xrightarrow{\mu} = \bigcup_{m \in M} \xrightarrow{\mu, m}$.

$$\forall s_1 \xrightarrow{\mu_{[k]}} s_{k+1} : \mu_i \neq \mu_{i+1} \iff \text{STBL}_{F_{\mu_i}}(s_{i+1}),$$

$$\text{given that } s_i \xrightarrow{\mu_i} s_{i+1}, 1 \leq i \leq k. \quad (6)$$

In ConEvs dynamics, changing the control is the only way to evolve the dynamics since a stable state is reached with the current instance of the Boolean network resulting from the application of a control input to the BCN. The trajectory described in (5) is ConEvs.

Contracted Control Sequence. The *contracted control sequence* keeps only one instance of the control input for each sub-sequence having identical control inputs. For ConEvs dynamics, the contracted control sequence can be somehow considered as the sequence making the dynamics evolve from stable states to stable states, and the initial control sequence can be easily retrieved by connecting the encountered stable states for each F_{μ_i} by a path and applying the same control for this path. In the case of example (5), the contracted sequence represented by the active controls is: $(\emptyset, \{u_1^1\}, \{u_1^0\})$.

Classes of Sequences. The control sequences can be categorized into families based on the evolution of the control between steps:

1. *Total Control Sequence* (TCS): all the controls are triggered during the first phase for controlled variables and remain active all along the sequence, possibly changing the values to which the variables are frozen.
2. *Abiding Control Sequence* (ACS): once a control on a variable is triggered, the variable stays controlled but the freezing nature may possibly vary.
3. *Opened Control Sequence* (OCS): no constraints on control parameters are imposed. Therefore a control can be changed or released freely.

The sequence described in (5) is an ACS sequence and, starting with the state 001, it is a TCS sequence since only x_1 is controlled. The OCS family corresponds to the largest class of control sequences including ACS, which in its turn includes TCS. Therefore, the following inclusions between these families hold:

$$\text{TCS} \subset \text{ACS} \subset \text{OCS}.$$

The TCS class is mainly used for proofs with no specific biological application. The ACS class models the consequences of the disease as mutations forbidding the relaxation of the control definitively while enabling the possibility to change it or not according to the context of the biological process. The OCS class is the most general class that may represent the action of the drugs on molecular network potentially implying the modification and the relaxation of the actions.

3   Control Sequence Discovery

Finding a control sequence altering the dynamics to evolve towards an expected state is a major challenge that can be defined as a reachability problem stated as follows:

Let $S_\alpha, S_\omega \subseteq S_X$ be two set of states, can we find a control sequence:

$$\mu_{[k]} = (\mu_1, \ldots, \mu_k) \text{ such that there exists a path } s_1 \xrightarrow{\mu_{[k]}} s_{k+1},$$

with: $s_1 \in S_\alpha$ and $s_{k+1} \in S_\omega$?

We refer to this problem as *"Controlled Fate in Sequence"* (CoFaSe) problem. For the example (Fig. 1), the controlled variable is $C_X = \{x_1\}$ and the uncontrolled ones are $\bar{C}_X = \{x_2, x_3\}$. The set of initial states is $S_\alpha = \{000\}$ and the set of final states is $S_\omega = \{010, 110\}$, corresponding to the states where $x_2 = 1$ and $x_3 = 0$.

In biological modelling, the outcome of reprogramming can be formulated as a condition on the biomarkers checking whether the system has reached an expected signature. Therefore, by considering that the biomarkers are represented by the uncontrolled variables, S_ω will be defined by a predicate p formalizing the expected biological property as follows: $S_\omega = \{s \in S_X \mid p(s_{\downarrow \bar{C}_X})\}$. Notice that achieving a given state of controlled variables is trivial and consists in merely assigning their expected values by setting the appropriate control inputs. Therefore, the main problem lies on the way to indirectly influence the state variation of the uncontrolled variables by freezing actions on controlled variables.

A sequence $\mu_{[k]}$ is said *minimal* for the CoFaSe problem with respect to F_u, S_α, and S_ω if no control sequences $\nu_{[l]}$ satisfying the problem have a lower cardinality: $l < k$. A sequence $\mu_{[k]}$ is said *parsimonious* if the number of activated controls is minimal to achieve the expected transition $s_i \xrightarrow{\mu_i} s_{i+1}$ for each control input $\mu_i, 1 \leq i \leq k$. Applied to ConEvs dynamics, the problem also imposes that the states appearing S_ω should be stable. The contracted control sequence $(\{u_1^1\}, \{u_1^0\})$ of the example (Fig. 1) is minimal, parsimonious, and complies to the ConEvs condition for S_ω.

3.1 Complexity of CoFaSe

Finding a single parsimonious control is known to be NP-complete [2]. In this section we show that the inference of a control sequence satisfying CoFaSe is PSPACE-hard, which makes this problem even less tractable than finding single controls (assuming that PSPACE $\neq$ NP). Since the freezing to 0 and to 1 cannot be triggered simultaneously for a single variable, the cardinality of possible controlled transitions from a state is $3^{|X|} \cdot |M|$, meaning that finding the control sequence by exhaustively exploring these spaces is not practically tractable.

The problem of reachability in Boolean networks working in synchronous mode can actually be formalized as a CoFaSe problem. Indeed, reachability on a Boolean network is precisely the CoFaSe problem for a Boolean control network without controlled variables. Proof of Lemma 1 shows however that this reduction is not merely an artefact specific for such degenerate BCN. We can construct a network with a non-empty set of control variables and reduce the CoFaSe problem for this network to a reachability problem for a standard Boolean network.

Lemma 1. *Deciding whether a control sequence exists for the CoFaSe problem in the synchronous mode is at least as hard as reachability in (uncontrolled) Boolean networks in synchronous mode.*

The reachability hardness of CoFaSe is based on the result stating the PSPACE-completeness of reachability in 1-safe Petri nets [6], applied to Boolean networks with the *asynchronous* mode. In the appendix of this paper, we provide an extension of this result to the synchronous mode. As far as we know, no characterization of complexity of reachability in Boolean networks working in the synchronous mode has been established in the literature before.

Theorem 1. *Given a Boolean network F with the variables X, a set of starting states $S_\alpha \subseteq S_X$, and the set of target states $S_\omega \subseteq S_X \setminus S_\alpha$, it is PSPACE-complete to decide whether F can reach any of the states in S_ω from one of the states in S_α.*

This theorem, combined with Lemma 1, gives the following lower bound on the complexity of CoFaSe.

Theorem 2. *Deciding the existence of a control sequence for the CoFaSe problem in the synchronous mode is PSPACE-hard.*

Whether CoFaSe is in PSPACE remains an open question. Two levels of complexity can be considered in CoFaSe: reachability and control discovery, indicating that the upper bound on the complexity of this problem may be high.

3.2   Bounds on Sequence Size

The properties related to the equivalences of classes of sequences enable us to define an upper bound on the length of the minimal control sequence providing relevant insights for the resolution of the CoFaSe problem.

Proposition 1 states the observational equivalence between OCS and TCS classes, namely, that any OCS control sequence can be simulated by a TCS sequence having the same state trace.

Proposition 1. *For any control sequence $\mu_{[k]}$ there exists a total control sequence of the same size $\nu_{[k]} \in TCS$ generating the same state trace by using the synchronous mode.*

As a TCS sequence is an ACS sequence by definition, we can conclude from Proposition 1 that all the OCS sequences solving a given CoFaSe problem can be simulated with ACS/TCS sequences, and conversely, since OCS includes ACS which in turn includes TCS. Therefore, the three categories of sequences have the same expressive capabilities to characterize the CoFaSe problem solutions. Hence, the use of a particular class of sequences does not interfere with the capability of solving the CoFaSe problem.

The necessity to find an exact solution precludes the use of approximation heuristic. Thereby, we are looking for factors that would significantly reduce the search space in practice. Theorem 3 defines an upper bound of the size of minimal sequence that only depends on the number of uncontrolled variables.

Theorem 3. *The size of the minimal control sequence $\mu_{[k]}$ solving CoFaSe problem is bounded by $2^{|\bar{C}_X|}$ ($|\mu_{[k]}| \leq 2^{|\bar{C}_X|}$) for the synchronous mode.*

Theorem 3 shows the critical role of the uncontrolled variables for the definition of the control sequence. In practice, the number of uncontrolled variables standing for biomarkers is still very low compared to the controlled variables that represent the other molecules of a network [2,4,5]. Moreover by the definition of the CoFaSe problem, the evolution of the uncontrolled variables guides the discovery of the control since the objective is to reach an expected final state characterized by a property defined on the uncontrolled variables. Thus, the algorithm will be based on the exploration of the states of uncontrolled variables to incrementally construct a control sequence.

3.3   Bounds on Sequence Size for ConEvs Dynamics

Finding a minimal parsimonious contracted OCS solving the CoFaSe problem under the ConEvs dynamics appears relevant for biological applications. Indeed, this framework models either the different mutational steps where a mutation rewires the network reaching another fate as it is the case for the Vogelstein sequence [9], or a therapeutic regimen where the drug administering depends on the therapeutic evaluation modelled by a stable state assessment [11].

In fact, determining an upper bound on the size of such sequences provides relevant insights to design an efficient algorithm for the control sequence inference based on the intelligent exploration of the sequence space. Theorem 3 is proved for the case where the control is changed at any state, thus this upper bound is not directly applicable in the ConEvs case, where the control changes are allowed at stable states only.

A similar upper bound for ConEvs is established that allows us to design an algorithm for solving the CoFaSe problem under the ConEvs dynamics with a reasonable running time for biological applications with a limited number of biomarkers (see Subsect. 3.4).

Theorem 4. *The size of the minimal contracted control sequence $\mu_{[k]} \in OCS$ solving the CoFaSe problem (F, S_α, S_ω) for the ConEvs model of dynamics under the synchronous mode is at most $2^{|\bar{C}_X|+1}$:*

$$|\mu_{[k]}| \leq 2^{|\bar{C}_X|+1}.$$

Theorem 4 implies the possibility of the occurrence of states with the same $\bar{C}_X$ part in the contracted trace, called *duplicates*. The proof of the theorem also entails that duplicates appear in two successive states at most, except for the first state of the trace. Intuitively, if duplicate $\bar{C}_X$ profiles appear in non-successive steps i and j, the whole evolution between i and j can be skipped by applying an appropriate control input.

Corollary 1. *Consider a minimal contracted control sequence $\mu_{[k]} \in OCS$ solving the CoFaSe problem (F, S_α, S_ω) for the ConEvs model of dynamics under the* *synchronous mode. Take the sequence $\tau = (s_i)_{1 \le i \le k+1}$ of states induced by $\mu_{[k]}$, with $s_1 \in S_\alpha$, $s_{k+1} \in S_\omega$, and the states s_i, $1 < i < k+1$, being the stable states at which the control is changed. If there exist two indices $1 < i < j < k+1$ such that $s_{i \downarrow \bar{C}_X} = s_{j \downarrow \bar{C}_X}$, then $j = i+1$.*

By setting $S_\alpha = \{010\}$, $\bar{C}_X = \{x_1\}$, and $S_w = \{1\star\star\}$ as CoFaSe parameters, the following trajectory for example Fig. 1, controlled by the minimal parsimonious contracted control sequence $(\{u_2^0\}, \{u_2^1\})$, contains a duplicate, $x_1 = 0$, occurring in the initial state 010 and the stable state 001.

$$010 \xrightarrow{\{u_2^0\}} 000 \xrightarrow{\{u_2^0\}} \mathbf{001} \xrightarrow{\{u_2^1\}} 011 \xrightarrow{\{u_2^1\}} \mathbf{110}.$$

3.4 Inference of Minimal Parsimonious Contracted Control Sequences

The algorithm infers all minimal parsimonious control sequences solving the CoFaSe problem for the ConEvs model of dynamics. Hence, the algorithm will find a sequence of controls evolving from stable states to stable states. By convention motivated by the biological application, we assume the property on the expected final states to only concern the uncontrolled variables.

As the number of uncontrolled variables is in practice markedly lower than the number of controlled variables (*e.g.*, [2,4,5]), the exhaustive exploration of all possible profiles for these variables constitutes an efficient approach for control sequence computation. Furthermore, our algorithm is optimized to avoid redundant operations. Informally the algorithm builds a tree describing the possible paths from the initial states reaching a state of S_w, where a node corresponds to a set of states having the same $\bar{C}_X$–profiles. The shortest paths/trajectories are found in the tree, from which the minimal parsimonious control sequences are directly derived by keeping their control inputs.

Phases of Algorithm. Algorithm 1 comprises two major phases. The first phase (steps 1 and 2) corresponds to the search for a control allowing to directly attain a state of S_w. The second phase (steps 3 and 4) corresponds to searching a trajectory visiting the intermediary states. Moreover, at each step the parsimonious control input is inferred with the method presented in [2,3]. A detailed version of the algorithm is given in Appendix. The evolution of the main steps is detailed in Fig. 2.

Data Structures. The algorithm relies on the following data structures:

1. *the list Δ* of partial states over $\bar{C}_X$ induced by the candidate controlled sequences;
2. *the exploration tree G* with nodes labelled by sets of stable states and edges labelled by controls;
3. *the sets Γ_l, Γ_{l+1}, and Γ_{l+2}* of unexplored nodes of the tree for the current level of depth l, and the next two levels, respectively.

At the beginning, Δ is initialized to contain the partial states that do not appear in $S_{\omega \downarrow \bar{C}_X}$, Γ_l contains the root node $\{S_\alpha\}$ of the exploration tree, and all the other data structures are empty.

Algorithm 1. Inference of minimal parsimonious contracted control sequences

---

1. *Direct reachability of S_ω:* For all $\gamma \in \Gamma_l$, infer the control μ taking the BCN from γ to some of the target $\bar{C}_X$−profiles appearing in $S_{\omega \downarrow \bar{C}_X}$. If such a μ exists, add the arc labelled by μ to G and go to step 6.

2. *Reachability of S_ω via a duplicate:* For all $\gamma \in \Gamma_l$, infer a pair of controls (μ, μ') such that μ takes the BCN to some states having $\bar{C}_X$−profiles from $\gamma_{\downarrow \bar{C}_X}$, and μ' takes the BCN from there to some of the target $\bar{C}_X$−profiles. If such a pair of controls exists, add two chained arcs labelled by μ and μ' to G and go to step 6.

3. *Direct reachability of Δ:* For every $\gamma \in \Gamma_l$, infer a set of controls $\mathcal{U}$ taking the BCN from γ to some of the $\bar{C}_X$−profiles appearing in Δ. If $\mathcal{U}$ is non-empty, add the arcs labelled by the controls from $\mathcal{U}$ to G, and store the sets of stable states they allow to reach in Γ_{l+1}.

4. *Reachability of Δ via a duplicate:* For every $\gamma \in \Gamma_l$, infer a set of pairs of controls $\mathcal{D} = \{(\mu, \mu') \mid \mu, \mu' \in S_U\}$ such that μ takes the BCN to some states having $\bar{C}_X$−profiles from $\gamma_{\downarrow \bar{C}_X}$, and μ' takes the BCN from there to some of the profiles in $\Delta' \subset \Delta$ which we could not be directly reached at the previous step. If $\mathcal{D}$ is not empty, add chained arcs labelled by the pairs of controls from $\mathcal{D}$ to G, and store the sets of stable states they allow to reach in Γ_{l+2}.

5. *Continue if states left:* If one of Γ_l, Γ_{l+1}, or Γ_{l+2} is non-empty, go to step 1.

6. *Produce the result:* Find the sequence of controls by backtracking G from a leaf found in steps 1 or 2 to the root S_α. If no such leaf was found, return $\emptyset$.

Duplicated States. A specific treatment is applied to take into account the case where a trajectory passes through duplicates with the same $\bar{C}_X$−profiles. Hence, only the states of the controlled variables are modified without necessary freezing all of them. Therefore their profile varies and must be assessed by the algorithm.

The algorithm first infers the set of parsimonious control inputs validating the following equation where γ is the initial set of states:

$$\exists s' : s \xrightarrow{\mu}{}^* s' \land \mathrm{STBL}_{F_\mu}(s') \land s'_{\downarrow \bar{C}_X} \subseteq \gamma_{\downarrow \bar{C}_X} \land s' \notin \gamma. \tag{7}$$

From the situation formalized in (7), the new set of stable states γ', where the C_X−profile is modified, represents the stable states of F_μ reachable from γ. Subsequently, a parsimonious control input μ' is inferred such that the set of stable states $S_{\mu'}$ of $F_{\mu'}$ reachable from γ' contains some elements satisfying the property p. We prove that at most 2 successive duplicates may occur in any given sequence, except for the initial state, in which case the number of repetitions may reach 3 (Corollary 1).

Correctness. Algorithm 1 closely follows the proofs of Theorem 4 and of Corollary 1, which guarantees the correctness, and the minimal parsimony of the result. In other words, the sequences found by Algorithm 1 solve the CoFaSe problem for the ConEvs model of dynamics under the synchronous mode and they are minimal and parsimonious.

Theorem 5. *Algorithm 1 returns all minimal parsimonious control sequences $\mu_{[k]}$ solving the CoFaSe problem for the ConEvs model of dynamics under the synchronous mode.*

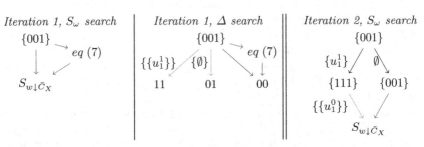

Iteration 1, S_ω search	Iteration 1, Δ search	Iteration 2, S_ω search

Legend : Construction of the tree built by Algorithm 1 to infer the control sequence allowing to reach $S_\omega = \{\star10\}$ for the Boolean control network F_U of Figure 1 from the set of initial states $S_\alpha = \{001\}$. The only controlled variable is $C_X = \{x_1\}$ and the uncontrolled ones are $\bar{C}_X = \{x_2, x_3\}$.

On the left and in the middle are the two phases of the algorithm exploring for $\Gamma_l = \{S_\alpha\}$ the reachability of S_ω and of the partial states of Δ respectively. On the right is the first phase of the second iteration for $\Gamma_l = \{\{111\}, \{001\}\}$ and the final step of the algorithm where it finds the sequential control $\{\{u_1^1\}, \{u_1^0\}\}$ reaching S_ω.

Green edges represent the reachability of the target property (corresponding to S_ω) and are annotated by sets of inferred parsimonious controls. Red edges correspond to a failure of the algorithm for finding a control inputs leading to the target profile. A black edge is a branch of the tree G connecting the previous initial state to the new set of initial states created by the application of the annotated control to F_U.

Fig. 2. Iterations of Algorithm 1 on the Boolean network of Fig. 1.

Example. Figure 2 shows the evolution of the control sequence computation. At the beginning, we have: $\Gamma_l = \{S_\alpha\}, \Gamma_{l+1} = \emptyset, \Gamma_{l+2} = \emptyset$, and $\Delta = \{11, 01, 00\}$. On the left, we seek to reach a state from S_ω. $S_{\omega \downarrow \bar{C}_X}$ being not reachable from S_α in one shot, we search for the existence of a possible trajectory via a duplicate. Since no states validate (7), we know that such a trajectory does not exist.

As no sequences were found, we arrive at the second phase. Here, all states of Δ will be used as the new targets. In the middle, we see that $11 \in \Delta$ is reachable from s_α with the control $\{u_1^1\}$ and $01 \in \Delta$ is reachable from s_α with the control $\emptyset$, but $00 \in \Delta$ is not reachable. We add to Γ_{l+1} the stable states γ'

of $F_{\{u_1^1\}}$ validating the search state $11 \in \Delta$ and γ'' of $F_\emptyset$ validating the search state $01 \in \Delta$, and update Δ by deleting 11. After creating the new edges of the tree, $s_\alpha \xrightarrow{\{u_1^1\}} \gamma'$ and $s_\alpha \xrightarrow{\emptyset} \gamma''$, we test for new trajectories which would attain a state of Δ via a duplicate. Since no states validate (7), we know that such a trajectory does not exist.

On the right, after updating all variables, we start the second iteration of the loop with $\Gamma_l = \{\{111\}, \{001\}\}, \Gamma_{l+1} = \emptyset, \Gamma_{l+2} = \emptyset$, and $\Delta = \{00, 01\}$. We observe that $S_{\omega \downarrow \bar{C}_X}$ is reachable from the initial state $\{111\}$ by the control $\{u_1^0\}$. After creating the new edge, the algorithm breaks the loop and returns the found sequence $\{\{u_1^1\}, \{u_1^0\}\}$ by backtracking the resulting tree.

Parsimony and Optimality. Minimality is proved by considering the parsimony condition for the control inputs at each step. However, it is worth noticing that it may be possible to find shorter control sequences by relaxing the parsimony constraint. Let us consider the following Boolean network:

$$F = \begin{cases} x_1 = x_1 \vee (x_2 \wedge x_3 \wedge \neg x_4) \\ x_2 = (\neg x_1 \wedge x_2) \vee (\neg x_1 \wedge x_4) \vee (x_2 \wedge \neg x_3) \vee (x_2 \wedge x_4) \\ x_3 = (x_1 \wedge x_3) \vee (\neg x_2 \wedge x_3) \vee (x_3 \wedge x_4) \\ x_4 = x_4 \end{cases},$$

with $\bar{C}_X = \{x_1, x_2\}, C_X = \{x_3, x_4\}, S_\alpha = \{0000\}$, and $S_\omega = \{11\star\star\}$ as CoFaSe parameters. The minimal parsimonious control sequence of size 3 inferred by Algorithm 1 is $\mu = \{\{u_4^1\}, \{u_3^1\}, \{u_4^0\}\}$. However, the following sequence of size 2: $\nu = \{\{u_3^1, u_4^1\}, \{u_4^0\}\}$ also converges to the same final state 1100. Their trajectories are respectively:

$$T_\mu = 0000 \xrightarrow{\{u_4^1\}} 0001 \xrightarrow{\{u_4^1\}} 0101 \xrightarrow{\{u_3^1\}} 0111 \xrightarrow{\{u_4^0\}} 0110 \xrightarrow{\{u_4^0\}} 1100.$$
$$T_\nu = 0000 \xrightarrow{\{u_3^1, u_4^1\}} 0011 \xrightarrow{\{u_3^1, u_4^1\}} 0111 \xrightarrow{\{u_4^0\}} 0110 \xrightarrow{\{u_4^0\}} 1100.$$

As the parsimony condition is related to the target property on $\bar{C}_X$-profiles, Algorithm 1 infers a minimal control sequence passing through some intermediary stable states since $\bar{C}_X$-profiles 11 (S_ω) cannot be directly attained. Let us consider $\{x_1 = 0, x_2 = 1\}$ as the intermediary $\bar{C}_X$-profile to reach. The stable state 0101 is then first reached by the single parsimonious freezing of x_4 to 1 before reaching stable state 0111 whereby state 1100 can be finally attained. $\bar{C}_X$-profile 01 is then duplicated suggesting that the sequence does not evolve in the $\bar{C}_X$-profile exploration but moves to a more appropriate stable state. By contrast, the application of a non-parsimonious control $\{u_3^1, u_4^1\}$ gathering the two first controls, directly reaches this state and reduce the size of the control sequence (ν) to 2. As the variations between these states occur for the controlled variables only (x_3), the sequence is here shortened by gathering the controls between duplicated $\bar{C}_X$-profiles.

4 Conclusion

In this article, we study the sequential control applied to Boolean networks. We propose a theoretical framework aiming at discovering minimal control sequences evolving from stable states to stable states (ConEvs model of dynamics) for parsimonious control actions at each step. Algorithm 1 infers the minimal parsimonious control sequence and runs in time which is exponential in the number of $\bar{C}_X$-variables ($\mathcal{O}(2^{|\bar{C}_X|})$). The inference of control sequences can be used in precision medicine for the causal analysis of such complex diseases as cancer [9], stressing the evolution of the tumorigenesis from benign to malignant tumoral states. The control sequence inference can also be used to determine efficacious order of drug administering for chemotherapy [11].

Moreover, our analysis emphasizes non-obvious complex features of the sequence, such as the occurrence of duplicates where only the controlled variables evolve without changing the states of uncontrolled variables. Such occurrences are interpreted as the need to evolve to different stable states with the same profiles for uncontrolled variables, but which could not be reached previously.

The perspective is twofold: applying the method to biological cases for investigating complex treatment schemes, more specifically for cancer, and investigating an algorithm guaranteeing the minimality in a broader context by notably relaxing the parsimony requirement and by considering other modes as the asynchronous one. The synchronous mode enables an efficient computation of the reachability condition in practice by a symbolic composition since the transition is a function. For the asynchronous mode we need to tackle the fact that the transition becomes a relation inducing non-determinism that should not be exhaustively explored for ensuring the efficiency of the algorithm.

Appendix

Proofs

Proof (Lemma 1). Take an n-variable Boolean network F and construct a Boolean control network F' by adding to F the single control variable x_0 and defining the update functions f_i' of F' in terms of the update functions f_i of F in the following way:

$$f_i' = f_i \wedge x_0, 1 \le i \le n,$$
$$f_0' = 0,$$

where f_0' is the update function for x_0.

Consider the controls $\mu_1 = d_0^1$ and $\mu_0 = d_0^0$ controlling x_0 to 1 and 0 respectively. The previous two properties ensure that the state graph of F_{μ_1}' is that of F, with $x_0 = 1$ added to each state, and that the state graph F_{μ_0}' only contains transitions to the state $\mathbf{0}$, which is the state in which all variables are 0.

Let X be the set of variables of F. The set of variables of F' is thus $X' = X \cup \{x_0\}$. Fix a set of starting states $S_\alpha \subseteq S_{X'}$ and a set of target states $S_\omega \subseteq S_{X'} \backslash (S_\alpha \cup \{\mathbf{0}\})$, such that the states in both sets satisfy $x_0 = 1$. The CoFaSe problem for the tuple (F', S_α, S_ω) has a solution if and only if the states in $S_{\omega \downarrow X}$ are reachable from $S_{\alpha \downarrow X}$ in F. Indeed, by construction of F' and since $\mathbf{0} \notin S_\omega$, the control sequence for this instance of CoFaSe may only be the singleton control sequence consisting of μ_1, and it must ensure the reachability of S_ω from S_α in F_{μ_1}, whose state graph is trivially isomorphic to that of F. Therefore, an oracle for CoFaSe would allow to solve reachability in Boolean networks working in the synchronous mode with at most polynomial overhead, which proves the statement of the lemma. $\qquad\square$

Lemma 2. *Given a Boolean network F with the variables X, a set of starting states $S_\alpha \subseteq S_X$, the set of target states $S_\omega \subseteq S_X \setminus S_\alpha$, it is PSPACE-hard to decide whether F can reach any of the states in S_ω from a state in S_α.*

Proof. The proof idea is to polynomial-time reduce the acceptance problem of a deterministic linear bounded automaton (a DLBA) to reachability for Boolean networks working in asynchronous mode. An LBA is a Turing machine which is only allowed to use at most $f(n)$ contiguous tape cells, where n is the size of the input and f is a linear function. Deciding whether a DLBA accepts a given input string is a PSPACE-complete problem (e.g., [10]).

Take a DLBA M and construct the Boolean network F simulating M in the following way. Define the Boolean variables $A_{i,j}$ and $Q_{i,k}$, where i indexes the tape cells of M, j indexes the symbols in the tape alphabet of M, and k indexes the states of M. The situation in which the i-th tape cell contains the j-th symbol is represented by setting $A_{i,j}$ to 1. The situation in which M is in the k-th state and the head is on the i-th tape cell is represented by setting $Q_{i,k}$ to 1. F operates by stepwise simulating the evolution of M: rewriting the j_1-th symbol to the j_2-th symbol in the i-th tape cell is done by setting A_{i,j_1} to 0 and A_{i,j_0} to 1, while moving the head from cell i_1 to i_2 and changing the state from k_1 to k_2 is simulated by setting Q_{i_1,k_1} to 0 and Q_{i_2,k_2} to 1. The synchronous dynamics of F can therefore faithfully simulate M, because M is deterministic.

For any input word w, the DLBA M reaches a configuration in the set of accepting configurations C_A if and only if F can reach the encoding of one of the configurations C_A from the encoding of the initial configuration of M. The statement of the lemma follows from the facts that the procedure of constructing F from M is polynomial and that acceptance for DLBA is PSPACE-complete. $\qquad\square$

Lemma 3. *Given a Boolean network F with the variables X, a set of starting states $S_\alpha \subseteq S_X$, the set of target states $S_\omega \subseteq S_X \setminus S_\alpha$, it is in PSPACE to decide whether F can reach any of the states in S_ω from one of the states in S_α.*

Proof. The proof idea is to construct a DLBA M which accepts the input if and only if the Boolean network F can reach a state in S_ω from a state in S_α. The initial configuration of M consists of the following three segments:

1. the list of binary vectors representing the states in S_α, each vector written in two copies;

2. the list of binary vectors representing the states in S_ω, each vector written in one copy;
3. an $|X|$-bit binary counter initialised to 0, where $|X|$ is the number of binary variables of F.

In the remainder of the proof we implicitly assume that the states of F are represented as Boolean vectors.

Consider a state $s \in S_\alpha$. The initial configuration of M contains a substring ss. M starts by simulating the transitions of F on one copy of s and replacing the other copy by the new state $s' = F(s)$, thereby yielding the new substring ss'. The subsequent operation of M is divided into macrosteps, during which it carries out the following actions:

1. calculate the new state for each pair of states in segment (1);
2. compare each new state with the states written in segment (2); if one of these comparisons is successful, M accepts, otherwise its continues to the following substep;
3. check if all the bits of the binary counter in segment (3) are 1; if yes, reject, otherwise, commence the next macrostep.

Intuitively, M simulates the deterministic synchronous dynamics of F on every state in segment (1), accepts if it sees a target state from S_ω, or rejects after $2^{|X|}$ steps. Counting to $2^{|X|} = |S_X|$ ensures that the entire state graph of F reachable from S_α is visited. Therefore, M accepts if and only if F can reach at least one state in S_ω from at least one state in S_α. Constructing M from the triple (F, S_α, S_ω) is a polynomial-time procedure, meaning that an oracle for DLBA acceptance would allow deciding reachability for Boolean networks working in the synchronous mode with polynomial overhead. This proves the statement of the theorem.   □

Proof (Proposition 1). Take a control sequence $\mu_{[k]}$ and an initial state. For a transition $s_i \xrightarrow{\mu_i} s_{i+1}, 1 \leq i \leq k$, two cases may occur for the control parameters of the controlled variables, $x_j \in C_X$.:

1. If one of the two control parameters u_j^0, u_j^1 is already activated then the configuration remains the same for ν.
2. If the control parameters are both idle ($u_j^0 = 1, u_j^1 = 1$) then we directly fix the expected final state value by setting the control appropriately, namely:
   $\nu_i(u_j^0) = 0, \nu_i(u_j^1) = 1$ if $s_{i+1}(x_j) = 0$ and,
   $\nu_i(u_j^0) = 1, \nu_i(u_j^1) = 0$ if $s_{i+1}(x_j) = 1$.

As the update is synchronous then all the value of the controlled variables, x_j, leads to the state $s_{i+1}(x_j)$ in a controlled way. For uncontrolled variables $x_j \in \bar{C}_X$, we have: $(f_\nu)_j = (f_\mu)_j$ since no modifications occur, then the update is the same.

Since transition only depends on the previous state that can be obtained by application of a TCS control input ν_i, we can define a TCS control input for each step finally leading to a total controlled sequence $\nu_{[k]}$ simulating the trajectory controlled by $\mu_{[k]}$ from a s_1.   □

Proof (Theorem 3). Assume that $\mu_{[k]}$ is a minimal sequence solving CoFaSe problem with regards to $F_u, S\alpha, S_\omega$ and consider the total control sequence $\nu_{[k]}$ simulating $\mu_{[k]}$ from s_1 (Proposition 1). As $\mu_{[k]}$ is minimal by assumption then $\nu_{[k]}$ is also minimal, yielding the following path: $s_1 \xrightarrow{\nu_1} s_2 \ldots s_k \xrightarrow{\nu_k} s_{k+1}$, with $s_1 \in S_\alpha$ and $s_{k+1} \in S_\omega$.

Now, assume that $k > 2^{|\bar{C}_X|}$, and consider the projection of the state trace onto $\bar{C}_X$: $T = (s_{i\downarrow\bar{C}_X})_{1\leq i\leq k+1}$. As there exists $k+1$ states, we deduce that $|T| > 2^{|\bar{C}_X|}$. Hence the cardinality of T exceeds the cardinality of the set of states of $\bar{C}_X$, $S_{\bar{C}_X}$, thus leading to the existence of two states in the trace $s_i, s_j, 1 < i < j \leq k+1$ such that $s_{i\downarrow\bar{C}_X} = s_{j\downarrow\bar{C}_X}$. Based on this conclusion, we can observe that it is possible to skip all the controls in between step $i-1$ to step $j-1$, thus leading to the following trajectory:

$$s_1 \xrightarrow{\nu_1} s_2 \ldots s_{i-1} \xrightarrow{\nu_{j-1}} s_i' \xrightarrow{\nu_j} s_{i+1}' \xrightarrow{\nu_{j+1}} s_{i+2}' \ldots s_k' \xrightarrow{\nu_k} s_{k+1}'.$$

As the variables in C_X are fully controlled with the TCS sequence ν, we deduce that the application of ν_{j-1} at state s_{i-1} with respect to the synchronous mode gives rise to the same state than its application at state s_{j-1} for these variables. Hence, we have the following equivalence:

$$s_{j\downarrow C_X} = s_{i\downarrow C_X}'.$$

Since no control variations affect the uncontrolled variable by definition, we deduce that: $s_{i\downarrow\bar{C}_X} = s_{i\downarrow\bar{C}_X}'$. Moreover, by hypothesis on i, j we know that $s_{j\downarrow\bar{C}_X} = s_{i\downarrow\bar{C}_X}$, thus leading to the following equivalence by approximating the two former equalities:

$$s_{j\downarrow\bar{C}_X} = s_{i\downarrow\bar{C}_X}'.$$

Hence, from these equalities, we deduce that $s_i' = s_j$ because the state of controlled and uncontrolled variables is the same for s_i', s_j. From s_i' the sequence we conclude that s_{k+1} will be finally reached as this was done from s_j for ν (resp. μ). The control sequence we have constructed then validates CoFaSe problem since it reaches s_{k+1} and $s_{k+1\downarrow\bar{C}_X} \in S_{\omega\downarrow\bar{C}_X}$ by hypothesis. The size of this sequence is $k - (j - i) < k$, meaning that $\nu_{[k]}$ (resp. $\mu_{[k]}$) is not minimal, that contradicts the original assumption and proving the statement of the theorem. $\qquad\square$

Lemma 4. *Consider a control sequence $\mu_{[k]} \in OCS$ solving the CoFaSe problem (F, S_α, S_ω) for the ConEvs model of dynamics under the synchronous mode. Take the trajectory induced by $\mu_{[k]}$: $T = s_1 \xrightarrow{\mu_1}{}^* s_2 \ldots s_k \xrightarrow{\mu_k}{}^* s_{k+1}$, where $s_1 \in S_\alpha$, $s_{k+1} \in S_\omega$, and the states $s_i, 1 < i < k+1$, are the stable states at which the control is changed. Denote by s_i' the states immediately following s_i in T: $s_i \xrightarrow{\mu_i} s_i'$. The following statement holds:*

$$s_{i\downarrow\bar{C}_X} = s_{i\downarrow\bar{C}_X}', \quad 1 < i < k+1.$$

Proof. Suppose for the sake of contradiction that there exists such an index i, $1 < i < k+1$, for which $s_{i\downarrow\bar{C}_X} \neq s'_{i\downarrow\bar{C}_X}$. Since the controls do not affect the variables in $\bar{C}_X$, the evolution of these variables between s_i and s'_i does not depend on the control applied at s_i. But then s_i cannot be stable, because the partial state of uncontrolled variables evolves from $s_{i\downarrow\bar{C}_X}$ to $s'_{i\downarrow\bar{C}_X}$, whatever the control applied at step i. It must be therefore that $s_{i\downarrow\bar{C}_X} = s'_{i\downarrow\bar{C}_X}$ for all $1 < i < k+1$, which proves the lemma.   □

Notice that the previous lemma does not consider the very first pair of states $s_1 \xrightarrow{\mu_1} s'_1$. Indeed, since we do not require the initial states to be stable, we may have: $s_{1\downarrow\bar{C}_X} \neq s'_{1\downarrow\bar{C}_X}$.

Proof (Theorem 4). Consider the CoFaSe problem $(F_u, S_\alpha, S_\omega)$ and assume that $\mu_{[k]} \in$ OCS is a minimal contracted control sequence solving it for the ConEvs model of dynamics. This control sequence gives rise to the trajectory $T = s_1 \xrightarrow{\mu_1}{}^* s_2 \ldots s_k \xrightarrow{\mu_k}{}^* s_{k+1}$, with $s_1 \in S_\alpha$, $s_{k+1} \in S_\omega$, and the states s_i, $1 < i < k+1$, being the stable states at which the control is changed. We will use the symbol τ to refer to the sequence of stable states, plus the initial and the final states: $\tau = (s_i)_{1 \leq i \leq k+1}$.

Now assume that $k > 2^{|\bar{C}_X|+1}$ and consider the sequence of partial states over $\bar{C}_X$: $\tau_{\downarrow\bar{C}_X} = (s_{i\downarrow\bar{C}_X})_{1 \leq i \leq k+1}$. Since k is greater than the double of the number of all states over $\bar{C}_X$, it must be that $\tau_{\downarrow\bar{C}_X}$ contains an element appearing three times at different positions. Suppose that these three positions are $1 \leq h < i < j \leq k+1$, i.e., $s_{h\downarrow\bar{C}_X} = s_{i\downarrow\bar{C}_X} = s_{j\downarrow\bar{C}_X}$. Note that, at least two different controls must appear between the states s_h and s_j because there is at least the stable state s_i in between the two.

In what follows, we will distinguish two cases based on whether s_h is the first state in the trajectory.

Case $h > 1$: Suppose that, instead of applying μ_h at s_h, we apply the total control $\hat{\mu}_{j-1}$ which sets the values of the controlled variables to $s_{j\downarrow C_X}$: $s_h \xrightarrow{\hat{\mu}_{j-1}} \hat{s}_j$. By construction, $\hat{s}_{j\downarrow C_X} = s_{j\downarrow C_X}$. Furthermore, according to Lemma 4, $\hat{s}_{j\downarrow\bar{C}_X} = s_{h\downarrow\bar{C}_X} = s_{j\downarrow\bar{C}_X}$. But then $\hat{s}_j = s_j$, meaning that we can replace the controls between s_h and s_j (at least two controls) by a single control $\hat{\mu}_{j-1}$, and get a shorter control sequence which still solves the same CoFaSe problem under ConEvs. This contradicts our initial assumption that $\mu_{[k]}$ is minimal.

Case $h = 1$: If the state s'_h immediately following s_h in T ($s_h \xrightarrow{\mu_h} s'_h$) has the same values for uncontrolled variables as s_h, $s_{h\downarrow\bar{C}_X} = s'_{h\downarrow\bar{C}_X}$, we may apply the same contraction as in the case $h > 1$.

Suppose now that $s_{h\downarrow\bar{C}_X} \neq s'_{h\downarrow\bar{C}_X}$. In this case we cannot directly jump from s_h to s_j, because the uncontrolled variables evolve. Nevertheless, one of the following will hold:

1. There is another element $s_{\downarrow\bar{C}_X} \neq s_{h\downarrow\bar{C}_X}$ appearing three times in τ, and we can then apply the contraction from the case $h > 1$ to its repetitions.

Algorithm 2. Inference of minimal parsimonious contracted control sequences

1. For all $\gamma \in \Gamma_l$ do :
   (a) Infer a parsimonious control parameter μ such that the set of stable states S_μ of F_μ, reachable from the set of states γ, has the property $S_{\omega \downarrow \bar{C}_X} \cap S_{\mu \downarrow \bar{C}_X} \neq \emptyset$.
   (b) If such a μ exists, a sequence of control is found, add to G the new arc $\gamma \xrightarrow{\mu} \gamma'$, where $\gamma' \subseteq S_\mu$ are the stable states of F_μ reachable from γ satisfying $\gamma'_{\downarrow \bar{C}_X} \subseteq S_{\omega \downarrow \bar{C}_X}$. Go to 8.
   Otherwise continue.

2. The states of S_ω not being attainable from Γ_l in one shot, we must test the existence of successive duplicate.
   For all $\gamma \in \Gamma_l$ do :
   (a) Infer a parsimonious pair of control parameter (μ, μ') such that the set of stable states S_μ of F_μ, reachable from the set of states γ, has the property $\gamma_{\downarrow \bar{C}_X} \cap S_{\mu \downarrow \bar{C}_X} \neq \emptyset$ and the set of stable states $S_{\mu'}$ of $F_{\mu'}$, reachable from the set of states γ', has the property $S_{\omega \downarrow \bar{C}_X} \cap S_{\mu' \downarrow \bar{C}_X} \neq \emptyset$, where $\gamma' \subseteq S_\mu$ are the stable states of F_μ reachable from γ satisfying $\gamma'_{\downarrow \bar{C}_X} \subseteq \gamma_{\downarrow \bar{C}_X}$.
   (b) If such a pair exists, a sequence of control is found, add to G the new arc $\gamma \xrightarrow{\mu} \gamma' \xrightarrow{\mu'} \gamma''$, where $\gamma'' \subseteq S_{\mu'}$ are the stable states of $F_{\mu'}$ reachable from γ' satisfying $\gamma''_{\downarrow \bar{C}_X} \subseteq S_{\omega \downarrow \bar{C}_X}$. Go to 8.
   Otherwise continue.

3. The states of S_ω not being attainable from Γ_l, we must explore the states of Δ.
   For all $\gamma \in \Gamma_l$ do : For all $\delta \in \Delta$ do :
   (a) Infer the list of parsimonious control parameters $\mathcal{U}$ such that, for every $\mu \in \mathcal{U}$, the set of stable states S_μ of F_μ which can be reached from γ has the property $\delta \in S_{\mu \downarrow \bar{C}_X}$.
   (b) If a control validating the property exists ($\mathcal{U} \neq \emptyset$), for each $\mu \in \mathcal{U}$, add γ' to Γ_{l+1} and create a new arc $\gamma \xrightarrow{\mu} \gamma'$ to G, where $\gamma' \subseteq S_\mu$ are the stable states of F_μ reachable from γ satisfying $\gamma'_{\downarrow \bar{C}_X} \subseteq \delta$.

4. For every $\gamma \in \Gamma_{l+1}$, if $\gamma_{\downarrow \bar{C}_X} \subseteq \Delta$, delete $\gamma_{\downarrow \bar{C}_X}$ from Δ .

5. For all $\delta \in \Delta$ not being attainable from Γ_l, we must test the existence of a successive duplicates.
   For all $\gamma \in \Gamma_l$ do : For all $\delta \in \Delta$ do :
   (a) Infer the list of parsimonious pair of control parameters $\mathcal{D}$ such that, for every $(\mu, \mu') \in \mathcal{D}$, the set of stable states S_μ of F_μ, which can be reached from γ, has the property $\gamma_{\downarrow \bar{C}_X} \cap S_{\mu \downarrow \bar{C}_X} \neq \emptyset$ and the set of stable states $S_{\mu'}$ of $F_{\mu'}$, reachable from the set of states γ', has the property $\delta \in S_{\mu' \downarrow \bar{C}_X}$, where $\gamma' \subseteq S_\mu$ are the stable states of F_μ reachable from γ satisfying $\gamma'_{\downarrow \bar{C}_X} \subseteq \gamma_{\downarrow \bar{C}_X}$.
   (b) If a control validating the property exists ($\mathcal{D} \neq \emptyset$), for each $(\mu, \mu') \in \mathcal{D}$, add γ'' to Γ_{l+2} and create a new arc $\gamma \xrightarrow{\mu} \gamma' \xrightarrow{\mu'} \gamma''$ to G, where $\gamma'' \subseteq S_{\mu'}$ are the stable states of $F_{\mu'}$ reachable from γ' satisfying $\gamma''_{\downarrow \bar{C}_X} \subseteq \delta$.

6. Prepares the next exploration layer. $\Gamma_l = \Gamma_{l+1}$, $\Gamma_{l+1} = \Gamma_{l+2}$ and $\Gamma_{l+2} = \{\}$.

7. If the lists of unexplored nodes Γ_l, Γ_{l+1} and Γ_{l+2} are not empty go to 1.

8. Find the sequence of control by backtracking G, the created tree, from a leaf γ' found in 1b or 2b to the root S_α. If this leaf does not exists, then return $\emptyset$.

---

2. The only element appearing three times in τ is $s_{h\downarrow\bar{C}_X}$. But then, since $|\tau| > 2^{|\bar{C}_X|+1}$, all partial states from $S_{\bar{C}_X} \setminus \{s_{h\downarrow\bar{C}_X}\}$ must appear exactly two times in τ. In particular, there exist two indices $1 < i' < j' < k+1$ such that $s'_{h\downarrow\bar{C}_X} = s_{i'\downarrow\bar{C}_X} = s_{j'\downarrow\bar{C}_X}$. If we now set the values of the controlled variables to $s_{j'\downarrow C_X}$ by applying the total control $\hat{\mu}_{j'-1}$ at s_h, we will end up in the state $s_{j'}$, contracting the sequence $\mu_{[k]}$ just like in the case $h > 1$.   □

It follows from this discussion that any control sequence solving the CoFaSe problem for the ConEvs model of dynamics and having more than $2^{|\bar{C}_X|+1}$ elements is not minimal, that proves the statement of the theorem.   □

Proof (Corollary 1). It follows from the proof of Theorem 4 that, whenever $j > i + 1$, we can use the control $\hat{\mu}_{j-1}$ to force the values of all controllable variables of s_i to $s_{j\downarrow C_X}$ and thus reach s_j from s_i in one step.   □

Proof (Theorem 5). The compliance of $\mu_{[k]}$ with the ConEvs model of dynamics and its minimal parsimony are guaranteed by the use of the parsimonious one-shot inference algorithm from [2,3]. This algorithm yields parsimonious controls allowing to reach stable states satisfying certain properties. The fact that $\mu_{[k]}$ solves the given CoFaSe problem follows from the end condition in step 6.   □

Complete Inference Algorithm
We give here the complete version of the algorithm presented in Sect. 3.4. First off, we provide a detailed presentation of the data structures used by the algorithm together with the values to which they are initialised.

1. $\Delta := S_{X\downarrow\bar{C}_X} \setminus S_{\omega\downarrow\bar{C}_X}$ is a list of potential intermediate partial states, between S_ω and S_α, in the contracted controlled sequence.
2. $G := \emptyset$ is the list of edges of the exploration tree constructed by the algorithm. Each node of this tree is labeled by set of stable states $S_i \in S_X$.
3. $\Gamma_l := \{S_\alpha\}$ is a set of unexplored nodes of the tree at the depth l and is initialize with the root node S_α for the depth 0. With $\Gamma_{l+1} := \{\}$ and $\Gamma_{l+2} := \{\}$ being sets of unexplored nodes of the tree respectively at the depth $l+1$ and the depth $l+2$.

References

1. Barabási, A.-L., Gulbahce, N., Loscalzo, J.: Network medicine: a network-based approach to human disease. Nat. Rev. Genet. **12**, 56–68 (2011)
2. Biane, C., Delaplace, F.: Abduction based drug target discovery using Boolean control network. In: Feret, J., Koeppl, H. (eds.) CMSB 2017. LNCS, vol. 10545, pp. 57–73. Springer, Cham (2017). https://doi.org/10.1007/978-3-319-67471-1_4
3. Biane, C., Delaplace, F.: Causal reasoning on Boolean control networks based on abduction: theory and application to cancer drug discovery. IEEE/ACM Trans. Comput. Biol. Bioinf. (2018). (Epub ahead of print)

4. Burga, L.N., et al.: Loss of BRCA1 leads to an increase in epidermal growth factor receptor expression in mammary epithelial cells, and epidermal growth factor receptor inhibition prevents estrogen receptor-negative cancers in BRCA1-mutant mice. Breast Cancer Res. **13**(2), R30 (2011)
5. Chau, C.H., Rixe, O., McLeod, H., Figg, W.D.: Validation of analytic methods for biomarkers used in drug development. Clin. Cancer Res. **14**(19), 5967–5976 (2008)
6. Cheng, A., Esparza, J., Palsberg, J.: Complexity results for 1-safe nets. Theoret. Comput. Sci. **147**(1), 117–136 (1995)
7. Creixell, P., et al.: Kinome-wide decoding of network-attacking mutations rewiring cancer signaling. Cell **163**(1), 202–217 (2015)
8. Csermely, P., Korcsmàros, T., Kiss, H.J.M., London, G., Nussinov, R.: Structure and dynamics of molecular networks: a novel paradigm of drug discovery: a comprehensive review. Pharmacol. Ther. **138**(3), 333–408 (2013)
9. Fearon, E.R., Vogelstein, B.: A genetic model for colorectal tumorigenesis. Cell **61**(5), 759–767 (1990)
10. Garey, M.R., Johnson, D.S.: Computers and Intractability; A Guide to the Theory of NP-Completeness. W. H. Freeman & Co., New York (1990)
11. Lee, M., et al.: Sequential application of anti-cancer drugs enhances cell death by re-wiring apoptotic signaling networks. Cell **149**, 780–794 (2012)
12. Lin, P.-C.K., Khatri, S.P.: Application of Max-SAT-based ATPG to optimal cancer therapy design. BMC Genom. **13 Suppl 6**(Suppl. 6), S5 (2012)
13. Mandon, H., Haar, S., Paulevé, L.: Temporal reprogramming of Boolean networks. In: Feret, J., Koeppl, H. (eds.) CMSB 2017. LNCS, vol. 10545, pp. 179–195. Springer, Cham (2017). https://doi.org/10.1007/978-3-319-67471-1_11
14. Mandon, H., Su, C., Pang, J., Paul, S., Haar, S., Paulevé, L.: Algorithms for the sequential reprogramming of Boolean networks. IEEE/ACM Trans. Comput. Biol. Bioinf. (2019 to appear)
15. Murrugarra, D., Veliz-Cuba, A., Aguilar, B., Laubenbacher, R.: Identification of control targets in Boolean molecular network models via computational algebra. BMC Syst. Biol. **10**(1), 94 (2016)
16. Sahni, N., et al.: Edgotype: a fundamental link between genotype and phenotype. Curr. Opin. Genet. Dev. **23**(6), 649–657 (2013)
17. Strimbu, K., Tavel, J.A.: What are Biomarkers? Curr. Opin. HIV AIDS **5**(6), 463–466 (2011)
18. Vogel, G.: Reprogramming cells. Science **322**(5909), 1766–1767 (2008)
19. Zanudo, J.G.T., Albert, R.: Cell fate reprogramming by control of intracellular network dynamics. PLoS Comput. Biol. **11**(4), e1004193 (2015)
20. Zhong, Q., et al.: Edgetic perturbation models of human inherited disorders. Mol. Syst. Biol. **5**(321), 321 (2009)

Control Variates for Stochastic Simulation
of Chemical Reaction Networks

Michael Backenköhler[1]([✉]), Luca Bortolussi[1,2], and Verena Wolf[1]

[1] Saarland University, Saarbrücken, Germany
`michael.backenkoehler@uni-saarland.de`
[2] University of Trieste, Trieste, Italy

Abstract. Stochastic simulation is a widely used method for estimating quantities in models of chemical reaction networks where uncertainty plays a crucial role. However, reducing the statistical uncertainty of the corresponding estimators requires the generation of a large number of simulation runs, which is computationally expensive. To reduce the number of necessary runs, we propose a variance reduction technique based on control variates. We exploit constraints on the statistical moments of the stochastic process to reduce the estimators' variances. We develop an algorithm that selects appropriate control variates in an on-line fashion and demonstrate the efficiency of our approach on several case studies.

Keywords: Chemical reaction network · Chemical master equation ·
Stochastic simulation algorithm · Moment equations ·
Control variates · Variance reduction

1 Introduction

Chemical reaction networks that are used to describe cellular processes are often subject to inherent stochasticity. The dynamics of gene expression, for instance, is influenced by single random events (e.g. transcription factor binding) and hence, models that take this randomness into account must monitor discrete molecular counts and reaction events that change these counts. Discrete-state continuous-time Markov chains have successfully been used to describe networks of chemical reactions over time that correspond to the basic events of such processes. The time-evolution of the corresponding probability distribution is given by the chemical master equation, whose numerical solution is extremely challenging because of the enormous size of the underlying state-space.

Analysis approaches based on sampling, such as the Stochastic Simulation Algorithm (SSA) [18], can be applied independent of the size of the model's state-space. However, statistical approaches are costly since a large number of simulation runs is necessary to reduce the statistical inaccuracy of estimators. This problem is particularly severe if reactions occur on multiple time scales

Electronic supplementary material The online version of this chapter (https://doi.org/10.1007/978-3-030-31304-3_3) contains supplementary material, which is available to authorized users.

© Springer Nature Switzerland AG 2019
L. Bortolussi and G. Sanguinetti (Eds.): CMSB 2019, LNBI 11773, pp. 42–59, 2019.
https://doi.org/10.1007/978-3-030-31304-3_3

or if the event of interest is rare. A particularly popular technique to speed up simulations is τ-leaping which applies multiple reactions in one step of the simulation. However, such multi-step simulations rely on certain assumptions about the number of reactions in a certain time interval. These assumptions are typically only approximately fulfilled and therefore introduce approximation errors on top of the statistical uncertainty of the considered point estimators.

Moment-based techniques offer a fast approximation of the statistical moments of the model. The exact moment dynamics can be expressed as an infinite-dimensional system of ODEs, which cannot be directly integrated for a transient analysis. Hence, ad-hoc approximations need to be introduced, expressing higher order moments as functions of lower-order ones [1,13]. However, moment-based approaches rely on assumptions about the dynamics that are often not even approximately fulfilled and may lead to high approximation errors. Recently, equations expressing the moment dynamics have also been used as constraints for parameter estimation [5] and for computing moment bounds using semi-definite programming [12,15].

In this work, we propose a combination of such moment constraints with the SSA approach. Specifically, we interpret these constraints as random variables that are correlated with the estimators of interest usually given as functions of chemical population variables. These constraints can be used as (linear) control variates in order to improve the final estimate and reduce its variance [23,34]. The method is easy on an intuitive level: If a control variate is positively correlated with the function to be estimated then we can use the estimate of the variate to adjust the target estimate.

The incorporation of control variates into the SSA introduces additional simulation costs for the calculation of the constraint values. These values are integrals over time, which we accumulate based on the piece-wise constant trajectories. This introduces a trade-off between the variance reduction that is achieved by using control variates versus the increased simulation cost. This trade-off is expressed as the product of the variance reduction ratio and the cost increase ratio.

For a good trade-off, it is crucial to find an appropriate set of control variates. Here we propose a class of constraints which is parameterized by a moment vector and a weighting parameter, resulting in infinitely many choices. We present an algorithm that samples from the set of all constraints and proceeds to remove constraints that are either only weakly correlated with the target function or are redundant in combination with other constraints.

In a case study, we explore different variants of this algorithm both in terms of generating the initial constraint set and of removing weak or redundant constraints. We find that the algorithm's efficiency is superior to a standard estimation procedure using stochastic simulation alone in almost all cases.

Although in this work we focus on estimating first order moments at fixed time points, the proposed approach can in principle deal with any property that can be expressed in terms of expected values such as probabilities of complex path properties. Another advantage of our technique is that an increased efficiency is achieved without the price of an additional approximation error as it is the case for methods based on moment approximations or multi-step simulations.

This paper is structured as follows. In Sect. 2 we give a brief survey of methods and tools related to efficient stochastic simulation and moment techniques. In Sect. 3 we introduce the common stochastic semantics of chemical reaction networks. From these semantics we show in Sect. 4 how to derive constraints on the moments of the transient distribution. The variance reduction technique of control variates is described in Sect. 5. We show the design of an algorithm using moment constraints to reduce sample variance in Sect. 6. The efficiency and other characteristics of this algorithm are evaluated on four non-trivial case studies in Sect. 7. Finally, we discuss the findings and give possibilities for further work in Sect. 8.

2   Related Work

Much research has been directed at the efficient analysis of stochastic chemical reaction networks. Usually research focuses on improving efficiency by making certain approximations.

If the state-space is finite and small enough one can deal with the underlying Markov chain directly. But there are also cases where the transient distribution has an infinitely large support and one can still deal with explicit state probabilities. To this end, one can fix a finite state-space, that should contain most of the probability [27]. Refinements of the method work dynamically and adjust the state-space according to the transient distributions [3,20,26].

On the other end of the spectrum there are mean-field approximations, which model the mean densities faithfully in the system size limit [6]. In between there are techniques such as moment closure [32], that not only consider the mean, but also the variance and other higher order moments. These methods depend on ad-hoc approximations of higher order moments to close the ODE system given by the moment equations. Yet another class of methods approximate molecular counts continuously and approximate the dynamics in such a continuous space, e.g. the system size expansion [35] and the chemical Langevin equation [16].

While the moment closure method uses ad-hoc approximations for high order moments to facilitate numerical integration, they can be avoided in some contexts. For the equilibrium distribution, for example, the time-derivative of all moments is equal to zero. This directly yields constraints that have been used for parameter estimation at steady-state [5] and bounding moments of the equilibrium distribution using semi-definite programming [14,15,21]. The latter technique of bounding moments has been successfully adapted in the context of transient analysis [12,29,30]. We adapt the constraints proposed in these works to improve statistical estimations via stochastic simulation (cf. Sect. 4).

While the above techniques give a deterministic output, stochastic simulation generates single executions of the stochastic process [18]. This necessitates accumulating large numbers of simulation runs to estimate quantities. This adds a significant computational burden. Consequently, some effort has been directed at lowering this cost. A prominent technique is τ-leaping [17], which in one step performs multiple instead of only a single reaction. Another approach is to find approximations that are specific to the problem at hand, such as approximations based on time-scale separations [7,8].

Recently, multilevel Monte Carlo methods have been applied in to time-inhomogenous CRNs [2]. In this techniques estimates are combined using estimates of different approximation levels.

The most prominent application of a variance reduction technique in the context of stochastic reaction networks is importance sampling [22]. This technique relies on an alteration of the process and then weighting samples using the likelihood-ratio between the original and the altered process.

3   Stochastic Chemical Kinetics

A chemical reaction network (CRN) describes the interactions between a set of species $S_1, \ldots, S_{n_S}$ in a well-stirred reactor. Since we assume that all reactant molecules are spatially uniformly distributed, we just keep track of the overall amount of each molecule. Therefore the state-space is given by $\mathcal{S} \subseteq \mathbb{N}^{n_S}$. These interactions are expressed a set of *reactions* with a certain inputs and outputs, given by the vectors v_j^- and v_j^+ for reaction $j = 1, \ldots, n_R$, respectively. Such reactions are denoted as

$$\sum_{i=1}^{n_S} v_{ji}^- S_i \xrightarrow{c_j} \sum_{i=1}^{n_S} v_{ji}^+ S_i \,. \tag{1}$$

The reaction rate constant $c_j > 0$ gives us information on the propensity of the reaction. If just a constant is given, *mass-action* propensities are assumed. In a stochastic setting for some state $x \in \mathcal{S}$ these are

$$\alpha_j(x) = c_j \prod_{i=1}^{n_S} \binom{x_i}{v_{ji}^-} \,. \tag{2}$$

The system's behavior is described by a stochastic process $\{X_t\}_{t \geq 0}$. The propensity function gives the infinitesimal probability of a reaction occurring, given a state x. That is, for a small time step $\delta t > 0$

$$\Pr(X_{t+\delta t} = x + v_j \mid X_t = x) = \alpha_j(x)\delta t + o(\delta t) \,. \tag{3}$$

This induces a corresponding continuous-time Markov chain (CTMC) on $\mathcal{S}$ with generator matrix[1]

$$Q_{x,y} = \begin{cases} \sum_{j:x+v_j=y} \alpha_j(x), & \text{if } x \neq y \\ -\sum_{j=1}^{n_R} \alpha_j(x), & \text{otherwise.} \end{cases} \tag{4}$$

---

[1] Assuming a fixed enumeration of the state space.

Accordingly, the time-evolution of the process' distribution, given an initial distribution π_0, is given by the Kolmogorov forward equation, i.e. $\frac{d\pi_t}{dt} = Q\pi_t$, where $\pi_t(x) = \Pr(X_t = x)$. For a single state, it is commonly referred to as the *chemical master equation* (CME)

$$\frac{d}{dt}\pi_t(x) = \sum_{j=1}^{n_R} (\alpha_j(x - v_j)\pi_t(x - v_j) - \alpha_j(x)\pi_t(x)). \tag{5}$$

A direct solution of (5) is usually not possible. If the state-space with non-negligible probability is suitably small, a state space truncation could be performed. That is, (5) is integrated on a possibly time-dependent subset $\hat{\mathcal{S}}_t \subseteq \mathcal{S}$ [20,27,33]. Instead of directly analyzing (5), one often resorts to simulating trajectories. A trajectory $\tau = x_0 t_1 x_1 t_1 \ldots t_n x_n$ over the interval $[0, T]$ is a sequence of states x_i and corresponding jump times t_i, $i = 1, \ldots, n$ and $t_n = T$. We can sample trajectories of X by using stochastic simulation [18].

Consider the birth-death model below as an example.

Model 1 (Birth-death process). *A single species* A *has a constant production and a decay that is linear in the current amount of molecules. Therefore the model consists of two mass-action reactions*

$$\varnothing \xrightarrow{\gamma} A, \quad A \xrightarrow{\delta} \varnothing,$$

where $\varnothing$ *denotes no reactant or no product, respectively.*

For Model 1 the change of probability mass in a single state $x > 0$ is described by expanding (5) and

$$\frac{d}{dt}\pi_t(x) = \gamma\pi_t(x - 1) + \delta\pi_t(x + 1) - (\gamma + \delta)\pi_t(x).$$

We can generate trajectories of this model by choosing either reaction, with a probability that is proportional to its rate given the current state x_i. The jump time $t_i - t_{i+1}$ is determined by sampling from an exponential distribution with rate $\gamma + x_i\delta$.

4   Moment Constraints

The time-evolution of $\mathbb{E}(f(X_t))$ for some function f can be directly derived from (5) by computing the sum $\sum_{x \in \mathcal{S}} f(x)\frac{d}{dt}\pi_t(x)$, which yields

$$\frac{d}{dt}\mathbb{E}(f(X_t)) = \sum_{j=1}^{n_R} \mathbb{E}((f(X_t + v_j) - f(X_t))\alpha_j(X_t)). \tag{6}$$

While many choices of f are possible, for this work we will restrict ourselves to monomial functions $f(x) = x^m$, $m \in \mathbb{N}^{n_S}$ i.e. the *non-central moments* of the process. The *order* $|m|$ of a moment $\mathbb{E}(X^m)$ is the sum over the exponents, i.e. $|m| = \sum_i m_i$. The integration of (6) with such functions f is well-known in the context of moment approximations of CRN models. For most models the arising ODE system is infinitely large, because the time-derivative of low order moments usually depends on the values of higher order moments. To close this system, *moment closures*, i.e. ad-hoc approximations of higher order moments are applied [31]. The main drawback of this kind of analysis is that it is not known whether the chosen closure gives an accurate approximation for the case at hand. Here, such approximations are not necessary, since we will apply the moment dynamics in the context of stochastic sampling instead of trying to integrate (6).

Apart from integration strategies, setting (6) to zero has been used as a constraint for parameter estimation at steady-state [5] and bounding moments at steady-state [11,15,21]. The extension of the latter has recently lead to the adaption of these constraints to a transient setting [12,30]. These two transient constraint variants are analogously derived by multiplying (6) by a time-dependent, differentiable weighting function $w(t)$ and integrating:

Multiplying with $w(t)$ and integrating on $[t_0, T]$ yields [12,30]

$$
w(T)\mathbb{E}(f(X_T)) - w(t_0)\mathbb{E}(f(X_{t_0})) - \int_{t_0}^{T} \frac{dw(t)}{dt} \mathbb{E}(f(X_t))\, dt
$$
$$
= \sum_{j=1}^{n_R} \int_{t_0}^{T} w(t)\mathbb{E}((f(X_t + v_j) - f(X_t))\, \alpha_j(X_t))\, dt
\tag{7}
$$

In the context of computing moment bounds via semi-definite programming the choices $w(t) = t^s$ [30] and $w(t) = e^{\lambda(T-t)}$ [12] have been proposed. While both choices proved to be effective in different case studies, relying solely on the latter choice, i.e. $w(t) = e^{\lambda(T-t)}$ was sufficient.

By expanding the rate functions and f in (7) and substituting the exponential weight function we can re-write (7) as

$$
0 = \mathbb{E}(f(X_T)) - e^{\lambda T}\mathbb{E}(f(X_{t_0})) + \sum_k c_k \int_{t_0}^{T} e^{\lambda(T-t)}\mathbb{E}(X_t^{m_k})\, dt
\tag{8}
$$

with coefficients c_k and vectors m_k defined accordingly. Assuming the moments remain finite on $[0, T]$, we can define the random variable

$$
Z = f(X_T) - e^{\lambda T} f(X_{t_0}) + \sum_k c_k \int_{t_0}^{T} e^{\lambda(T-t)} X_t^{m_k}\, dt
\tag{9}
$$

with $\mathbb{E}(Z) = 0$.

Note, that a realization of Z depends on the whole trajectory $\tau = x_0 t_1 x_1 t_1 \ldots t_n x_n$ over $[t_0, T]$. Thus, for the integral terms in (9) we have to compute sums

$$
\frac{1}{\lambda} \sum_{i=1}^{n} \left(e^{\lambda(T-t_{i+1})} - e^{\lambda(T-t_i)} \right) x_i^{m_k},
\tag{10}
$$

over a given trajectory. This accumulation is best done during the simulation to avoid storing the whole trajectory. Still, the cost of a simulation run increases. For the method to be efficient, the variance reduction (Sect. 5) needs to overcompensate for this increased cost of a simulation run.

For Model 1 the moment equation for $f(x) = x$ becomes

$$\frac{d}{dt}\mathbb{E}\left(X_t\right) = \gamma - \delta\mathbb{E}\left(X_t\right).$$

The corresponding constraint (8) with $\lambda = 0$ gives

$$0 = \mathbb{E}\left(X_T\right) - \mathbb{E}\left(X_0\right) - \gamma T + \delta \int_0^T \mathbb{E}\left(X_t\right) dt.$$

In this instance the constraint leads to an explicit function of the moment over time. If $X_0 = 0$ w.p. 1, then (8) becomes

$$\mathbb{E}\left(X_T\right) = \frac{\gamma}{\delta}\left(1 - e^{-\delta T}\right) \tag{11}$$

when choosing $\lambda = -\delta$.

5   Control Variates

Now, we are interested in the estimation of some quantity $\mathbb{E}\left(V\right)$ by stochastic simulation. Let $V_1, \ldots, V_n$ be independent samples of V. Then the sample mean $\hat{V}_n = \frac{1}{n}\sum_{i=1}^n V_k$ is an estimate of $\mathbb{E}\left(V\right)$. By the central limit theorem

$$\sqrt{n}\hat{V}_n \xrightarrow{d} N(\mathbb{E}\left(V\right), \sigma_V^2).$$

Now suppose, we know of a random variable Z with $0 = \mathbb{E}\left(Z\right)$. The variable Z is called a *control variate*. If a control variate Z is correlated with V, we can use it to reduce the variance of $\hat{V}_n$ [19,28,34,36]. For example, consider we are running a set of simulations and consider a single constraint. If the estimated value of this constraint is larger than zero and we estimate a positive correlation between the constraint Z and V, we would, intuitively, like to decrease our estimate $\hat{V}_n$ accordingly. This results in an estimation of the mean of the random variable

$$Y_\beta = V - \beta Z$$

instead of V. The variance

$$\sigma_{Y_\beta}^2 = \sigma_V^2 - 2\beta\mathrm{Cov}(V, Z) + \beta^2\sigma_Z^2.$$

The optimal choice β can be computed by considering the minimum of $\sigma_{Y_\beta}^2$. Then

$$\beta^* = \mathrm{Cov}(V, Z)/\sigma_Z^2.$$

Therefore $\sigma_{Y_{\beta^*}} = \sigma_Z^2(1 - \rho_{VZ}^2)$, where ρ_{VZ} is the correlation of Z and V.

If we have multiple control variates, we can proceed in a similar fashion. Now, let Z denote a vector of d control variates and let

$$\Sigma = \begin{bmatrix} \Sigma_Z & \Sigma_{VZ} \\ \Sigma_{ZV} & \sigma_V^2 \end{bmatrix}$$

be the covariance matrix of (Z, V). As above, we estimate the mean of $Y_\beta = V - \beta^\top Z$. The ideal choice of β is the result of an ordinary least squares regression between V and Z_i, $i = 1, \ldots, n$. Specifically, $\beta^* = \Sigma_Z^{-1} \Sigma_{ZV}$. Then, asymptotically the variance of this estimator is [34],

$$\sigma_{\hat{Y}_{\beta^*}}^2 = (1 - R_{ZV}^2)\sigma_V^2, \quad R_{ZV}^2 = \Sigma_{ZV} \Sigma_Z^{-1} \Sigma_{ZV} / \sigma_V^2. \tag{12}$$

In practice, however, β^* is unknown and needs to be replaced by an estimate $\hat{\beta}$. This leads to an increase in the estimator's variance. Under the assumption of Z and V having a multivariate normal distribution [10,23], the variance of the estimator is $\hat{Y}_{\hat{\beta}} = \hat{V} - \hat{\beta}^\top \hat{Z}$

$$\sigma_{\hat{Y}_{\hat{\beta}}}^2 = \frac{n-2}{n-2-d}(1 - R_{ZV}^2)\sigma_{\hat{V}}^2. \tag{13}$$

Clearly, a control variate is "good" if it is highly correlated with V. The constraint in (11) is an example of the extreme case. When we use this constraint as a control variate for the estimation of the mean at some time point t, it has a correlation of ± 1 since it describes the mean at that time precisely. Therefore the variance is reduced to zero. We thus aim to pick control variates that are highly correlated with V.

Consider, for example, the above case of the birth-death process. If we choose (11) as a constraint, it would always yield the exact difference of the exact mean to the sample mean and therefore have a perfect correlation. Clearly, $\hat{\beta}$ reduces to 1 and $\hat{Y}_1 = \mathbb{E}(X_t)$.

6 Moment-Based Variance Reduction

We propose an adaptive estimation algorithm (Algorithm 1) that starts out with an initial set of control variates and periodically removes potentially inefficient variates. The "accumulator set" A represents the time-integral terms (10). The size of A has the most significant impact on the overall speed of the algorithm since it represents the only factor incurring a direct cost increase in the SSA itself (line 5).

The algorithm consists of a main loop which performs n simulation runs (line 4). Between each run the mean and covariance estimates of $[Z, V]$ are updated (line 6). Every $d < n$ iterations, the control variates are checked for *efficiency* and *redundancy* (lines 7–12).

Checking both conditions is based on the correlation ρ_{ij} between the i-th and j-th control variate and the correlation ρ_{iv} of a control variate i to V. The first

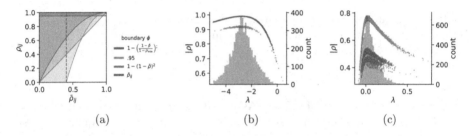

Fig. 1. (a) Different decision functions used in the redundant control variate removal. The weaker of any two control variates is removed if the pair $(\bar{\rho}_{ij}, \rho_{ij})$ belongs to the shaded area of the considered function. The vertical dashed line indicates $\rho_{\min}$. (b, c) The absolute correlation of different constraints to V arising from different choices of λ. The blue dots represent constraints based on first order moments, while the orange refers to control variates derived from second order moments. In both cases 10,000 samples were used with 30 initial samples for λ from $N(0, 1)$ and $k_{\min} = 2$. A quadratic decision bound was used for the redundancy removal. Furthermore, a histogram of control variates selected by Algorithm 1 is given. In (b) $\mathbb{E}\left(X_2^A\right)$ in the dimerization model was estimated. In (c) $\mathbb{E}\left(X_{50}^X\right)$ in the processive modification model was estimated. (Color figure online)

condition is a simple lower threshold $\rho_{\min}$ for a correlation ρ_{iv}. This condition aims to remove those variates from the control variate set that are only weakly correlated to V (line 9). The rationale is that, if variate i has a low correlation with the variable of interest V, its computation may not be worth the costs. Here, we propose to set $\rho_{\min}$ heuristically as

$$\rho_{\min} = \min\left(0.1, \frac{\max_i \rho_{iv}}{k_{\min}}\right),$$

where $k_{\min} > 1$ is an algorithm parameter.

The second condition aims to remove redundant conditions. This is not only beneficial for the efficiency of the estimator, but also necessary for the matrix inversion (12) because perfectly and highly correlated constraints will make the covariance matrix estimate $\hat{\Sigma}_Z$ (quasi-) singular. For all considered criteria we iterate over all tuples $(i, j) \in \{1, \ldots, k\}^2$, $i \neq j$, removing the weaker of the two, i.e. arg $\min_{k \in \{i,j\}} \rho_{kv}$, if the two control variates are considered redundant (line 10).

There are many ways to define such a redundancy criterion. Here, we focus on criteria that are defined in terms of the average correlation $\bar{\rho}_{ij} = (\rho_{iv} + \rho_{jv})/2$. For two variates i and j we then check if their mutual correlation ρ_{ij} exceeds a some function ϕ of $\bar{\rho}_{ij}$, i.e. we check the inequality

$$\phi(\bar{\rho}_{ij}) \leq \rho_{ij}.$$

If this inequality holds, constraint arg $\min_{k \in \{i,j\}} \rho_{kv}$ is removed. Naturally, there are many possible choices for the above decision boundary ϕ (cf. Fig. 1a).

Algorithm 1. Estimate the mean of species i at time T

1: **procedure** ESTIMATEMEAN($n, d, n_{\max}, n_\lambda, k_{\min}$)
2:      $L = \{\lambda_i \sim \pi_\lambda \mid 1 \le i < n_\lambda\} \cup \{0\}$
3:      $P \leftarrow \{(m, \lambda) \mid 1 \le |m| \le n_{\max}, \lambda \in L\}$
4:      **for** $i = 1, \ldots, n$ **do**
5:          $\tau \leftarrow \mathrm{SSA}(\pi_0, T, A)$
6:          compute constraint values using A and update $\hat{\Sigma}$ and $\hat{V}_i$
7:          **if** $i \bmod d = 0$ **then**
8:              $\rho_{\min} \leftarrow \min(0.1, \max_i \rho_{iv}/k_{\min})$
9:              $P \leftarrow P \setminus \{(m_k, \lambda_k) \mid \rho_{kv} < \rho_{\min}\}$
10:            $P \leftarrow P \setminus \{(m_k, \lambda_k) \mid \exists i, j.i \ne j, \phi(\bar{\rho}_{ij}) < \rho_{ij}, k = \arg\min_{k \in \{i,j\}} \rho_{kv}\}$
11:          **end if**
12:          remove unneeded accumulators from A
13:      **end for**
14:      **return** $\hat{V}_n - \left(\hat{\Sigma}_Z^{-1} \hat{\Sigma}_{ZV}\right)^\top \hat{Z}_n$
15: **end procedure**

The simplest choice is to ignore $\bar{\rho}_{ij}$ and just fix a constant close to 1 as a threshold, e.g. $\phi_c(\bar{\rho}_{ij}) = .99$. While this often leads to the strongest variance reduction and avoids numerical issues in the control variate computation, it turns out that the computational overhead is not as well-compensated as by other choices of ϕ (see Sect. 7).

Another option is to fix a simple linear function, i.e. $\phi_\ell(\bar{\rho}_{ij}) = \bar{\rho}_{ij}$. For this choice the intuition is, that one of two constraints is removed if their mutual correlation exceeds their average correlation with V.

Here, we also assess two quadratic choices for ϕ. The first choice of $\phi_q(\bar{\rho}) = 1 - (1 - \bar{\rho})^2$ is more tolerant than the linear function and more strict than a threshold function, except for highly correlated control variates. Another variant of ϕ is given by including the lower bound $\rho_{\min}$ and scaling the quadratic function accordingly: $\phi_{sq}(\bar{\rho}) = 1 - ((1 - \bar{\rho})/(1 - \rho_{\min}))^2$. The different choices of ϕ considered here are plotted in Fig. 1a.

Now, we discuss the choice of the initial control variates. We identify control variate k by a tuple (m_k, λ_k) of a moment vector m_k and a time-weighting parameter λ_k. That is, we use $w(t) = e^{\lambda_k(T-t)}$ and $f(x) = x^{m_k}$ in (7). For a given set of parameters L, we use all moments up to some fixed order $n_{\max}$ (line 3). The ideal set of parameters L is generally not known. For certain choices the correlation of the control variates and the variable of interest is higher then for others. To illustrate this, consider the above example of the birth-death process. Choosing $\lambda = -\delta$ leads to a control variate that has a correlation of ± 1 with V. Therefore, the ideal choice of initial values for would be $L = \{-\delta\}$. This, however, is generally not known. Therefore, we sample a set of λ's from some fixed distribution π_λ (line 2).

7   Case Studies

We first define a criterion of *efficiency* in order to estimate whether the reduction in variance is worth the increased cost. A natural baseline of a variance reduction is, that it is more efficient to pay for the overhead of the reduction than to generate more samples to achieve a similar reduction of variance. Let σ_Y^2 be the variance of Y. The *efficiency* of the method is the ratio of the necessary cost to achieve a similar reduction with the CV estimate Y_{CV} compared to the standard estimate Y [24], i.e.

$$E = \frac{c_0 \sigma_Y^2}{c_1 \sigma_{Y_{CV}}^2}. \tag{14}$$

That ratio c_0/c_1 depends on both the specific implementation and the technical setup. The cost increase is mainly due to the computation of the integrals in (8). But the repeated checking of control variates for efficiency also increases the cost. The accumulation over the trajectory directly increases the cost of a single simulation which is the critical part of the estimation. To estimate the baseline cost c_0, 2000 estimations were performed without considering any control variates.

The simulation is implemented in the Rust programming language[2]. The model description is parsed from a high level specification. Rate functions are compiled to stack programs for fast evaluation. Code is made available online [4].

We consider four non-trivial case studies. Three models exhibit complex multi-modal behaviour. We now describe the models and the estimated quantities in detail.

The first model is a simple dimerization on a countably infinite state-space.

Model 2 (Dimerization). *We first examine a simple dimerization model on an unbounded state-space*

$$\varnothing \xrightarrow{10} M, \quad 2M \xrightarrow{0.1} D$$

with initial condition $X_0^M = 0$.

Despite the models simplicity, the moment equations are not closed for this system due to the second reaction which is non-linear. Therefore a direct analysis of the expected value would require a closure. For this model we will estimate $\mathbb{E}\left(X_2^M\right)$.

The following two models are bimodal, i.e. they each posses two stable regimes among which they can switch stochastically. For both models we choose the initial conditions such that the process will move towards either attracting region with equal probability.

---

[2] https://www.rust-lang.org.

Model 3 (Distributive Modification). *The distributive modification model was introduced in* [9]. *It consists of the reactions*

$$X + Y \xrightarrow{.001} B + Y, \quad B + Y \xrightarrow{.001} 2Y,$$
$$Y + X \xrightarrow{.001} B + X, \quad B + X \xrightarrow{.001} 2X$$

with initial conditions $X_0^X = X_0^Y = X_0^B = 100$.

Model 4 (Exclusive Switch). *The exclusive switch model consists of 5 species, 3 of which are typically binary (activity states of the genes)* [25].

$$P_1 \to \varnothing \quad G \to G + P_2 \quad G + P_1 \to G_1 \quad G_1 \to G + P_1 \quad G_1 \to G_1 + P_1$$
$$P_2 \to \varnothing \quad G \to G + P_1 \quad G + P_2 \to G_2 \quad G_2 \to G + P_2 \quad G_2 \to G_2 + P_2$$

with initial conditions $X_0^G = 1$ *and* $X_0^{G_1} = X_0^{G_2} = X_0^{P_1} = X_0^{P_2} = 0$.

We evaluate the influence of algorithm parameters, choices of distributions to sample λ from, and the influence of the sample size on the efficiency of the proposed method. Note that the implementation does not simplify the constraint representations or the state space according to stoichiometric invariants or limited state spaces. Model 3, for example has the invariant $X_t^X + X_t^Y + X_t^B = \text{const.}$, $\forall t \geq 0$, which could be used to reduce the state-space dimensionality to two. In Model 4 the invariant $\forall t \geq 0. X_t^G, X_t^{G_1}, X_t^{G_2} \in \{0, 1\}$ could be used to optimize the algorithm by eliminating redundant moments, e.g. $\mathbb{E}((X^G)^2) = \mathbb{E}(X^G)$. Such an optimization would further increase the efficiency of the algorithm.

We first turn to the choice of the λ sampling distribution. Here we consider two choices:

1. a standard normal distribution $N(0, 1)$,
2. a uniform distribution on $[-5, 5]$.

We deterministically include $\lambda = 0$ in the constraint set, as this parameter corresponds to a uniform weighting function. We performed estimations on the case studies using different valuations of the algorithm parameters of the minimum threshold $k_{\min}$ and the number of λ-samples n_λ. We used samples size $n = 10{,}000$ and checked the control variates every $d = 100$ iterations for the defined criteria. For each valuation 1000 estimations were performed. In Fig. 2b, we summarize the efficiencies for the arising parameter combinations on the three case studies. Most strikingly, we can note that the efficiency was consistently larger than one in all cases. Generally, the normal sampling distribution out-performed the alternative uniform distribution, except in case of the dimerization. The reason for this becomes apparent, when examining Fig. 1b,c: In case of the dimerization model the most efficient constraints are found for $\lambda \approx -3$, while in case of the distributive modification they are located just above 0 (we observe a similar pattern for the exclusive switch case study). Therefore the sampling of efficient λ values is more likely using a uniform distribution for the dimerization case

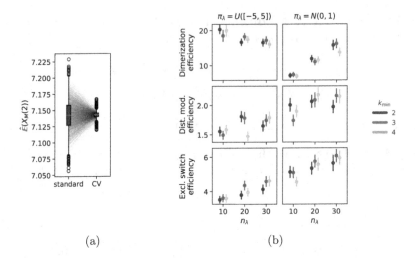

Fig. 2. (a) The effect of including control variates (CV) on the mean estimates $\hat{\mathbb{E}}(X_2^M)$ in the dimerization case study. Parameters were $\pi_\lambda = N(0,1)$, $n_\lambda = 30$, $k_{\min} = 4$, $\phi(\bar{\rho}) = 1 - (1 - \bar{\rho})^2$. (b) The efficiencies for different valuations of n_λ and $k_{\min}$ and choices of π_λ. The sample size was $n = 10{,}000$ in all cases with $d = 100$. The bars give the bootstrapped (1000 iterations) standard deviations.

study, than it is for the others. Given that larger absolute values for λ seem unreasonable due their exponential influence on the weighting function and the problem of fixing a suitable interval for a uniform sampling scheme, the choice of a standard normal distribution for π_λ seems superior.

In Fig. 3 we compare efficiencies for different maximum orders of constraints $n_{\max}$. This comparison is performed for different choices of the redundancy rule and initial λ sample sizes n_λ. Again, for each parameter valuation 1000 estimations were performed. With respect to the maximum constraints order $n_{\max}$ we see a clear tendency, that the inclusion of second order constraints lessens the efficiency of the method. In case of a constant redundancy threshold it even dips below break-even for the distributive modification case study. This is not surprising, since the inclusion of second order moments increases the number of initial constraints quadratically and the incurred cost, especially of the first iterations, lessens efficiency.

Figure 4 depicts the trade-off between the variance reduction σ_0^2/σ_1^2 versus the cost ratio c_0/c_1. Comparing the redundancy criterion based on a constant threshold ϕ_c to the others, we observe both a larger variance reduction and an increased cost. This is due to the fact, that more control variates are included throughout the simulations (cf. See Supplementary Material Tables 1 and 2). Depending on the sample distribution π_λ and the case study, this permissive strategy may pay off. In case of the dimerization, for example, it pays off, while in case of the distributive modification it leads to a lower efficiency ratio. In the latter case the model is more complex, and therefore the set of initial control

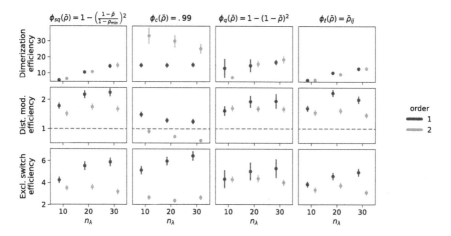

Fig. 3. The efficiency for different redundancy policies ϕ and maximal moment orders n_{max}. The sample size was $n = 10{,}000$ in all cases with $d = 100$. Furthermore, $k_{min} = 3$, $\pi_\lambda = N(0,1)$, and $n_{max} = 1$. The bars give the bootstrapped (1000 iterations) standard deviations.

variates is larger. With a more permissive redundancy strategy, more control variates are kept (ca. 10 when using ϕ_c vs. ca. 2–3 for the others). The other redundancy boundaries move the results further in the direction of less variance reduction while keeping the cost increase low. On the opposite end is the linear ϕ_ℓ. The quadratic versions ϕ_q and ϕ_{sq} can be found in the middle of this spectrum.

We also observe, that an increase of n_λ is particularly beneficial, if the sampling distribution π_λ does not capture the parameter region of the highest correlations well. This can be seen for the Dimerization case study, where the variance reduction increases strongly with increasing sample size (See Supplementary Material Tables 1 and 2). Since $\pi_\lambda = N(0,1)$, more samples are needed to sample efficient λ-values (cf. Fig. 1b).

Finally, we discuss the effect of the sample size n on the efficiency E. In Fig. 5 we give both the efficiencies and the slowdown for different sample sizes. As a redundancy rule we used the unscaled quadratic function, 30 initial values of λ, and $k_{min} = 3$. With increasing sample size, the efficiency usually approaches an upper limit. This is due to the fact that most control variates are dropped early on and the control variates often remain the same for the rest of the simulations. If we assume there are no helpful control variates in the initial set and all would be removed at iteration 100, the efficiency would approach 1 with $n \to \infty$.

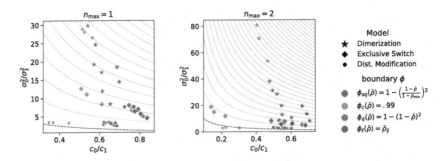

Fig. 4. A visualisation of the trade-off between variance reduction σ_0^2/σ_1^2 and cost ratio c_0/c_1. Isolines for efficiencies are given in grey. The break-even is marked by the dashed red line. Markers of the same kind differ in n_λ and shift with increasing value upwards in variance reduction and lower in c_0/c_1, i.e. the shift is to the left and upwards with increasing n_λ. The sample size was $n = 10{,}000$ in all cases with $d = 100$. Furthermore, $k_{\min} = 3$ and $\pi_\lambda = N(0,1)$. (Color figure online)

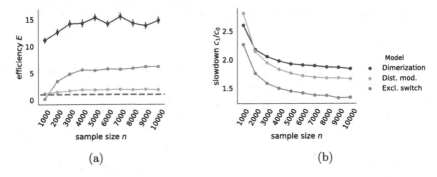

Fig. 5. (a) The effect of sample size on the efficiency E in the different case studies. The break-even $E = 1$ is marked by the dashed red line. (b) The cost increase due to the variance reduction over different sample sizes. (Color figure online)

8   Conclusion

In this work we have shown that known constraints on the moment dynamics can be successfully leveraged in simulation-based estimation of expected values. The empirical results indicate that the supplementing a standard SSA estimation with moment information can drastically reduce the estimators' variance. This reduction is paid for by accumulating information on the trajectory during simulation. However, the reduction is able to compensate for this increase. This means that for fixed costs, using estimates with control variates is more beneficial than using estimates without control variates.

While a variety of algorithmic variants was evaluated, many aspects remain subject to further study. In particular different choices of f and w in (7) may improve efficiency further. These choices become particularly interesting when moving from the estimation of simple first order moments to more complex queries such as behavioural probabilities of trajectories. In such cases, one might even attempt to find efficient control variate functions using machine learning methods.

Another open question regarding this work is its performance when multiple quantities instead of a single quantity are to be estimated. In such a case, constraints would be particularly beneficial, if they lead to improvements as many estimation targets as possible.

Furthermore the identification of the best weighting parameters λ could be done in a more adaptive fashion. The presented scheme of a sampling from π_λ could be extended into a Bayesian-like procedure, wherein the values for λ are periodically re-sampled from a distribution that is adjusted according to the best-performing constraints up to that point.

Acknowledgement. This work was supported by the DFG project MULTIMODE.

References

1. Ale, A., Kirk, P., Stumpf, M.P.: A general moment expansion method for stochastic kinetic models. J. Chem. Phys. **138**(17), 174101 (2013)
2. Anderson, D.F., Yuan, C.: Low variance couplings for stochastic models of intracellular processes with time-dependent rate functions. Bull. Math. Biol. **81**, 1–29 (2018)
3. Andreychenko, A., Mikeev, L., Spieler, D., Wolf, V.: Parameter identification for Markov models of biochemical reactions. In: Gopalakrishnan, G., Qadeer, S. (eds.) CAV 2011. LNCS, vol. 6806, pp. 83–98. Springer, Heidelberg (2011). https://doi.org/10.1007/978-3-642-22110-1_8
4. Backenköhler, M.: CME stochastic simulation code (2019). https://github.com/mbackenkoehler/cme-simulation
5. Backenköhler, M., Bortolussi, L., Wolf, V.: Moment-based parameter estimation for stochastic reaction networks in equilibrium. IEEE/ACM Trans. Comput. Biol. Bioinform. (TCBB) **15**(4), 1180–1192 (2018)
6. Bortolussi, L., Hillston, J., Latella, D., Massink, M.: Continuous approximation of collective system behaviour: a tutorial. Perform. Eval. **70**(5), 317–349 (2013)
7. Bortolussi, L., Milios, D., Sanguinetti, G.: Efficient stochastic simulation of systems with multiple time scales via statistical abstraction. In: Roux, O., Bourdon, J. (eds.) CMSB 2015. LNCS, vol. 9308, pp. 40–51. Springer, Cham (2015). https://doi.org/10.1007/978-3-319-23401-4_5
8. Cao, Y., Gillespie, D.T., Petzold, L.R.: The slow-scale stochastic simulation algorithm. J. Chem. Phys. **122**(1), 014116 (2005)
9. Cardelli, L., Csikász-Nagy, A.: The cell cycle switch computes approximate majority. Sci. Rep. **2**, 656 (2012)
10. Cheng, R.C.: Analysis of simulation experiments under normality assumptions. J. Oper. Res. Soc. **29**(5), 493–497 (1978)
11. Dowdy, G.R., Barton, P.I.: Bounds on stochastic chemical kinetic systems at steady state. J. Chem. Phys. **148**(8), 084106 (2018)
12. Dowdy, G.R., Barton, P.I.: Dynamic bounds on stochastic chemical kinetic systems using semidefinite programming. J. Chem. Phys. **149**(7), 074103 (2018)

13. Engblom, S.: Computing the moments of high dimensional solutions of the master equation. Appl. Math. Comput. **180**(2), 498–515 (2006)
14. Ghusinga, K.R., Lamperski, A., Singh, A.: Estimating stationary characteristic functions of stochastic systems via semidefinite programming. In: 2018 European Control Conference (ECC), pp. 2720–2725. IEEE (2018)
15. Ghusinga, K.R., Vargas-Garcia, C.A., Lamperski, A., Singh, A.: Exact lower and upper bounds on stationary moments in stochastic biochemical systems. Phys. Biol. **14**(4), 04LT01 (2017)
16. Gillespie, D.T.: The chemical Langevin equation. J. Chem. Phys. **113**(1), 297–306 (2000)
17. Gillespie, D.T.: Approximate accelerated stochastic simulation of chemically reacting systems. J. Chem. Phys. **115**(4), 1716–1733 (2001)
18. Gillespie, D.: Exact stochastic simulation of coupled chemical reactions. J. Phys. Chem. **81**(25), 2340–2361 (1977)
19. Glasserman, P., Yu, B.: Large sample properties of weighted Monte Carlo estimators. Oper. Res. **53**(2), 298–312 (2005)
20. Henzinger, T.A., Mateescu, M., Wolf, V.: Sliding window abstraction for infinite Markov chains. In: Bouajjani, A., Maler, O. (eds.) CAV 2009. LNCS, vol. 5643, pp. 337–352. Springer, Heidelberg (2009). https://doi.org/10.1007/978-3-642-02658-4_27
21. Kuntz, J., Thomas, P., Stan, G.B., Barahona, M.: Rigorous bounds on the stationary distributions of the chemical master equation via mathematical programming. arXiv preprint arXiv:1702.05468 (2017)
22. Kuwahara, H., Mura, I.: An efficient and exact stochastic simulation method to analyze rare events in biochemical systems. J. Chem. Phys. **129**(16), 10B619 (2008)
23. Lavenberg, S.S., Moeller, T.L., Welch, P.D.: Statistical results on control variables with application to queueing network simulation. Oper. Res. **30**(1), 182–202 (1982)
24. L'Ecuyer, P.: Efficiency improvement and variance reduction. In: Proceedings of the 26th conference on Winter simulation, pp. 122–132. Society for Computer Simulation International (1994)
25. Loinger, A., Lipshtat, A., Balaban, N.Q., Biham, O.: Stochastic simulations of genetic switch systems. Phys. Rev. E **75**(2), 021904 (2007)
26. Mateescu, M., Wolf, V., Didier, F., Henzinger, T.: Fast adaptive uniformisation of the chemical master equation. IET Syst. Biol. **4**(6), 441–452 (2010)
27. Munsky, B., Khammash, M.: The finite state projection algorithm for the solution of the chemical master equation. J. Chem. Phys. **124**(4), 044104 (2006)
28. Nelson, B.L.: Control variate remedies. Oper. Res. **38**(6), 974–992 (1990)
29. Sakurai, Y., Hori, Y.: A convex approach to steady state moment analysis for stochastic chemical reactions. In: 2017 IEEE 56th Annual Conference on Decision and Control (CDC), pp. 1206–1211. IEEE (2017)
30. Sakurai, Y., Hori, Y.: Bounding transient moments of stochastic chemical reactions. IEEE Control. Syst. Lett. **3**(2), 290–295 (2019)
31. Schnoerr, D., Sanguinetti, G., Grima, R.: Comparison of different moment-closure approximations for stochastic chemical kinetics. J. Chem. Phys. **143**(18), 11B610_1 (2015)
32. Singh, A., Hespanha, J.P.: Lognormal moment closures for biochemical reactions. In: Proceedings of the 45th IEEE Conference on Decision and Control, pp. 2063–2068. IEEE (2006)
33. Spieler, D.: Numerical analysis of long-run properties for Markov population models. Ph.D. thesis, Saarland University (2014)

34. Szechtman, R.: Control variate techniques for Monte Carlo simulation: control variates techniques for Monte Carlo simulation. In: Proceedings of the 35th Conference on Winter Simulation: Driving Innovation, pp. 144–149. Winter Simulation Conference (2003)
35. Van Kampen, N.G.: Stochastic Processes in Physics and Chemistry, vol. 1. Elsevier, Amsterdam (1992)
36. Wilson, J.R.: Variance reduction techniques for digital simulation. Am. J. Math. Manag. Sci. 4(3–4), 277–312 (1984)

Effective Computational Methods
for Hybrid Stochastic Gene Networks

Guilherme C. P. Innocentini[1], Fernando Antoneli[2], Arran Hodgkinson[3],
and Ovidiu Radulescu[3(⊠)]

[1] Federal University of ABC, Santo André, Brazil
[2] Escola Paulista de Medicina, Universidade Federal de São Paulo, São Paulo, Brazil
[3] LPHI UMR CNRS 5235, University of Montpellier, Montpellier, France
`ovidiu.radulescu@umontpellier.fr`

Abstract. At the scale of the individual cell, protein production is
a stochastic process with multiple time scales, combining quick and
slow random steps with discontinuous and smooth variation. Hybrid
stochastic processes, in particular piecewise-deterministic Markov processes (PDMP), are well adapted for describing such situations. PDMPs
approximate the jump Markov processes traditionally used as models
for stochastic chemical reaction networks. Although hybrid modelling
is now well established in biology, these models remain computationally challenging. We propose several improved methods for computing
time dependent multivariate probability distributions (MPD) of PDMP
models of gene networks. In these models, the promoter dynamics is
described by a finite state, continuous time Markov process, whereas the
mRNA and protein levels follow ordinary differential equations (ODEs).
The Monte-Carlo method combines direct simulation of the PDMP with
analytic solutions of the ODEs. The push-forward method numerically
computes the probability measure advected by the deterministic ODE
flow, through the use of analytic expressions of the corresponding semigroup. Compared to earlier versions of this method, the probability of
the promoter states sequence is computed beyond the naïve mean field
theory and adapted for non-linear regulation functions.

Availability. The algorithms described in this paper were implemented
in MATLAB. The code is available at Zenodo. https://doi.org/10.5281/
zenodo.3251708.

1 Introduction

In PDMP models of gene networks, each gene promoter is described as a finite
state Markov process [3,10,12,13]. The promoter triggers synthesis of gene
products (mRNAs and proteins) with intensities depending on its state. The
promoter can exhibit two state (ON-OFF) dynamics, but also dynamics with
more than two states and arbitrarily complex transitions [11,19]. The transition
rates between the states of the promoter depend on the expression levels of proteins expressed by the same or by other promoters. In PDMP models, the gene

© Springer Nature Switzerland AG 2019
L. Bortolussi and G. Sanguinetti (Eds.): CMSB 2019, LNBI 11773, pp. 60–77, 2019.
https://doi.org/10.1007/978-3-030-31304-3_4

products are considered in sufficiently large copy numbers and are represented as continuous variables following ordinary differential equations (ODEs). The sources of noise in these models are thus the discrete transitions between the promoter states.

In single cell experimental settings the quantities of mRNA [1,14,17,18] and proteins [5,6] can be determined for each cell. By double or multiple- fluorophore fluorescence techniques products from several genes can be quantified simultaneously and one can have access to multivariate probability distributions (MPD) of mRNA or proteins. The stochastic dynamics of promoters and gene networks can have important consequences for fundamental biology [4] but also for HIV [15] and cancer research [7]. For this reason we aim to develop effective methods for computing time-dependent MPDs for PDMP models. Our main objective is the reduction of computation time which is prerequisite for parameter scans and machine learning applications [8].

PDMPs already represent a gain with respect to the chemical Markov equation from which they are derived by various limit theorems [2]. A gene network PDMP model can be simulated by numerical integration of ODEs coupled with a driven inhomogeneous Poisson process for the successive transitions of the promoters [3,13,16,20]. The simulation becomes particularly effective when analytic solutions of the ODEs are available [10].

However, very little has been done to further improve the computational power by optimising simulation and analysis of PDMP models.

Numerical integration of the PDE satisfied by MPD is an interesting option combining precision and speed for small models. Finite difference methods, however, are of limited use in this context as they can not cope with many RNA and protein variables (extant examples are restricted to the dimension 2, corresponding to a single promoter, with or without self-regulation see [10,12]).

Another interesting method for computing time dependent MPDs is the push-forward method. For gene networks, this method has been first introduced in [9] and further adapted for continuous mRNA variables in [10]. It is based on the idea to compute the MPD as the push-forward measure of the semigroup defined by the ODEs. This method is approximate, as one has to consider that the discrete PDMP variables are piecewise constant on a deterministic time partition. Furthermore, the transition rates between promoter states were computed in a mean field approximation. In this paper we replace the mean field approximation by the next order approximation taking into account the moments of the protein distribution.

2   Methods

2.1   PDMP Models of Gene Networks

The state of a PDMP gene network model is defined by the elements of the set $E = \mathbb{R}^{2N} \times \{0, \ldots, s_{max} - 1\}^N$ where N is the number of genes and s_{max} is a positive integer representing the maximum number of states of a gene promoter. It is a process $\zeta_t = (x_t, y_t, s_t)$, where x_t, y_t, are vectors in $\mathbb{R}^N$ and their components x_t^i, y_t^i $(1 \leq i \leq N)$ encode the dynamics of protein and mRNA densities, respectively, associated with gene i. The vector s_t is an element of $S := \{0, \ldots, s_{max} - 1\}^N$ and the components s_t^i $(0 \leq s^i \leq s_{max} - 1)$ describe the jump Markov process between states of the gene i, where s_t^i can assume integer values in $\{0, \ldots, s_{max} - 1\}$. The process $\zeta_t = (x_t, y_t, s_t)$ is determined by three characteristics:

(1) For all fixed $s \in S$ a vector field $F_s : \mathbb{R}^{2N} \to \mathbb{R}^{2N}$ determining a unique global flow $\Phi_s(t, x, y)$ in $\mathbb{R}^{2N}$, the space of all protein ($x \in \mathbb{R}^N$) and mRNA ($y \in \mathbb{R}^N$) values such that, for $t > 0$,

$$\frac{d\Phi_s(t, x, y)}{dt} = F_s(\Phi_s(t, x, y)), \quad \Phi_s(0, x, y) = (x, y). \tag{1}$$

On coordinates, this reads

$$\frac{d\Phi_i^x}{dt} = b_i \Phi_i^y - a_i \Phi_i^x,$$
$$\frac{d\Phi_i^y}{dt} = k_i(s^i) - \rho_i \Phi_i^y, \quad 1 \leq i \leq N, \tag{2}$$

where b_i, k_i, a_i, ρ_i are translation efficiencies, transcription rates, protein degradation coefficients and mRNA degradation coefficients of the ith gene, respectively. Note that transcription rates depend on the relevant promoter states.

The flow $\Phi_s(t, x, y)$ represents a one parameter semigroup fulfilling the properties
(i) $\Phi_s(0, x_0, y_0) = (x_0, y_0)$,
(ii) $\Phi_s(t + t', x_0, y_0) = \Phi_s(t', \Phi_s^x(t, x_0, y_0), \Phi_s^y(t, x_0, y_0))$.

(2) A transition rate matrix for the promoter states $H : \mathbb{R}^{2N} \to M_{(Ns_{max}) \times (Ns_{max})}(\mathbb{R})$, such that $H_{r,s}(x, y) \geq 0$ is the (r, s) element of the matrix H. If $(s \neq r)$ this is the rate of probability to jump to the state r from the state s. Furthermore, $H_{s,s}(x, y) = -\sum_{r \neq s} H_{r,s}(x, y)$ for all $s, r \in S, s \neq r$ and for all $(x, y) \in \mathbb{R}^{2N}$.

(3) A jump rate $\lambda : E \to \mathbb{R}^+$. The jump rate can be obtained from the transition rate matrix

$$\lambda(x, y, s) = \sum_{r \neq s} H_{r,s}(x, y) = -H_{s,s}(x, y). \tag{3}$$

From these characteristics, right-continuous sample paths $\{(x_t, y_t) : t > 0\}$ starting at $\zeta_0 = (x_0, y_0, s_0) \in E$ can be constructed as follows. Define

$$x_t(\omega) := \Phi_{s_0}(t, x_0, y_0) \text{ for } 0 \leq t \leq T_1(\omega), \tag{4}$$

where $T_1(\omega)$ is a realisation of the first jump time of s_t, with the distribution

$$F(t) = \mathbb{P}[T_1 > t] = \exp(-\int_0^t \lambda(\Phi_{s_0}(u, x_0, y_0))du), \ t > 0, \tag{5}$$

and ω is the element of the probability space for which the particular realisation of the process is given. The pre-jump state is $\boldsymbol{\zeta}_{T_1^-(\omega)}(\omega) = (\boldsymbol{\Phi}_{s_0}(T_1(\omega)), \boldsymbol{x}_0, \boldsymbol{y}_0), s_0)$ and the post-jump state is $\boldsymbol{\zeta}_{T_1(\omega)}(\omega) = (\boldsymbol{\Phi}_{s_0}(T_1(\omega)), \boldsymbol{x}_0, \boldsymbol{y}_0), \boldsymbol{s})$, where $\boldsymbol{s}$ has the distribution

$$\mathbb{P}[\boldsymbol{s} = \boldsymbol{r}] = \frac{H_{r,s_0}(\boldsymbol{\Phi}_{s_0}(T_1(\omega)), \boldsymbol{x}_0, \boldsymbol{y}_0), s_0)}{\lambda(\boldsymbol{\Phi}_{s_0}(T_1(\omega)), \boldsymbol{x}_0, \boldsymbol{y}_0), s_0)}, \quad \text{for all } \boldsymbol{r} \neq s_0. \tag{6}$$

We then restart the process $\boldsymbol{\zeta}_{T_1(\omega)}$ and recursively apply the same procedure at jump times $T_2(\omega)$, etc.

Note that between each two consecutive jumps $(\boldsymbol{x}_t, \boldsymbol{y}_t)$ follow deterministic ODE dynamics defined by the vector field $\boldsymbol{F}_s$. At the jumps, the protein and mRNA values $(\boldsymbol{x}_t, \boldsymbol{y}_t)$ are continuous.

The calculation of the flow between two jumps and of the jump time can be gathered in the same set of differential equations

$$\frac{d\Phi_i^x}{dt} = b_i \Phi_i^y - a_i \Phi_i^x,$$
$$\frac{d\Phi_i^y}{dt} = k_i(s^i) - \rho_i \Phi_i^y, \ 1 \leq i \leq N,$$
$$\frac{d \log F}{dt} = -\lambda(\boldsymbol{x}, \boldsymbol{y}, s_0), \tag{7}$$

that has to be integrated with the stopping condition $F(T_1) = U$, where U is a random variable, uniformly distributed on $[0, 1]$.

We define multivariate probability density functions $p_s(t, \boldsymbol{x}, \boldsymbol{y})$. These functions satisfy the Liouville-master equation which is a system of partial differential equations:

$$\frac{\partial p_s(t, \boldsymbol{x}, \boldsymbol{y})}{\partial t} = -\nabla_{x,y} \cdot (\boldsymbol{F}_s(\boldsymbol{x}, \boldsymbol{y}) p_s(t, \boldsymbol{x}, \boldsymbol{y})) + \sum_r H_{s,r}(\boldsymbol{s}, \boldsymbol{x}, \boldsymbol{y}) p_r(t, \boldsymbol{x}, \boldsymbol{y}). \tag{8}$$

2.2   ON/OFF Gene Networks

In this paper, for the purpose of illustration only, all the examples are constituted by ON/OFF gene networks. For an ON/OFF gene each component s^i has two possible values 0 for OFF and 1 for ON.

As a first example that we denote as model M_1, let us consider a two genes network; the expression of the first gene being constitutive and the expression of the second gene being activated by the first. We consider that the transcription activation rate of the second gene is proportional to the concentration of the first protein $f_2 x^1$. All the other rates are constant f_1, h_1, h_2, representing the transcription activation rate of the first gene, and the transcription inactivation rates of gene one and gene two, respectively. For simplicity, we consider that the two genes have identical protein and mRNA parameters $b_1 = b_2 = b$, $a_1 = a_2 = a$, $\rho_1 = \rho_2 = \rho$. We further consider that $k_i(s^i) = k_0$ if the gene i is OFF and $k_i(s^i) = k_1$ if the gene i is ON.

The gene network has four discrete states, in order $(0,0)$, $(1,0)$, $(0,1)$, and $(1,1)$. Then, the transition rate matrix for the model M_1 is

$$
\begin{bmatrix}
-(f_1 + f_2 x^1) & h_1 & h_2 & 0 \\
f_1 & -(h_1 + f_2 x^1) & 0 & h_2 \\
f_2 x^1 & 0 & -(f_1 + h_2) & h_1 \\
0 & f_2 x^1 & f_1 & -(h_1 + h_2)
\end{bmatrix}. \tag{9}
$$

The model M_2 differs from the model M_1 by the form of the activation function. Instead of a linear transcription rate $f_2 x^1$ we use a Michaelis-Menten model $f_2 x^1/(K_1 + x^1)$. This model is more realistic as it takes into account that the protein x^1 has to attach to specific promoter sites which become saturated when the concentration of this protein is high.

The transition rate matrix for the model M_2 is

$$
\begin{bmatrix}
-(f_1 + f_2 x^1/(K_1 + x^1)) & h_1 & h_2 & 0 \\
f_1 & -(h_1 + f_2 x^1/(K_1 + x^1)) & 0 & h_2 \\
f_2 x^1/(K_1 + x^1) & 0 & -(f_1 + h_2) & h_1 \\
0 & f_2 x^1/(K_1 + x^1) & f_1 & -(h_1 + h_2)
\end{bmatrix}. \tag{10}
$$

2.3   Monte-Carlo Method

The Monte-Carlo method utilizes the direct simulation of the PDMP based on Eq. 7. A larger number M of sample paths is generated and the values of (x_t, y_t) are stored at selected times. Multivariate probability distributions are then estimated from this data.

The direct simulation of PDMPs needs the solutions of (7) which can be obtained by numerical integration. This is not always computationally easy. Problems may arise for fast switching promoters when the ODEs have to be integrated many times on small intervals between successive jumps. Alternatively, the numerical integration of the ODEs can be replaced by analytic solutions or quadratures. Analytic expressions are always available for the gene network flow (2) and read

$$
\Phi_i^x(t, x_0, y_0) = x_0 \exp(-a_i t) + b_i \left[\left(y_0 - \frac{k_i(s^i)}{\rho_i} \right) \frac{\exp(-\rho_i t) - 1}{a_i - \rho_i} + \frac{k_i(s^i)}{\rho_i} \frac{1 - \exp(-a_i t)}{a_i} \right],
$$
$$
\Phi_i^y(t, x_0, y_0) = (y_0 - k_i/\rho_i) \exp(-\rho_i t) + k_i/\rho_i. \tag{11}
$$

Let us consider the following general expression of the jump intensity function

$$
\lambda(\boldsymbol{x}, \boldsymbol{y}, \boldsymbol{s}) = c_0(\boldsymbol{s}) + \sum_i^N c_i(\boldsymbol{s}) x^i + \sum_i^N d_i(\boldsymbol{s}) f_i(x^i),
$$

where f_i are non-linear functions, for instance Michaelis-Menten $f_i(x^i) = x^i/(K_i + x^i)$ or Hill functions $f_i(x^i) = (x^i)^{n_i}/(K_i^{n_i} + (x^i)^{n_i})$. If $d_i = 0$ for all $1 \le i \le N$, the cumulative distribution function of the waiting time T_1 can be solved analytically [10], otherwise it can be obtained by quadratures. For example, for the model M_2 one has

$$\lambda(x, y, s) = \left(f_1 + f_2 \frac{x^1}{K_1 + x^1}\right)\delta_{s,1} + \left(h_1 + f_2 \frac{x^1}{K_1 + x^1}\right)\delta_{s,2} + (h_2 + f_1)\delta_{s,3} + (h_2 + h_1)\delta_{s,4},$$

where $\delta_{i,j}$ is Kronecker's delta. In this case, the waiting time T_1 is obtained as the unique solution of the equation

$$
-\log(U) = \left[(f_1 + f_2)T_1 + f_2 \int_0^{T_1} \frac{1}{K_1 + \Phi_1^x(t', x_0, y_0)} dt'\right]\delta_{s_0,1} + \left[(h_1 + f_2)T_1 + \right.
$$
$$
\left. f_2 \int_0^{T_1} \frac{1}{K_1 + \Phi_1^x(t', x_0, y_0)} dt'\right]\delta_{s_0,2} + (h_2 + f_1)T_1\delta_{s_0,3} + (h_2 + h_1)T_1\delta_{s_0,4},
$$
$$(12)$$

where U is a random variable, uniformly distributed on $[0, 1]$. In our implementation of the algorithm we solve (12) numerically, using the bisection method.

2.4 Push-Forward Method

This method allows one to compute the MPD of proteins and mRNAs at a time τ given the MPD of proteins and mRNAs at time 0.

In order to achieve this we use a deterministic partition $\tau_0 = 0 < \tau_1 < \ldots < \tau_M = \tau$ of the interval $[0, \tau]$ such that $\Delta_M = \max_{j \in [1,M]}(\tau_j - \tau_{j-1})$ is small. The main approximation of this method is to consider that s_t, for $t \in [0, \tau]$, is piecewise constant on this partition, more precisely, that $s_t = s_j := s_{\tau_j}$, for $t \in [\tau_j, \tau_{j+1})$, $0 \le j \le M - 1$. This is rigorously true for intervals $[\tau_j, \tau_{j+1})$ completely contained between two successive random jump times of s_t. This case becomes very frequent for a very fine partition (large M). Thus, the error generated by the approximation vanishes in the limit $M \to \infty$ (for a rigorous result see the Theorem 1 in Sect. 3.1 and the Lemma 1 in the Appendix C).

For each path realization $S_M := (s_0, s_1, \ldots, s_{M-1}) \in \Omega := \{0, 1, \ldots, 2^N - 1\}^M$ of the promoter states, we can compute (see Appendix B) the protein and mRNA levels x_t, y_t of all genes $i \in [1, N]$, at $t = \tau$:

$$y_\tau^i = y_0^i e^{-\rho\tau} + \frac{k_0}{\rho}(1 - e^{-\rho\tau}) + \frac{k_1 - k_0}{\rho}\sum_{j=1}^{M-1} e^{-\rho\tau}(e^{-\rho\tau_{j+1}} - e^{-\rho\tau_j})s_j^i \quad (13)$$

$$x_\tau^i = x_0^i e^{-a\tau} + \frac{by_0^i}{a - \rho}(e^{-\rho\tau} - e^{-a\tau}) + \frac{bk_0}{\rho}\left(\frac{1 - e^{-a\tau}}{a} + \frac{e^{-a\tau} - e^{-\rho\tau}}{a - \rho}\right) +$$

$$+ \frac{b(k_1 - k_0)}{\rho}e^{-a\tau}\sum_{j=1}^M s_{j-1}^i w_j, \, i \in [1, N] \quad (14)$$

where $w_j = \frac{e^{(a-\rho)\tau} - e^{(a-\rho)\tau_j}}{a-\rho}(e^{\rho\tau_j} - e^{\rho\tau_{j-1}}) - \frac{e^{(a-\rho)\tau_j} - e^{(a-\rho)\tau_{j-1}}}{a-\rho}e^{\rho\tau_{j-1}} + \frac{e^{a\tau_j} - e^{a\tau_{j-1}}}{a}$
and $s_j^i := 0$ if promoter i is OFF for $t \in [\tau_j, \tau_{j+1})$ and $s_j^i := 1$ if promoter i is ON for $t \in [\tau_j, \tau_{j+1})$.

In order to compute the MPD at time τ one has to sum the contributions of all solutions (13) and (14), obtained for the 2^{NM} realisations of promoter state paths with weights given by the probabilities of the paths.

Equations 13 and 14 can straightforwardly be adapted to compute $\boldsymbol{x}_t, \boldsymbol{y}_t$ for all $t \in [0, \tau]$. To this aim, τ should be replaced by t and M should be replaced by M_t defined by the relation $t \in [\tau_{M_t}, \tau_{M_t+1}]$.

Suppose that we want to estimate the MPD of all mRNAs and proteins of the gene network, using a multivariate histogram with bin centers $(x_0^{l_i}, y_0^{m_i})$, $1 \leq i \leq N, 1 \leq l_i \leq n_x, 1 \leq m_i \leq n_y$ where n_x, n_y are the numbers of bins in the protein and mRNA directions for each gene, respectively. Typically $x_0^{l_i} = b/(a\rho)(k_0 + (k_1 - k_0)(l_i - 1/2)), 1 \leq i \leq N, 1 \leq l_i \leq n_x$, $y_0^{m_i} = 1/\rho(k_0 + (k_1 - k_0)(m_i - 1/2)), 1 \leq i \leq N, 1 \leq m_i \leq n_y$. The initial MPD at time $t = 0$ is given by the bin probabilities $p_0^{l_i, m_i}, 1 \leq i \leq N, 1 \leq l_i \leq n_x, 1 \leq m_i \leq n_y$. Let $(x^{l_i, m_i}, y^{l_i, m_i})$ be the solutions, with $x_0^i = x_0^{l_i}$ and $y_0^i = y_0^{m_i}$. The many-to-one application $(l_i', m_i') = \psi(l_i, m_i)$ provides the histogram bin (l_i', m_i') in which falls the vector $(x^{l_i, m_i}, y^{l_i, m_i}))$. The push forward MPD at time $t = \tau$ is defined by the bin probabilities $p^{l_i, m_i}, 1 \leq i \leq N, 1 \leq l_i \leq n_x, 1 \leq m_i \leq n_y$ that are computed as

$$p^{l_i, m_i} = \sum_{S_M \in \Omega} \sum_{\psi(l_i', m_i') = (l_i, m_i)} p_0^{l_i', m_i'} \mathbb{P}[S_M]. \tag{15}$$

In order to compute $\mathbb{P}[S_M]$ we can use the fact that, given $\boldsymbol{x}_t$, $\boldsymbol{s}_t$ is a finite state Markov process, therefore

$$\mathbb{P}[S_M] = \Pi_{s_{N-1}, s_{N-2}}(\tau_{N-2}, \tau_{N-1}) \ldots \Pi_{s_1, s_0}(\tau_0, \tau_1) P_0^S(s_0), \tag{16}$$

where $P_0^S : \{0, 1, \ldots, 2^N - 1\} \to [0, 1]$ is the initial distribution of the promoter state,

$$\boldsymbol{\Pi}(\tau_j, \tau_{j+1}) = \exp\left(\int_{\tau_j}^{\tau_{j+1}} \boldsymbol{H}(\boldsymbol{x}_t)\, dt\right), \tag{17}$$

and $\boldsymbol{x}_t$ is given by (14).

The push-forward method can be applied recursively to compute the MPD for times $\tau, 2\tau, \ldots, n_t\tau$. The complexity of the calculation scales as $n_t(n_x)^N(n_y)^N 2^{NM}$ which is exponential in the number of genes N. The exponential complexity comes from considering all the 2^{NM} possible paths S_M. However, many of these paths have almost the same probability and impose very similar trajectories to the variables $(\boldsymbol{x}_t, \boldsymbol{y}_t)$. In fact, a convenient approximation is to consider that different genes are switching between ON and OFF states according to Markov processes with rates given by the mean values of regulatory proteins (mean field approximation, [10]). This approximation consists in applying the push-forward procedure for each gene separately, using averaged transition probabilities. Thus, the 2^N states transition matrix $\boldsymbol{H}$ has to be replaced by N, 2×2 state transition matrices for each gene. This approximation reduces the complexity of the calculations to $n_t n_x n_y N 2^M$ which is linear in the number of genes.

In [10] we have replaced the regulation term $f_2 x_t^1$ occurring in the transition matrix by its mean $f_2 \mathbb{E}[x_t^1]$. In this case both H and Π can be computed analytically, which leads to a drastic reduction in the execution time. This approach is suitable for the model M_1, which contains only linear regulation terms. For nonlinear regulation terms, Π can not generally be computed analytically. Furthermore, the mean field approximation introduces biases. For instance, in the case of the model M_2, the approximation $f_2 x_t^1/(K_1 + x_t^1) \approx f_2 \mathbb{E}[x_t^1]/(K_1 + \mathbb{E}[x_t^1])$ is poor. A better approximation in this case is to replace $f_2 x_t^1/(K_1 + x_t^1)$ by its mean and use

$$\mathbb{E}\left[\frac{f_2 x_t^1}{K_1 + x_t^1}\right] \approx \frac{f_2 \mathbb{E}[x_t^1]}{(K_1 + \mathbb{E}[x_t^1])} - \frac{f_2}{(K_1 + \mathbb{E}[x_t^1])^3} Var(x_t^1), \qquad (18)$$

in order to correct the bias. Here Var indicates the variance.

As in [10] we can use analytic expressions for $\mathbb{E}[x_t^1]$, but also for $Var(x_t^1)$. These expressions can be found in Appendix A. Although the elements of matrix H have analytic expressions, the elements of the matrix Π contain integrals that must be computed numerically. For the model M_2, we have

$$\Pi_1(\tau, \tau') = \begin{bmatrix} (1 - p_{1,on}) + p_{1,on} e^{-\epsilon_1(\tau'-\tau)}, & (1 - p_{1,on}) - (1 - p_{1,on}) e^{-\epsilon_1(\tau'-\tau)} \\ p_{1,on} - p_{1,on} e^{-\epsilon_1(\tau'-\tau)}, & p_{1,on} + (1 - p_{1,on}) e^{-\epsilon_1(\tau'-\tau)} \end{bmatrix}, \qquad (19)$$

for the transition rates of the first gene, where $p_{1,on} = f_1/(f_1 + h_1)$, $\epsilon_1 = (f_1 + h_1)/\rho$, and

$$\Pi_2(\tau, \tau') =$$
$$\begin{bmatrix} e^{-\int_\tau^{\tau'}(h_2 + F_2(t))\,dt} + h_2 \int_\tau^{\tau'} e^{-\int_t^{\tau'}(h_2 + F_2(t'))\,dt'}\,dt, & h_2 \int_\tau^{\tau'} e^{-\int_t^{\tau'}(h_2 + F_2(t'))\,dt'}\,dt \\ 1 - e^{-\int_\tau^{\tau'}(h_2 + F_2(t))\,dt} - h_2 \int_\tau^{\tau'} e^{-\int_t^{\tau'}(h_2 + F_2(t'))\,dt'}\,dt, & 1 - h_2 \int_\tau^{\tau'} e^{-\int_t^{\tau'}(h_2 + F_2(t'))\,dt'}\,dt \end{bmatrix}, \qquad (20)$$

for the transitions of the second gene, where $F_2(t) = f_2 \mathbb{E}\left[\frac{x_t^1}{K_1 + x_t^1}\right]$

3   Results

3.1   Convergence of the Push-Forward Method

The probability distribution obtained with the push-forward method converges to the exact PDMP distribution in the limit $M \to \infty$. This is a consequence of the following theorem.

Theorem 1. *Let $\Phi_{S_M}(t, \boldsymbol{x}, \boldsymbol{y})$ be the flow defined by the formulas (14) and (13), such that $(\boldsymbol{x}_t, \boldsymbol{y}_t) = \Phi_{S_M}(t, \boldsymbol{x}(0), \boldsymbol{y}(0))$ for $t \in [0, \tau]$, and let $\mu_t^M : \mathcal{B}(\mathbb{R}^{2N}) \to \mathbb{R}_+$ be the probability measure defined as $\mu_t^M(A) = \sum_{S_M \in \Omega} \mathbb{P}[S_M] \mu_0(\Phi_{S_M}^{-1}(t, A))$, where $\mu_0 : \mathcal{B}(\mathbb{R}^{2N}) \to \mathbb{R}_+$ is the probability distribution of $(\boldsymbol{x}, \boldsymbol{y})$ at $t = 0$, $\mathbb{P}[S_M]$ are given by (16), and $\mathcal{B}(\mathbb{R}^{2N})$ are the Borel sets on $\mathbb{R}^{2N}$. Let μ_t, the exact distribution of $(\boldsymbol{x}_t, \boldsymbol{y}_t)$ for the PDMP defined by (1), (2) and (3), with initial values $(\boldsymbol{x}_0, \boldsymbol{y}_0, \boldsymbol{s}_0)$ distributed according to $\mu_0 \times P_0^S$. Assume that $|\tau_i - \tau_{i-1}| < C/M$ for all $i \in [1, M]$, where C is a positive constant. Then, for all $t \in [0, \tau]$, μ_t^M converges in distribution to μ_t, when $M \to \infty$.*

The proof of this theorem is given in the Appendix C.

3.2   Testing the Push-Forward Method

In order to test the push-forward method, we compared the resulting probability distributions with the ones obtained by the Monte Carlo method using the direct simulation of the PDMP. We considered the models M_1 and M_2 with the following parameters: $\rho = 1$, $p_1 = \frac{f}{f+h} = 1/2$, $a = 1/5$, $b = 4$, $k_0 = 4$, $k_1 = 40$ for the two genes. The parameter ϵ took two values $\epsilon = 0.5$ for slow genes and $\epsilon = \frac{f+h}{\rho} = 5.5$ for fast genes. We tested the slow-slow and the fast-fast combinations of parameters.

The initial distribution of the promoters states was $P_0^S((0,0)) = 1$ where the state $(0,0)$ means that both promoters are OFF. The initial probability measure μ_0 was a delta Dirac distribution centered at $x^1 = x^2 = 0$ and $y^1 = y^2 = 0$. This is obtained by always starting the direct simulation of the PDMP from $x_0^1 = x_0^2 = 0$, $y_0^1 = y_0^2 = 0$, and $s_0^1 = s_0^2 = 0$. The simulations were performed between $t_0 = 0$ and $t_{max} = 20$ for fast genes and between $t_0 = 0$ and $t_{max} = 90$ for slow genes. In order to estimate the distributions we have used $MC = 50000$ samples for the highest sampling.

The push-forward method was implemented with $M = 10$ equal length subintervals of $[0, \tau]$. The time step τ was chosen $\tau = 2$ for fast genes and $\tau = 15$ for slow genes. The procedure was iterated 10 times for fast genes (up to $t_{max} = 20$) and 6 times for slow genes (up to $t_{max} = 90$).

The execution times are provided in the Table 1. The comparison of the probability distributions are illustrated in the Figs. 1 and 2. In order to quantify the relative difference between methods we use the L^1 distance between distributions. More precisely, if $p(x)$ and $\tilde{p}(x)$ are probability density functions to be compared, the distance between distributions is

$$d = \int |p(x) - \tilde{p}(x)| \, dx. \tag{21}$$

Table 1. Execution times for different methods. All the methods were implemented in Matlab R2013b running on a single core (multi-threading inactivated) of a Intel i5-700u 2.5 GHz processor. The Monte-Carlo method computed the next jump waiting time using the analytical solution of Eq. 12 for M_1 and the numerical solution of Eq. 12 for M_2. The push-forward method used analytic solutions for mRNA and protein trajectories from (13) and (14), and numerical computation of the integrals in Eq. 20, for both models.

Model	Monte-Carlo high sampling [min]	Push-forward [s]
M_1 slow-slow	45	20
M_1 fast-fast	74	30
M_2 slow-slow	447	20
M_2 fast-fast	758	30

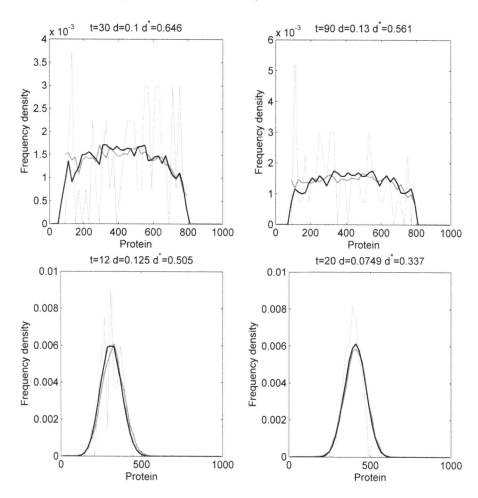

Fig. 1. Histograms of protein for the second gene, produced by the Monte-Carlo method (green lines) and by the push-forward method (black lines) for the model M_1. The green dotted line results from low sampling Monte-Carlo with similar execution time as the push-forward method, whereas the solid green line results from high sampling Monte-Carlo. The distances, defined by (21), are between low sampling and high sampling Monte-Carlo (d^*) and between push-forward and high sampling Monte-Carlo (d). (Color figure online)

This distance was computed between distributions resulting from the push-forward method and the Monte-Carlo method with the highest sampling. We have also used a reduced sampling Monte-Carlo scheme whose execution time is similar to the one of the push-forward method. The distributions resulting from low sampling and high sampling Monte-Carlo were compared using the same distance. Figures 1 and 2 clearly show that for the same execution time, the push-forward method outperforms the Monte-Carlo method.

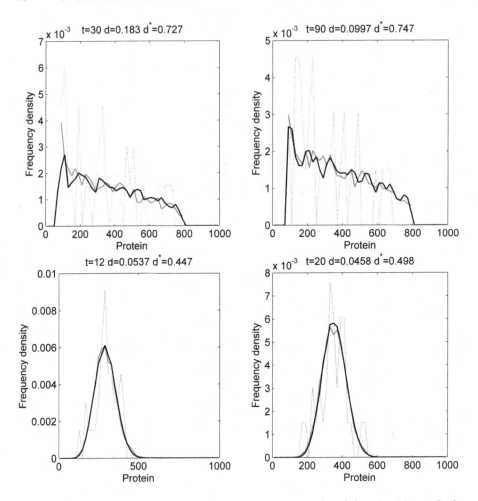

Fig. 2. Histograms of protein for the second gene, produced by the Monte-Carlo method (green lines) and by the Push-forward method (black lines) for the model M_2. The green dotted line results from low sampling Monte-Carlo with similar execution time as the push-forward method, whereas the solid green line results from high sampling Monte-Carlo. The distances, defined by (21), are between low sampling and high sampling Monte-Carlo (d^*) and between push-forward and high sampling Monte-Carlo (d). (Color figure online)

4   Discussion and Conclusion

Combining direct simulation of PDMP gene network models and analytic formulas for the ODE flow represents an effective, easy to implement method for computing time dependent MPD of these models. However, the precision of the Monte-Carlo estimates of the distributions increases like $\sqrt{MC}$, where MC is the number of Monte-Carlo samples. For this reason, the execution time of this method, although smaller compared to PDMP simulation methods that implement numerical resolution of the ODEs such as reported in [13] (data not shown), is large compared to deterministic methods such as the push-forward method.

The push-forward method represents an effective alternative to Monte-Carlo methods, ensuring reduced execution time. With respect to an earlier implementation of this method in [9] we used promoter states instead of mRNA copy numbers as discrete variables of the PDMP. As a consequence, the number of discrete states is lower and we can afford increasing the number M of time subdivisions. Compared to the similar work in [10] we used second moments of the protein distribution which took into account the correlation of the promoter states and lead to increased accuracy in the case of nonlinear regulation. We proved rigorously the convergence of the distributions calculated with the push-forward method to the exact distributions of the PDMP. However, the push-forward method is an approximate method, and its accuracy relies on the careful choice of the time and space steps, namely of the integers M, n_t, n_x, n_y. We will present elsewhere error estimates allowing an optimal choice of these parameters. Although the protein moments and the exponential transition rate matrix Π can be computed numerically, the effectiveness of the push-forward method is increased when analytic expressions are available for these quantities. In this paper, these expressions were computed for particular cases. In the future, we will provide expressions, as well as symbolic computation tools to compute these quantities in more general cases. We situate our findings in the broader effort of the community to produce new effective tools for computational biology by combining numerical and symbolic methods.

A   Appendix: Mean and Variance of the Protein

We compute here the mean and the variance of the protein synthesized by a constitutive promoter (gene 1 of models M_1 and M_2).

We start with

$$x_t = x_0 e^{-at} + b \int_0^t \left[y_0 e^{-\rho t'} + \int_0^{t'} \frac{k_0 + (k_1 - k_0)s_{t''}}{\rho} e^{\rho(t''-t')} dt'' \right] e^{a(t'-t)} dt',$$

(22)

where $s_t = 0$ if the promoter is OFF and $s_t = 1$ if the promoter is ON at the time t.

Equation 22 leads to

$$x_t = x_0 e^{-at} + b y_0 \frac{e^{-\rho t} - e^{-at}}{a - \rho} + \frac{b k_0}{\rho} \left(\frac{1 - e^{-at}}{a} - \frac{e^{-\rho t} - e^{-at}}{a - \rho} \right)$$
$$+ \frac{b(k_1 - k_0)}{\rho} \int_0^t \left[\int_0^{t'} s_{t''} e^{\rho t''} dt'' \right] e^{(a-\rho)t'} e^{-at} dt'.$$

(23)

From (23) it follows

$$
\mathbb{E}[x_t] = \mathbb{E}[x_0]\, e^{-at} + b\mathbb{E}[y_0] \frac{e^{-\rho t} - e^{-at}}{a - \rho} + \frac{bk_0}{\rho}\left(\frac{1 - e^{-at}}{a} - \frac{e^{-\rho t} - e^{-at}}{a - \rho}\right)
$$
$$
+ \frac{b(k_1 - k_0)}{\rho} \int_0^t \left[\int_0^{t'} \mathbb{E}[s_{t''}]\, e^{\rho t''}\, dt''\right] e^{(a-\rho)t'} e^{-at}\, dt'. \tag{24}
$$

The promoter state variable s_t follows the master equation

$$
\frac{d\mathbb{P}[s_t = 1]}{dt} = f(1 - \mathbb{P}[s_t = 1]) - (f + h)\mathbb{P}[s_t = 1], \tag{25}
$$

that has the solution

$$
\mathbb{E}[s_t] = \mathbb{P}[s_t = 1] = (p10 - p_1)e^{-\rho\epsilon t} + p_1, \tag{26}
$$

where $p10 = \mathbb{P}[s_0 = 1]$, $\epsilon = (f+h)/\rho$, and $p_1 = f/(f+h)$. Using straightforward algebra, we find

$$
\mathbb{E}[x_t] = M_0 + M_1 e^{-at} + M_2 e^{-\rho t} + M_3 e^{-\epsilon t}, \tag{27}
$$

where

$$
M_0 = \frac{b(k_0 + (k_1 - k_0)p_1)}{a}, \tag{28}
$$

$$
M_1 = \mathbb{E}[x_0] - \frac{b\mathbb{E}[y_0]}{a - \rho} + \frac{bk_0}{a(a - \rho)} + \frac{b(k_1 - k_0)(p_{10} - p_1)}{(a - \rho\epsilon)(a - \rho)} + \frac{b(k_1 - k_0)p_1}{a(a - \rho)}, \tag{29}
$$

$$
M_2 = \frac{b\mathbb{E}[y_0]}{a - \rho} - \frac{bk_0}{a - \rho} - \frac{b(k_1 - k_0)(p_{10} - p_1)}{\rho(1 - \epsilon)(a - \rho)} - \frac{b(k_1 - k_0)p_1}{a - \rho}, \tag{30}
$$

$$
M_3 = \frac{b(k_1 - k_0)(p_{10} - p_1)}{\rho(1 - \epsilon)(a - \epsilon\rho)}. \tag{31}
$$

From (23) we find also

$$
Var(x_t) = Var(x_0)e^{-2at} + b^2 Var(y_0)\left(\frac{e^{-\rho t} - e^{-at}}{a - \rho}\right)^2 + \left(\frac{b(k_1 - k_0)}{\rho}\right)^2 e^{-2at}
$$
$$
\times \int_0^t \int_0^t dt_2 dt_4 \left[\int_0^{t_2} \int_0^{t_4} (\mathbb{E}[s_{t_1} s_{t_3}] - \mathbb{E}[s_{t_1}]\mathbb{E}[s_{t_3}])e^{\rho t_1} e^{\rho t_3}\, dt_1 dt_3\right] e^{(a-\rho)t_2} e^{(a-\rho)t_4}. \tag{32}
$$

We have considered here that x_0, y_0 are uncorrelated, but more general expressions can be obtained.

In order to compute the two times covariance $\mathbb{E}[s_{t_1} s_{t_3}] - \mathbb{E}[s_{t_1}]\mathbb{E}[s_{t_3}]$ we combine the tower property of the conditional expectation with the Markov property satisfied by s_t. More precisely, for $t_1 \geq t_3$ we find $\mathbb{E}[s_{t_1} s_{t_3}] = \mathbb{E}[\mathbb{E}[s_{t_1} s_{t_3}|s_{t_3}]] = \mathbb{E}[((s_{t_3} - p_1)e^{-\rho\epsilon(t_1-t_3)} + p_1)s_{t_3}]$ and $\mathbb{E}[s_{t_1}]\,\mathbb{E}[s_{t_3}] = \mathbb{E}[\mathbb{E}[s_{t_1}|s_{t_3}]]\,\mathbb{E}[s_{t_3}] = ((\mathbb{E}[s_{t_3}] - p_1)e^{-\rho\epsilon(t_1-t_3)} + p_1)\mathbb{E}[s_{t_3}]$. Then, it follows

$$\mathbb{E}[s_{t_1} s_{t_3}] - \mathbb{E}[s_{t_1}]\,\mathbb{E}[s_{t_3}] = Var[s_{t_3}]e^{-\rho\epsilon(t_1-t_3)}. \tag{33}$$

s_{t_3} is a Bernoulli variable, therefore $Var[s_{t_3}] = \mathbb{E}[s_{t_3}]\,(1 - \mathbb{E}[s_{t_3}])$. From (33) and (26) it follows

$$\mathbb{E}[s_{t_1} s_{t_3}] - \mathbb{E}[s_{t_1}]\,\mathbb{E}[s_{t_3}] = p_1(1 - p_1)e^{-\rho\epsilon(t_1-t_3)} + (1 - 2p_1)(p10 - p_1)e^{-\rho\epsilon t_1}$$
$$- (p10 - p_1)^2 e^{-\rho\epsilon(t_1+t_3)}, \quad \text{for } t_1 \geq t_3. \tag{34}$$

Similarly, one gets

$$\mathbb{E}[s_{t_1} s_{t_3}] - \mathbb{E}[s_{t_1}]\,\mathbb{E}[s_{t_3}] = p_1(1 - p_1)e^{-\rho\epsilon(t_3-t_1)} + (1 - 2p_1)(p10 - p_1)e^{-\rho\epsilon t_3}$$
$$- (p10 - p_1)^2 e^{-\rho\epsilon(t_1+t_3)}, \quad \text{for } t_3 \geq t_1. \tag{35}$$

The domain of the multiple integral in (32) should be split in two sub-domains corresponding to $t_2 < t_4$ and to $t_2 > t_4$. Each of these sub-domains should be subdivided into two smaller sub-domains corresponding to $t_3 > t_1$ and $t_1 < t_3$. Symmetry arguments imply that the integrals on $t_2 < t_4$ and on $t_4 < t_2$ are equal, which allows us to perform the calculation of the integral on only two sub-domains, instead of four. After some algebra we find

$$Var(x_t) = Var(x_0)e^{-2at} + b^2 Var(y_0)\left(\frac{e^{-\rho t} - e^{-at}}{a - \rho}\right)^2$$
$$- \left[\frac{(p10 - p_1)(k_1 - k_0)b}{\rho(1 - \epsilon)}\left(\frac{e^{-\rho\epsilon t} - e^{-at}}{a - \rho\epsilon} - \frac{e^{-\epsilon t} - e^{-at}}{a - \rho\epsilon}\right)\right]^2$$
$$+ \frac{p_1(1 - p_1)(k_1 - k_0)^2 b^2}{\rho^2}(V_0 + V_1 e^{-(a+\rho\epsilon)t} + V_2 e^{-\rho(1+\epsilon)t} + V_3 e^{-2at} + V_4 e^{-(a+\rho)t} + V_5 e^{-2\rho t})$$
$$+ \frac{(1 - 2p_1)(p10 - p_1)(k_1 - k_0)^2 b^2}{\rho^2}(V_6 e^{-\rho\epsilon t} + V_7 e^{-(a-\rho\epsilon)t} + V_8 e^{-\rho(\epsilon+1)t}$$
$$+ V_9 e^{-(a+\rho\epsilon)t} + V_{10} e^{-\rho t} + V_{11} e^{-2\rho t}), \tag{36}$$

where

$$V_0 = \frac{a + (\epsilon + 1)\rho}{a(a + \rho\epsilon)(a + \rho)(\epsilon + 1)}, V_1 = -\frac{2}{(a^2 - \rho^2\epsilon^2)(a - \rho)(\epsilon - 1)},$$

$$V_2 = \frac{2}{(a - \rho\epsilon)(a - \rho)(\epsilon^2 - 1)}, V_3 = \frac{1}{a(a - \rho\epsilon)(a - \rho)^2},$$

$$V_4 = \frac{2(a + (1 - 2\epsilon)\rho)}{(a - \rho\epsilon)(a - \rho)^2(a + \rho)(\epsilon - 1)}, V_5 = -\frac{1}{(\epsilon - 1)(a - \rho)^2},$$

$$V_6 = -\frac{2(2a + (2 - \epsilon)\rho)}{a(\epsilon - 2)(2a - \rho\epsilon)(a + (1 - \epsilon)\rho)}, V_7 = \frac{2}{(1 - \epsilon)a(a - \rho)(a - \rho\epsilon)},$$

$$V_8 = \frac{2}{(a - \rho\epsilon)(a - \rho)(\epsilon - 1)}, V_9 = \frac{2(a + (1 - 2\epsilon)\rho)}{(a - \rho)^2(\epsilon - 1)(a - \rho\epsilon)(a + (1 - \epsilon))},$$

$$V_{10} = \frac{2}{(a - \rho)^2(2 - \epsilon)(1 - \epsilon)}, V_{11} = \frac{2}{(a - \rho)^2(2a - \rho\epsilon)(a - \rho\epsilon)}. \tag{37}$$

B   Appendix: Details of the Derivation of (13) and (14)

x_t and y_t satisfy the following system of equations

$$\frac{dx}{dt} = by - ax$$
$$\frac{dy}{dt} = k_0 + (k_1 - k_0)s - \rho y \tag{38}$$

For simplification, we rescale variables and parameters $t \to t\rho$, $k_i \to k_i/\rho$, $a \to a/\rho$, $b \to b/\rho$ and obtain

$$\frac{dx}{dt} = by - ax$$
$$\frac{dy}{dt} = k_0 + (k_1 - k_0)s - y \tag{39}$$

From (39) it follows $y_\tau = y_0 e^{-\tau} + \int_0^\tau d\tau' e^{-(\tau-\tau')}[k_0 + (k_1 - k_0)s(\tau')] = y_0 e^{-\tau} + \sum_{j=1}^{M-1} \int_{\tau_j}^{\tau_{j+1}} d\tau' e^{-(\tau-\tau')}[k_0 + (k_1 - k_0)s(\tau_j)] = y_0 e^{-\tau} + k_0(1 - e^{-\tau}) + (k_1 - k_0) \sum_{j=0}^{M-1} e^{-\tau}(e^{\tau_{j+1}} - e^{\tau_j})s(\tau_j)$ and hence (13).

From (39) we also obtain $x_\tau = x_0 e^{-a\tau} + b \int_0^\tau e^{a(\tau'-\tau)}y(\tau')d\tau' = x_0 e^{-a\tau} + \frac{by_0}{a-1}(e^{-\tau} - e^{-a\tau}) + b \int_0^\tau d\tau' e^{a(\tau'-\tau)} \int_0^{\tau'} d\tau'' e^{-(\tau'-\tau'')}(k_0 + (k_1-k_0)s_{\tau''}) = x_0 e^{-a\tau} + \frac{by_0}{a-1}(e^{-\tau} - e^{-a\tau}) + bk_0 \left(\frac{1-e^{-a\tau}}{a} + \frac{e^{-a(\tau-\tau_0)}-e^{-(\tau-\tau_0)}}{a-1} \right) + b(k_1 - k_0)I$, where

$$I = \int_0^\tau d\tau' e^{a(\tau'-\tau)} \int_0^{\tau'} d\tau'' e^{-(\tau'-\tau'')} s_{\tau''}.$$

In order to compute the integral I we decompose the triangular integration domain into $M - 1$ rectangles and M triangles on each of which $s_{\tau''}$ is constant, as in Fig. 3.

The contribution of each rectangle to the integral I is $\int_{\tau_i}^\tau d\tau' e^{a(\tau'-\tau)}$ $\int_{\tau_{i-1}}^{\tau_i} d\tau'' e^{-(\tau'-\tau'')} s_{\tau_{i-1}} = e^{-a\tau} \frac{e^{(a-1)\tau} - e^{(a-1)\tau_i}}{a-1}(e^{\tau_i} - e^{\tau_{i-1}})s_{\tau_{i-1}}$.

The contribution of each triangle to the integral I is $\int_{\tau_{i-1}}^{\tau_i} d\tau' e^{a(\tau'-\tau)} \int_{\tau_{i-1}}^{\tau'} d\tau''$ $e^{-(\tau'-\tau'')} s_{\tau_{i-1}} = e^{-a\tau} s_{\tau_{i-1}} \left(\frac{e^{a\tau_i} - e^{a\tau_{i-1}}}{a} - e^{\tau_{i-1}} \frac{e^{(a-1)\tau_i} - e^{(a-1)\tau_{i-1}}}{a} \right)$.

It follows that $I = \sum_{i=1}^{M-1} e^{-a\tau} \frac{e^{(a-1)\tau} - e^{(a-1)\tau_i}}{a-1}(e^{\tau_i} - e^{\tau_{i-1}})s_{\tau_{i-1}} + \sum_{i=1}^M e^{-a\tau}$ $s_{\tau_{i-1}} \left(\frac{e^{a\tau_i} - e^{a\tau_{i-1}}}{a} - e^{\tau_{i-1}} \frac{e^{(a-1)\tau_i} - e^{(a-1)\tau_{i-1}}}{a} \right)$.

Noting that the first sum in the expression of I can go to $i = M$ (the M-th term is zero) we obtain (14).

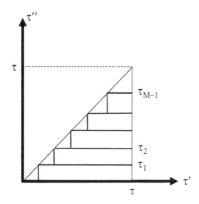

Fig. 3. Decomposition of the integration domain for computing the integral I.

C   Appendix: Proof of the Theorem 1

The proof the Theorem 1 relies on the following Lemma:

Lemma 1. *Let (x_t, y_t, s_t) be a realization of the PDMP such that $s(\tau_j) = s_j$ for all $j \in [0, M-1]$ and let (x_t^M, y_t^M, s_t^M) be the push-forward solutions computed with (13) and (14) considering that s_t^M is piecewise constant on the intervals $[\tau_j, \tau_{j+1}]$, $s_t^M = s_j$, $\tau_j \leq t < \tau_{j+1}$. Then, under the conditions of Theorem 1, $\mathbb{P}\big[||y_t - y_t^M|| > \delta\big] \to 0$ and $\mathbb{P}\big[||x_t - x_t^M|| > \delta\big] \to 0$ when $M \to \infty$, for any $\delta > 0$ and for all $t \in [0, \tau]$.*

We prove this Lemma for y_t. The proof for x_t follows the same principles.

Proof. For each gene i, x_t^i and y_t^i are solutions of the equations (38) with $s = s^i$. The box $\mathcal{B} = \{k_0/\rho \leq y^i \leq k_1/\rho, (bk_0)/(\rho a) \leq x^i \leq (bk_1)/(\rho a), 1 \leq i \leq N\}$ is invariant with respect to the gene network flow defined by (38). Precisely, if the initial data is in $\mathcal{B}$, then the entire trajectory is confined to $\mathcal{B}$. If the initial data is not in $\mathcal{B}$, then the solutions x_t^i, y_t^i, $1 \leq i \leq N$ converge monotonically to the unique point attractor $y^i = (k_0 + (k_1 - k_0)s^i)/\rho$, $x^i \leq (b(k_0 + (k_1 - k_0)s^i))/(\rho a)$, $i \in [1, N]$ in $\mathcal{B}$. From these elementary remarks it follows that the jump rate is bounded, $\lambda(x_t, y_t, s_t) < A$, for all $t \in [0, \tau]$. Therefore, the random variables η_i, representing the number of jumps of the promoter i in the interval $[0, \tau]$ has bounded mean $\mathbb{E}[\eta_i] < A\tau$.

Consider that for the ith gene there are η_i jumps such that s^i changes from ON to OFF or from OFF to ON, on the interval $[0, \tau]$.

The positions of these jumps are $\tau_{j_l}^* \in [\tau_{j_l}, \tau_{j_l+1})$, for $l \in [1, \eta_i]$. Using (13) it follows $|y_t^i - y_t^{i,M}| \leq e^{-\rho\tau} \frac{k_1 - k_0}{\rho} \sum_{l=1}^{\eta_i} |e^{\rho\tau_{j_l}^*} - e^{\rho\tau_{j_l}}| \leq (k_1 - k_0) \sum_{l=1}^{\eta_i} (\tau_{j_l+1} - \tau_{j_l}) < (k_1 - k_0)\frac{\eta_i C}{M}$.

Using Markov's inequality we find that $\mathbb{P}\big[(k_1 - k_0)\frac{\eta_i C}{M} > \delta\big] \leq \frac{AC\tau(k_1 - k_0))}{\delta M}$.

It follows that $\mathbb{P}\big[|y_t^i - y_t^{i,M}| > \delta\big] \to 0$ when $M \to \infty$, for any $\delta > 0$, $1 \leq i \leq N$.

The proof of the Theorem 1 follows from the Lemma 1 because by construction, the promoter states have the same distribution in the push-forward and PDMP schemes, and the convergence in probability of the mRNAs and of the proteins implies the convergence in distribution of these variables.

References

1. Cai, L., Friedman, N., Xie, X.S.: Stochastic protein expression in individual cells at the single molecule level. Nature **440**(7082), 358 (2006)
2. Crudu, A., Debussche, A., Muller, A., Radulescu, O.: Convergence of stochastic gene networks to hybrid piecewise deterministic processes. Ann. Appl. Probab. **22**, 1822–1859 (2012)
3. Crudu, A., Debussche, A., Radulescu, O.: Hybrid stochastic simplifications for multiscale gene networks. BMC Syst. Biol. **3**(1), 89 (2009)
4. Eldar, A., Elowitz, M.B.: Functional roles for noise in genetic circuits. Nature **467**(7312), 167 (2010)
5. Elowitz, M.B., Levine, A.J., Siggia, E.D., Swain, P.S.: Stochastic gene expression in a single cell. Science **297**(5584), 1183–1186 (2002)
6. Ferguson, M.L., et al.: Reconciling molecular regulatory mechanisms with noise patterns of bacterial metabolic promoters in induced and repressed states. Proc. Natl. Acad. Sci. USA **109**, 155 (2012)
7. Gupta, P.B.: Stochastic state transitions give rise to phenotypic equilibrium in populations of cancer cells. Cell **146**(4), 633–644 (2011)
8. Herbach, U., Bonnaffoux, A., Espinasse, T., Gandrillon, O.: Inferring gene regulatory networks from single-cell data: a mechanistic approach. BMC Syst. Biol. **11**(1), 105 (2017)
9. Innocentini, G.C.P., Forger, M., Radulescu, O., Antoneli, F.: Protein synthesis driven by dynamical stochastic transcription. Bull. Math. Biol. **78**(1), 110–131 (2016)
10. Innocentini, G.C.P., Hodgkinson, A., Radulescu, O.: Time dependent stochastic mRNA and protein synthesis in piecewise-deterministic models of gene networks. Front. Phys. **6**, 46 (2018)
11. da Costa Pereira Innocentini, G., Forger, M., Ramos, A.F., Radulescu, O., Hornos, J.E.M.: Multimodality and flexibility of stochastic gene expression. Bull. Math. Biol. **75**(12), 2600–2630 (2013)
12. Kurasov, P., Lück, A., Mugnolo, D., Wolf, V.: Stochastic hybrid models of gene regulatory networks-a PDE approach. Math. Biosci. **305**, 170–177 (2018)
13. Lin, Y.T., Buchler, N.E.: Efficient analysis of stochastic gene dynamics in the non-adiabatic regime using piecewise deterministic Markov processes. J. R. Soc. Interface **15**(138), 20170804 (2018)
14. Raj, A., Peskin, C.S., Tranchina, D., Vargas, D.Y., Tyagi, S.: Stochastic mRNA synthesis in mammalian cells. PLoS Biol. **4**(10), e309 (2006)
15. Razooky, B.S., Pai, A., Aull, K., Rouzine, I.M., Weinberger, L.S.: A hardwired HIV latency program. Cell **160**(5), 990–1001 (2015)
16. Riedler, M.G.: Almost sure convergence of numerical approximations for piecewise deterministic Markov processes. J. Comput. Appl. Math. **239**, 50–71 (2013)
17. Tantale, K., et al.: A single-molecule view of transcription reveals convoys of RNA polymerases and multi-scale bursting. Nat. Commun. **7**, 12248 (2016)

18. Thattai, M., Van Oudenaarden, A.: Stochastic gene expression in fluctuating environments. Genetics **167**(1), 523–530 (2004)
19. Thomas, P., Popović, N., Grima, R.: Phenotypic switching in gene regulatory networks. Proc. Natl. Acad. Sci. **111**(19), 6994–6999 (2014)
20. Zeiser, S., Franz, U., Wittich, O., Liebscher, V.: Simulation of genetic networks modelled by piecewise deterministic Markov processes. IET Syst. Biol. **2**(3), 113–135 (2008)

On Chemical Reaction Network Design by a Nested Evolution Algorithm

Elisabeth Degrand, Mathieu Hemery, and François Fages[(✉)]

Inria Saclay, Lifeware Group, Palaiseau, France
francois.fages@inria.fr

Abstract. One goal of synthetic biology is to implement useful functions with biochemical reactions, either by reprogramming living cells or programming artificial vesicles. In this perspective, we consider Chemical Reaction Networks (CRN) as a programming language, and investigate the CRN program synthesis problem. Recent work has shown that CRN interpreted by differential equations are Turing-complete and can be seen as analog computers where the molecular concentrations play the role of information carriers. Any real function that is computable by a Turing machine in arbitrary precision can thus be computed by a CRN over a finite set of molecular species. The proof of this result gives a numerical method to generate a finite CRN for implementing a real function presented as the solution of a Polynomial Initial Values Problem (PIVP). In this paper, we study an alternative method based on artificial evolution to build a CRN that approximates a real function given on finite sets of input values. We present a nested search algorithm that evolves the structure of the CRN and optimizes the kinetic parameters at each generation. We evaluate this algorithm on the Heaviside and Cosine functions both as functions of time and functions of input molecular species. We then compare the CRN obtained by artificial evolution both to the CRN generated by the numerical method from a PIVP definition of the function, and to the natural CRN found in the BioModels repository for switches and oscillators.

1 Introduction

One goal of Synthetic Biology is to implement useful functions using biochemical reactions, either by reprogramming living cells, or by programming artificial devices. While the former approach is mainstream in synthetic biology [1,12,17, 19,41,47], examples of the later approach are given by the whole field of DNA computing [2,8,11,15,38] or analog computing with enzymatic reactions [42], possibly encapsulated in artificial vesicles [13].

Chemical Reaction Networks (CRN) are used to describe systems of chemical reactions. In this article, we consider CRN as a programming language [20, 46]. We focus on their continuous semantics defined by Ordinary Differential Equations (ODE) on the molecular concentrations. We study the CRN program synthesis problem, i.e. the problem of designing a CRN for implementing a given

© Springer Nature Switzerland AG 2019
L. Bortolussi and G. Sanguinetti (Eds.): CMSB 2019, LNBI 11773, pp. 78–95, 2019.
https://doi.org/10.1007/978-3-030-31304-3_5

function, either a function of time (Problem 1) or a function of input molecular species (Problem 2):

Problem 1. Given a function $f : [0, a] \rightarrow \mathbb{R}$, construct a CRN such that one of the species has a concentration A implementing the function f. That is $A(t) = f(t), t \in [0, a]$.

Problem 2. Given a function $f : [0, a] \rightarrow \mathbb{R}$, construct a CRN such that, when initialized for a given input, one of the species converge to the desired result. That is $\forall x \in [0, a], X(t = 0) = P_I(x) \Rightarrow A(t = 1) = f(x)$ where X is the vector of species concentrations and P_I a polynomial of degree one.

A recent result proving the Turing completeness of continuous CRN [20], restricted to mass action law kinetics and with at most two reactants, has given rise to a numerical method for compiling computable real functions and programs in finite CRN. This approach is implemented in Biocham-4[1] [20] and CRN++[2] [46] with some different design choices. The theoretical result states that one can restrict to real functions presented as solutions of polynomial differential equations and polynomial initial values as functions of the inputs (PIVP). For a positive PIVP, the transformation can use a CRN inference algorithm from ODEs such as [29] or [21] initially dedicated to importing MatLab models in SBML [25]. For negative variables, the transformation can use the dual-rail encoding of a real variable by the difference between two positive variables for the positive and negative parts respectively [37]. In the generated CRN the molecular concentration play the role of information carriers, similarly to analog computation performed by protein complexes in cells [17,43].

In this article, we study an alternative CRN design method based on artificial evolution, and compare the results to the numerical method. In [15] a complete search method is described for scanning the space of CRN that may implement a given function with a limited number of species and reactions, and then optimizing the kinetic parameters using a metropolis like algorithm. This enumerative algorithm is limited to small size CRN with only a handful of species and reactions. Another method described in [6] consists in sketching a CRN in broad outline and letting an optimization algorithm complete the holes to perform the desired function. This method thus requires prior knowledge on the CRN. In the framework of gene regulatory networks, [33] gives a method to infer network parameters but with also prior knowledge on the structure of the network. In [18] a method to evolve reaction networks is presented using the DNA toolbox and one single genetic algorithm for both structure and parameters.

The method presented here does not require prior knowledge. It is based on artificial evolution with the hope of finding CRN more akin to comparison to natural CRN present in BioModels. The idea is to let an evolutionary algorithm (EA) evolve a population of CRN to approximate a real function given on a finite set of values. These data are either finite traces for approximating a function of

---

[1] http://lifeware.inria.fr/biocham4.
[2] https://github.com/marko-vasic/crnPlusPlus.

time (Problem 1), or dose-response diagrams for approximating a function of input (Problem 2). The general idea is to let evolve a population of potential CRN while selecting them according to a fitness that is the distance between their output and the desired target function. We take here as a distance the L2 norm evaluated in a specified set of points.

Such a CRN design method by artificial evolution can also be seen as a machine learning procedure [45] to learn a CRN from traces either observed in the systems biology perspective, or desired in the synthetic biology perspective. This method may be seen as a learning problem akin to those currently solved with neural networks or linear regression. The input is a set of points (x_i, y_i) called *data* and the goal is to find a function f from a certain class that minimizes the difference between the y_i and $f(x_i)$. The output of the approximation is the best function f. However, as the approximation function f is computed through a CRN, we also hope for a network that will gives us some explanatory power in systems biology, as well as some implementability in synthetic biology.

We present a nested evolution algorithm which combines a genetic algorithm for evolving the CRN structures of a population of CRN, with a black-box continuous optimization procedure, namely the covariance matrix adaptation evolutionary strategy (CMA-ES) [28], for optimizing the kinetic parameters of the CRN by evolving a population of parameter settings. This nesting of two levels of evolution has been proposed, see for instance [4] for mixed-variable optimization problems in the framework of non-linear optimization problems, refined in [5] to evolve a cell model with structure optimization and parameter optimization. The authors suggest to use a better genetic algorithm for parameter optimization and to parallelize the genetic algorithms. This is what we show here with the choice of CMA-ES [28] for parameter optimization, and the parallelization of both populations of the genetic algorithm and CMA-ES.

In the next section, we recall basic definitions of continuous CRN, their Turing completeness, and the automatic design method based on this approach. In Sect. 3 we present our two level evolutionary algorithm for learning CRN structure and kinetic parameters, and its parallel implementation. In Sect. 4 we evaluate the results obtained with this algorithm on both functions of time and input/output functions, by focusing in both cases, on the Heaviside function as an ideal sigmoid function, and on the cosine function as an ideal oscillator. On these examples, we compare the CRN obtained by artificial evolution to the CRN obtained by compilation. In Sect. 5, we compare the CRN obtained by our algorithm for the cosine function, to CRN present in BioModels using the subgraph epimorphism method [23,24], and discuss their relationship to circadian clock models.

2   CRN Design by PIVP Compilation

In [20] we have shown that any computable function over the reals can be implemented by a CRN over a finite set of molecular species. This result relies on previous results on analog computation and complexity showing the Turing completeness of polynomial initial value problems (PIVPs), i.e. numerical integration with arbitrary precision of polynomial ODEs from polynomial initial conditions [3]. More precisely, we showed

Theorem 1 ([20]). *A function over the reals is computable (resp. in polynomial time) if and only if it is computable by a CRN with mass action law kinetics using only synthesis reactions with at most two catalysts, of the form*

$$\text{- => z } or \text{ \_ =[x]=> z } or \text{ \_ =[x+y]=> z}$$

and annihilation reactions of the form

$$\text{x\_p + x\_m => \_}$$

(resp. with trajectories of polynomial length).

This form of reactions based solely on catalytic synthesis and degradation by annihilation is given by the proof of Turing completeness but is not very appealing from a biochemical point of view. It should be understood as an abstract form of reactions which can be realized with real reactions, for instance by replacing an annihilation reaction by a complexation reaction forming an inactive complex, a synthesis reaction by a transformation reaction from an inactive form, e.g. by a phosphorylation reaction where the catalyst is a kinase, etc. Such a realization of abstract CRN in CRN with real enzymes is described in [14] using the BRENDA database of enzymes[3] and has been used for instance in [13] to implement designed circuits with real enzymes in artificial vesicle.

Example 1. The cosine function of time can be specified by the PIVP

$$df/dt = z \quad dz/dt = -f \quad f(0) = 1 \quad z(0) = 0$$

The CRN generated by the PIVP method introduces variables for the positive and negative part of f and z together with annihilation reactions given with a sufficiently high value rate constant *fast*:

```
biocham: compile_from_expression(cos, time, f).
  _ = [z2_p] => f_p.
  _ = [z2_m] => f_m.
  _ = [f_m] => z2_p.
  _ = [f_p] => z2_m.
  fast*z2_m*z2_p for z2_m+z2_p => _.
  fast*f_m*f_p for f_m+f_p => _.
  present (f_p, 1).
biocham: list_ode.
  d(f_p)/dt = z2_p-fast*f_m*f_p
  d(f_m)/dt = z2_m-fast*f_m*f_p
  d(z2_p)/dt = f_m-fast*z2_m*z2_p
  d(z2_m)/dt = f_p-fast*z2_m*z2_p
```

The inference of general CRN has a low time complexity, linear in the number of variables and monomials for generating general reactions [21]. However, the transformation to reactions with at most two reactants, may take exponential time in the worst case [9].

---

[3] https://www.brenda-enzymes.org/.

3   CRN Design by Artificial Evolution

3.1   Nested Evolution Algorithms for Structure and Kinetics

The goal here is to learn a CRN that minimizes the error between the input trace and the simulation trace of the CRN. Thanks to Theorem 1 we know that we can restrict ourselves to elementary reactions with at most two reactants of some simple form and with mass action law kinetics, i.e. to PIVPs with monomials of degree at most 2.

While the structure of the CRN/PIVP defines what is possible, the value of the parameters (i.e. kinetic parameters and initial concentrations) is essential to the function and a bad exploration of the parameter space might lead to a wrong rejection of an actually good structure. To solve this difficulty we use two nested evolutionary algorithms where the first optimizes the structure and the second optimizes the parameter values, as summarized in Fig. 1.

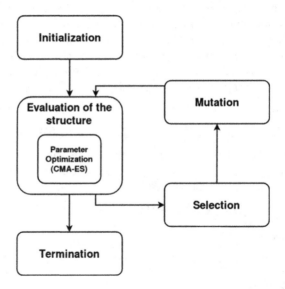

Fig. 1. Schematic representation of the nested evolution algorithm. The initialization corresponds to the creation of different random individuals. The evaluation of a given structure implies a sub-call to the CMA-ES algorithm with a limited time budget, and only the best set of parameters is kept to compute the fitness of the individual in the main algorithm. Elite selection is used where half the population is deleted while the second one is copied, this scheme allows for fast selection (no random number is needed) and keep a constant population size. The mutation scheme incorporates adding or removing variables or monomials, and switching signs. Termination is enforced by fixing the number of generations.

The Covariance Matrix Adaptation Evolution Strategy (CMA-ES) [28] is a black-box derivative-free continuous optimization algorithm used here to optimize the kinetic parameters at each step of the CRN structure evolution. This algorithm enjoys many invariance properties with respect to scales as well as linear, rotational or any order-preserving transformations. It is a state-of-the-art algorithm that shows best performances on the hardest non-convex even discontinuous problems [27]. Here the parameters are positive and are searched according to a logarithmic scale. CMA-ES has been used in Biocham for parameter search and robustness optimization with respect to quantitative temporal logic properties [22,39,40] with notable success in both systems biology [30] and synthetic biology [13] as well as external control of cell processes [44].

Algorithm 1. Structure Optimization

---

function EVOLUTION(f)
    population ← INITIALIZEPOPULATION(f)
    for g in generation **do**
        for p in population **do**
            STRUCTUREFITNESS(f,p)
        end for
        pop_new ← SELECT(population)
        population ← pop_new + MUTATE(pop_new)   ▷ Concatenate the two lists
    end for
    return SELECT_BEST(population)
end function

Algorithm 2. Parameter Optimization

---

procedure STRUCTUREFITNESS(f, pivp)
    u ← RANDOM(0, 1)
    if u == 1 **then**
        x0 ← RANDOMINITIALSTATE(pivp)
    else
        x0 ← INITIALSTATEFROMBEST(pivp)
    end if
    fit_value, fit_param ← CMAES(FITNESSFUNCTION(pivp),x0)
    UPDATEBESTPARAMETERS(pivp, fit_value, fit_param)
end procedure

The top-level genetic algorithm for learning the structure of the PIVP is given in Algorithm 1. Each monomial has a sign, $+$ or $-$. To handle the constraint of positivity of the concentration variables, a monomial in the ODE of one variable cannot have a negative sign if the variable is not present in the monomial. For example, the derivative of x can depend on $-xy$ but not on $-y$ which is equivalent to say that a degradation reaction must have the degraded species as reactant. This constraint ensures the positivity of the system (Prop. 1 in [21]). With this provision, there exists a canonical (yet not unique) way of associating a CRN to

---

Algorithm 3. Fitness function for functions of time

---

   function FITNESSFUNCTION(f, pivp, param)
      coeff, y0 ← PARAMTOCOEFF(param)
      sol ← INTEGRATE(pivp, coeff, y0)
      return LOSS(sol, f)
   end function

Algorithm 4. Fitness function for functions of input

---

   function FITNESSFUNCTION(f, pivp, param)
      coeff, y0 ← PARAMTOCOEFF(param)
      value ← 0
      for x_i in x **do**
         sol ← INTEGRATE(pivp, coeff, $y0(x_i)$)
         value ← value + LOSS(sol, f_i)
      end for
      return value/len(x)
   end function

---

a PIVP, by associating a possibly catalyzed synthesis reaction to each positive monomial, with mass action law kinetics with the monomial coefficient as rate constant, and similarly a degradation reaction to a negative monomial.

The population of possible solutions is initialized with random PIVPs. The number of variables is chosen in a given range with a uniform distribution. They have 1 or 2 monomials per variable with the same probability.

To evaluate the fitness score of a given structure we rely on CMA-ES to find a set of optimized kinetic parameters and initial concentration (Algorithm 2). As CMA-ES is a stochastic algorithm and its result may vary from one call to another even for the same structure, only the best parameter set found so far is used to affect the score (that is we do not allow the fitness to decrease). To keep the computation time tractable, CMA-ES is called with a time limit (of the order of the minute depending on the problem at hand). This also mean that, even with a proper initialization of the CMA-ES seed, the result of this optimization may vary according to the current state of the processor. However limiting the number of function evaluations may drastically slow down the algorithm as large PIVP are difficult to integrate.

As for selection, an *elitist* strategy is applied: the 50% best polynomials are kept and then mutated. It is worth noting that this criterion is only based on the ranking of the individual solutions according to their evaluation and is thus robust with respect to the precise fitness definition.

Several types of mutations are used:

Adding a monomial with a coefficient $\exp(v)$ where v is sampled according to the normal law $\mathcal{N}(0, 1)$.

Removing a monomial Note that if a variable has no more monomials, a random monomial is added to this variable (thus avoiding an empty ODE).

Adding a variable to avoid empty variable, two monomials are added with coefficients sampled as just explained, and the initial concentration is sampled with the same law.

Removing a variable all the monomials of this variable are deleted too.

Restarting That is replacing the PIVP by a new random one having the same number of variables and 1 or 2 monomials per variable (also called a restart, this enables the exploration of the search space)

At each generation, one of the mechanisms above can happen. The probability of adding or removing a monomial is 0.35. While the other mutations have a probability of 0.1. Moreover, at each generation, the sign of a monomial is changed (if it satisfies the positivity constraint). This high rate of mutation is tempered by the fact that only one half of the population is mutated while the other half is kept unchanged, thus ensuring that the best structures will never be lost through the mutation operator. Hence the term of elitist selection. During the mutation, the good parameters found during evaluation are kept. Note that no cross-overs are implemented in our mutation operator.

After a fixed number of generations, the best structure of the population is selected as the final result and CMA-ES is called a last time with an extended time limit to fine tuned the parameters.

3.2 Parallel Implementation

The previous algorithm has been implemented in Python 3.6.3 using the CMA-ES package `cma`, the numerical integrator LSODA of the `scipy` library, and the parallelization package `mpi4py` [16].

In our evolutionary algorithm, the most time consuming part is the evaluation part since it requires numerical integration to compare the simulation trace to the objective trace. However, since the evaluations of the different individuals in the population are independent, they can be parallelized. Furthermore, since our algorithm uses two nested evolutionary algorithms, it can benefit from two sources of parallelism. For each of them, we can achieve a linear speedup in the number of processors.

The first layer of parallelism is easy to implement, as there is a function `Scatter` that takes a list from a root process and scatter it to all processes. The function `Gather` can perform the inverse operation. Adding the second layer is more difficult, especially in our case where it is not possible to transform the two layers into one layer. To overcome this challenge, using the function parallelism is easy to implement, as there `Split`, new communicators are created. There are as many new communicators as individuals in the population. At each generation, the population is scattered between the different groups of processes. Each group is dedicated to one individual. Each group performs the evaluation of the individual. The population of CMA-ES is set here to 10 individual parameter sets. In a group, during the evaluation part of CMA-ES, each individual of CMA-ES is thus evaluated by a different process, through a scatter/gather mechanism.

4 Evolved CRN for Mathematical Functions

4.1 Functions of Time

In all this section, we use as fitness function the mean square error between our goal and the simulation trace evaluated on a finite set of time points.

Cosine. Interestingly, the algorithm evolves PIVPs that can be reduced to the following CRN of 3 species and 7 reactions[4]:

```
962*A*B for A =[B]=> _ .           present(A, 1).
4.78*C for _ =[C]=> A.             present(B, 0.0028).
544*A*B for B =[A]=> _ .           present(C, 0.539).
0.477*B for _ =[B]=> B.
1.5 for _ => B.
0.648*B for _ =[B]=> C.
0.346*C for C => _ .
```

On a trace of two periods, the obtained fitness value is $L_2(A, \cos) = 0.042$. This is an excellent fit comparable to the fitness value of $L_2(f_p, \cos) = 0.039$ obtained with the numerical method (Example 1) with a `fast` parameter set to 100.

In this CRN, the positive and negative parts of the cosine are the species A and B. It is worth noticing that these two species are subjected to a fast bi-degradation similar to the encoding of signed variables with two variables for the positive and negative parts shown in Example 1. However, this CRN uses the two parts in a non symmetric fashion that makes it unreachable for our compiler.

The oscillatory behavior of this CRN is the limit cycle of the ODE and is reached for a large space of initial conditions. We can also gather together the reactions that have similar monomials (reactions 1 and 3 for example) and even if this affect the precise fit of the cosine, we check that the oscillatory behavior is preserved.

Heaviside. The Heaviside function

$$\Theta(t) = \begin{cases} 0 & \text{if } t < 0.5, \\ 1 & \text{if } t \geq 0.5. \end{cases} \tag{1}$$

is a discontinuous (hence non-computable) function that cannot be defined by a PIVP. Because it is an ideal step function, it is interesting to see what CRN can emerge to best approximate that function. It takes nearly 10 times more generations to evolve a good approximation for Heaviside than it takes to evolve the cosine function. The best result found, with a fitness $\phi = L_2(A, \Theta) = 0.0078$, is of the following form:

---

[4] The evolved CRN of this section are available at https://lifeware.inria.fr/wiki/Main/Software#CMSB19b.

```
1.05e-4*C*C for _ =[C+C]=> A.
1.79*C*C for _ =[C+C]=> B.
3.07*C*C for _ =[C+C]=> C.
1*A*B for _ =[A+B]=> C.
1.38*B*C for C =[B]=> _.
```

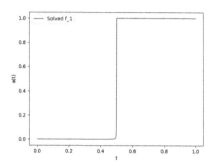

To understand the dynamics of this CRN, we see that the amplitude of the output species is tuned by the parameter of the first reaction. The fourth reaction (_ =[A+B]=> C) is actually unessential to the function even if it can result in a better approximation, its suppression does not significantly modify the trace. The core of the CRN is thus on the competition between species B and C, both starting at low concentration and increasing abruptly around the desired threshold. There, the last reaction imposes a slow-down that makes C disappear and B stabilize. The species of interest A is then only a linear scaling of B to respect the constraints imposed by the fitness function.

It is worth remarking that this solution is surprisingly close to the PIVP method to transform a function of time in a function of input by multiplying each monomial by a variable that decreases exponentially fast, hence halting the computation on the desired state [20]. Here an exploding variable is halted in its course to simulate non-linearity.

Another way our artificial evolutionary algorithm manages to produce Heaviside like functions is through the generation of sigmoidal functions. The CRN below

```
k*A*B for B =[A]=> A.                        present(A, C0).
                                             present(B, 1.0).
```

satisfies the relation $C_0 \simeq \exp(-\frac{k}{2})$ in order to have the jump around one half, and k large makes the jump as sharp as possible. Practically, this make the initial concentration of A vanishingly small.

This mechanism may have several variants depending on the evolutionary history followed by the algorithm. A common one is to implement the sigmoid function through two species with large concentration in order to produce a sharp exponential and to report the output on a third one that is a linear rescaling of a previous one as is the case in the previous example.

4.2 Functions of Input Variables

In this section, the performance of a CRN is measured by recording only the final value of its output species and using a mean square error as global loss function. Here, final value mean the value at a predefined time ($t = 1$), to ensure that this correspond to the steady state, the norm of the derivative is used as a penalty. Note that the choice $t = 1$ has nothing particular as the time may be rescaled.

Cosine. The PIVP structure that commonly emerges by artificial evolution is the following one:

$$\frac{da}{dt} = -2.9 \cdot 10^2 \cdot a - 1.8 \cdot 10^{-1} \cdot c \tag{2}$$

$$\frac{db}{dt} = -2.2 \cdot 10^9 \cdot ac - 4.2 \cdot 10^{-10} \cdot c \tag{3}$$

$$\frac{dc}{dt} = 3.8 \cdot 10^{-5} \cdot ab - 1.0 \cdot 10^0 \cdot c^2 \tag{4}$$

We choose to start from the PIVP to emphasize the importance of neutral transformation in the final result of our EA. Here, both time $2.9 \cdot 10^2 t \to t$ and the last variable $7.7 \cdot 10^6 c \to c$ can be rescaled, giving:

$$\frac{da}{dt} = -a - 8.1 \cdot 10^{-11} \cdot c \tag{5}$$

$$\frac{db}{dt} = -ac - 1.9 \cdot 10^{-19} \cdot c \tag{6}$$

$$\frac{dc}{dt} = ab - 4.5 \cdot 10^{-10} \cdot c^2 \tag{7}$$

where all the variable are of the order of unity. The final term in c may safely be ignored and this let us with the same PIVP as the one feed to our compiler to generate the cosine of input:

$$\frac{da}{dt} = -a \tag{8}$$

$$\frac{db}{dt} = -ac \tag{9}$$

$$\frac{dc}{dt} = ab \tag{10}$$

Here again, a plays the role of a halting species and b and c compute the cosine and sine functions that are halted on the desired value.

Sum of Two Inputs. The purpose here is to find a CRN to implement the sum function $f(x, y) = x + y$, as a first example of a case with two inputs. The best solution found by evolution is:

```
k1*A*A for A =[A]=> B.              present(A, input1).
k2*A*A for _ =[A+A]=> C.            present(B, input2).
k2*B*B for _ =[B+B]=> C.            present(C, 1.44).
k2*A*B for _ =[A+B]=> 2C.
k2*C*C for C =[C]=> _.
```

where we have labeled with the same name the rate constants found nearly equal by artificial evolution. On the other hand the values of k1 and k2 do not play any role. Actually, k1 may even be chosen to be null, highlighting a symmetry of the fitness function. The rest of the reaction implements the equation: $\dot{c} = (a+b)^2 - c^2$, where the square enforces a faster dynamics and thus a better convergence.

The general idea is however to balance the positive and negative monomials of variable c to stabilize it on the desired result. A direct consequence of this mechanism is that the initial concentration of C has no importance in the final output. While a human engineer would thus fixed it to 0, our evolutionary algorithm proposes a random value. We may however suspect that this is the mean of the output used to train our CRN as such a choice would accelerate the mean convergence time and decrease the final error.

That mechanism may be more clear if written without the squaring:

```
k1*A for _ =[A]=> C.              present(A, input1).
k2*B for _ =[B]=> C.              present(B, input2).
k3*C for C => _.
```

Interestingly, this also gives a generalization where different values for the rates allow us to compute the function:

$$f[k_1, k_2, k_3](x, y) = \frac{k_1}{k_3}x + \frac{k_2}{k_3}y. \tag{11}$$

Several remarks should be done on this CRN. First, it provides a CRN to compute online since a modification of the concentration of the input will be transmitted to the output. This online CRN algorithm will have the form of an exponential decay with a time rate $\frac{1}{k_3}$. A fast dynamics (high k_3) means that either the output is small, or the energetic cost will be high. Finally, it is known from the study of such process that on the stochastic regime, the distribution of species C will follow a Poisson distribution with a variance equal to its mean. In comparison, the solution found by our EA converges faster and present a higher signal to noise ratio in stochastic semantics.

Heaviside. When considered as a function of input, the non-linearity of the Heaviside function is no longer an issue as it can be computed with continuous function such as a bi-stable switch. For example, starting from the simplest bistable PIVP $\frac{da}{dt} = a(1-a)(2a-1)$, our compiler generates the following PIVP:

```
3*A*A for _ =[A+A]=> A.          present(A,input).
A for A => _.                     present(B,input2).
2*A*B for A =[B]=> _.
6*A*B for _ =[A+B]=> B.
4*B*B for B =[B]=> _.
2*B for B => _.
```

where input2 have to be the square of input. Interestingly, this switch CRN is an instance of the Approximate Majority distributed algorithm extensively studied and related to cell division-cycle progression models in [7].

Another solution would be to activate a production and degradation cycle based on the remainder of the input after a fast comparison to a predefined threshold:

```
fast*A*C for A+C => _.                    present(A,input).
 A   for _ =[A]=> B.                       present(C,0.5).
A*B for B =[A]=> _.
```

The best CRN found by evolution in our experiments is however very different from these two solutions[5], and plays upon the dynamics of the network and the precise point of evaluation defined by the input trace to fit, here at time $t = 1$. The CRN found displays a transient state of variable length τ_t where the output is near zero, and a single steady state where the output is near unity. The value of the input impacts the time to reach the steady state so that if the input is greater than one half then $\tau_t < 1$ and the output is near 1 at evaluation time, while in the other case, $\tau_t > 1$ and at the evaluation time $t = 1$ the output is null but just temporarily. This illustrates the typical problem of generalization in machine learning for an input/output function specified by a finite trace.

5   Comparison to Natural CRNs in BioModels

One first remark illustrated by the previous examples, is that when we usually end up with parameters value taking involved values it is often possible, by clever variable transformation, to recover more natural values thus highlighting several symmetries of our fitness function. This is typically the case when evolving the cosine function, the scale of the sine function is undefined and creates a whole variety of solutions that are strictly equivalent up to a rescaling of two parameters. This kind of symmetry of the fitness function is one of the plague of biology as it is often difficult to identify them and they may obfuscate a simpler design [26].

A second discussion may be raised by the two CRN for the cosine function of time obtained respectively by the numerical method and the artificial evolution algorithm. Figure 2 shows that those CRN can be compared to the circadian clocks present in eukaryote and prokaryote albeit with strong differences in their implementations. In prokaryotes, a circadian clock has been shown to be implemented through a cycle of phosphorylation and dephosphorylation, such as of the two sites S and T of a KaiC protein polymer found in cyanobacteria [35]. This mechanism is usually presented through a four step process:

$$ST \rightarrow SpT \rightarrow pSpT \rightarrow pST \rightarrow ST$$

where a p preceding a letter indicates that the corresponding site is phosphorylated. This mechanism is similar to the compiled version of our cosine function where 4 species activates one another in a circular fashion:

$$\cos^+ \rightarrow \sin^+ \rightarrow \cos^- \rightarrow \sin^- \rightarrow \cos^+$$

---

[5] This CRN is not shown in the main text but could be examined on the notebook available at: https://lifeware.inria.fr/wiki/Main/Software#CMSB19b.

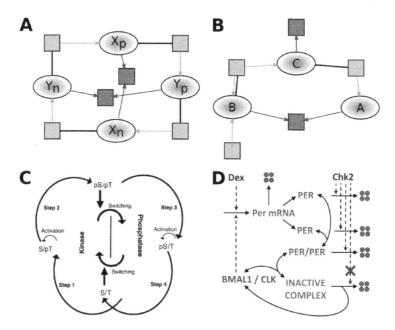

Fig. 2. Representation of our compiled (panel A) and evolved (panel B) CRN as reaction graphs for the cosine function of time. Ellipses represent species, squares indicate reactions in green for productions and red for degradations. Catalysts are linked to their reactions with a bold black line. Panel C shows the molecular oscillator of cyanobacteria as described in [35]. Panel D shows the genetic oscillator of mammal (similar in most eukaryotes) as proposed in [32]. (Color figure online)

where the upper signs denote the positive and negative part of the function.

The eukaryotes use a transcriptional mechanism where a CLOCK protein activates the transcription of its own inhibitor. The action of the inhibitor is however delayed through several steps that may vary from one organism to another. Usually this involves translation, dimerization and transportation to and from the nucleus. By using the graph matching method based on subgraph epimorphism (SEPI) [24] to compare our evolved CRN to models of the BioModels repository, we found a mapping from the mammalian circadian clock models such as [32] toward the simple oscillator in panel B of Fig. 2 presented above. The SEPI mapping makes B play the role of the transcription factor that activates the transcription of the messenger RNA C that then became the protein A that will bind to the transcription factor to form an inactive complex, thus inhibiting the transcription.

Table 1 shows general SEPI comparison results with models in BioModels [10]. We see that models presenting oscillations (circadian clock and cell cycle) are far more likely to exhibit a SEPI toward one of our evolved network than the MAPK models (the SEPIs found come from model 026 [36] and 010 [34]). The

Table 1. SEPI matchings from biological models in BioModels having between 5 and 15 species to the 10 best evolved CRNs (over 60 runs) for the cosine time function.

Model type	No. of models	Nb. SEPI matchings (min/mean/max)	% SEPI
Circadian clock	13	3/7.8/12	60%
Cell cycle	22	6/10.1/17	45%
Mapk	9	0/0.7/2	08%

evolved model that harbors nearly all SEPI from both cell cycle and circadian clock models is the one with the best fit presented in Sect. 4.1.

These connections between the result of our evolutionary algorithm and the actual mechanisms found in biology suggest a form of evolutionary convergence. It is well known in biology that organs in species that have evolved totally independently may in fact be similar. There might be actually few ways to implement a clock that are easily found by evolution so much that every solution that appears is actually a close repetition from the existing one. This kind of evolution convergence has already been emphasized in the case of the evolution of a biological logarithm were every successful run displays the same core mechanism [31]. This provides a tool to decipher the function and constraints of a biochemical network by comparing it to an idealized mathematical framework while still being able to make relevant connection with the biological case.

6   Conclusion

We have described a nested evolution algorithm to learn both the structure and the kinetic parameters of a CRN for approximating a function given by a finite trace of either time points for a function of time, or input values.

On a Heaviside function of time, the results obtained by artificial evolution lead to a remarkably simple CRN of 3 molecular species and 5 reactions with double catalysts which provide a very stiff transition although using mass action law kinetics. This solution is more economical than the CRN generated by the PIVP method for sigmoid functions [20]. On a Heaviside function of input, the CRN found by evolution are slightly more complicated than the bistable switch found in cell cycle CRN for instance, but much less complex than the MAPK signaling network that plays a similar role [34].

On the cosine function of time, the best CRN found by evolution contains an annihilation reaction similar to the CRN generated by the numerical method for positive and negative variables, but one less reaction thanks to an intriguing non symmetric use of the two variables which preserves the limit cycle. Interestingly, the evolved and the PIVP generated structures could be compared to prokaryote and eukaryote models of the circadian clock found in BioModels.

On the cosine function of input, a CRN surprisingly emerges with the structure of the CRN for cosine function of time, using the same trick as in [20] to stop time at the desired input value.

These results are encouraging to further develop artificial evolution methods for CRN design in both perspectives of systems biology with the study of evolution convergence, and synthetic biology in addition to rational design.

Acknowledgment. This work benefited from support from the ANR-MOST project BIOPSY "Biochemical Programming System"ANR-16-CE18-0029 and granted access to HPC resources with GENCI allocation 2018-AP011010715.

References

1. Batt, G., Yordanov, B., Weiss, R., Belta, C.: Robustness analysis and tuning of synthetic gene networks. Bioinformatics **23**(18), 2415–2422 (2007)
2. Benenson, Y., Adar, R., Paz-Elizur, T., Livneh, Z., Shapiro, E.: DNA molecule provides a computing machine with both data and fuel. Proc. Nat. Acad. Sci. **100**(5), 2191–2196 (2003)
3. Bournez, O., Graça, D.S., Pouly, A.: Polynomial time corresponds to solutions of polynomial ordinary differential equations of polynomial length. In: 43rd International Colloquium on Automata, Languages, and Programming, ICALP 2016, LIPIcs, 11–15 July 2016, Rome, Italy, vol. 55, pp. 109:1–109:15. Schloss Dagstuhl - Leibniz-Zentrum fuer Informatik (2016)
4. Cao, H., Kang, L., Chen, Y., Yu, J.: Evolutionary modeling of systems of ordinary differential equations with genetic programming. Genet. Program Evolvable Mach. **1**(4), 309–337 (2000)
5. Cao, H., Romero-Campero, F.J., Heeb, S., Cámara, M., Krasnogor, N.: Evolving cell models for systems and synthetic biology. Syst. Synth. Biol. **4**(1), 55–84 (2010)
6. Cardelli, L., et al.: Syntax-guided optimal synthesis for chemical reaction networks. In: Majumdar, R., Kunčak, V. (eds.) CAV 2017. LNCS, vol. 10427, pp. 375–395. Springer, Cham (2017). https://doi.org/10.1007/978-3-319-63390-9_20
7. Cardelli, L., Csikász-Nagy, A.: The cell cycle switch computes approximate majority. Sci. Rep. **2**, 656 (2012)
8. Cardelli, L., Kwiatkowska, M., Whitby, M.: Chemical reaction network designs for asynchronous logic circuits. In: Rondelez, Y., Woods, D. (eds.) DNA 2016. LNCS, vol. 9818, pp. 67–81. Springer, Cham (2016). https://doi.org/10.1007/978-3-319-43994-5_5
9. Carothers, D.C., Parker, G.E., Sochacki, J.S., Warne, P.G.: Some properties of solutions to polynomial systems of differential equations. Electron. J. Differ. Equ. **2005**(40), 1–17 (2005)
10. Chelliah, V., Laibe, C., Novère, N.: Biomodels database: a repository of mathematical models of biological processes. In: Schneider, M.V. (ed.) In Silico Systems Biology, Methods in Molecular Biology, vol. 1021, pp. 189–199. Humana Press (2013)
11. Chen, Y., et al.: Programmable chemical controllers made from DNA. Nat. Nanotechnol. **8**, 755–762 (2013)
12. Chen, Y., Smolke, C.D.: From DNA to targeted therrapeutics: bringing synthetic biology moving to the clinic. Sci. Trans. Med. **3**(106), 106ps42 (2011)
13. Courbet, A., Amar, P., Fages, F., Renard, E., Molina, F.: Computer-aided biochemical programming of synthetic microreactors as diagnostic devices. Mol. Syst. Biol. **14**(4), E7845 (2018)

14. Courbet, A., Molina, F., Amar, P.: Computing with synthetic protocells. Acta Biotheor. **63**(3), 309 (2015)
15. Dalchau, N., Murphy, N., Petersen, R., Yordanov, B.: Synthesizing and tuning chemical reaction networks with specified behaviours. In: Phillips, A., Yin, P. (eds.) DNA 2015. LNCS, vol. 9211, pp. 16–33. Springer, Cham (2015). https://doi.org/ 10.1007/978-3-319-21999-8_2
16. Dalcin, L.D., Paz, R.R., Kler, P.A., Cosimo, A.: Parallel distributed computing using Python. Adv. Water Resour. **34**(9), 1124–1139 (2011)
17. Daniel, R., Rubens, J.R., Sarpeshkar, R., Lu, T.K.: Synthetic analog computation in living cells. Nature **497**(7451), 619–623 (2013)
18. Dinh, H.Q., Aubert, N., Noman, N., Fujii, T., Rondelez, Y., Iba, H.: An effective method for evolving reaction networks in synthetic biochemical systems. IEEE Trans. Evol. Comput. **19**(3), 374–386 (2015)
19. Duportet, X., et al.: A platform for rapid prototyping of synthetic gene networks in mammalian cells. Nucl. Acids Res. **42**(21), 13440–13451 (2014)
20. Fages, F., Le Guludec, G., Bournez, O., Pouly, A.: Strong turing completeness of continuous chemical reaction networks and compilation of mixed analog-digital programs. In: Feret, J., Koeppl, H. (eds.) CMSB 2017. LNCS, vol. 10545, pp. 108–127. Springer, Cham (2017). https://doi.org/10.1007/978-3-319-67471-1_7
21. Fages, F., Gay, S., Soliman, S.: Inferring reaction systems from ordinary differential equations. Theor. Comput. Sci. **599**, 64–78 (2015)
22. Fages, F., Soliman, S.: On robustness computation and optimization in BIOCHAM-4. In: Češka, M., Šafránek, D. (eds.) CMSB 2018. LNCS, vol. 11095, pp. 292–299. Springer, Cham (2018). https://doi.org/10.1007/978-3-319-99429-1_18
23. Gay, S., Fages, F., Martinez, T., Soliman, S., Solnon, C.: On the subgraph epimorphism problem. Discrete Appl. Math. **162**, 214–228 (2014)
24. Gay, S., Soliman, S., Fages, F.: A graphical method for reducing and relating models in systems biology. Bioinformatics **26**(18), i575–i581 (2010). Special issue ECCB 2010
25. Gomez, H.F., Hucka, M., Keating, S.M., Nudelman, G., Iber, D., Sealfon, S.C.: MOCCASIN: converting MATLAB ODE models to SBML. Bioinformatics **21**(12), 1905–1906 (2016)
26. Gutenkunst, R.N., Waterfall, J.J., Casey, F.P., Brown, K.S., Myers, C.R., Sethna, J.P.: Universally sloppy parameter sensitivities in systems biology models. PLOS Comput. Biol. **3**(10), 1–8 (2007)
27. Hansen, N., Auger, A., Ros, R., Finck, S., Pošík, P.: Comparing results of 31 algorithms from the black-box optimization benchmarking BBOB-2009. In: Proceedings of the 12th Annual Conference Companion on Genetic and Evolutionary Computation - GECCO 2010, p. 1689. ACM Press, New York (2010)
28. Hansen, N., Ostermeier, A.: Completely derandomized self-adaptation in evolution strategies. Evol. Comput. **9**(2), 159–195 (2001)
29. Hárs, V., Tóth, J.: On the inverse problem of reaction kinetics. In: Farkas, M. (ed.) Colloquia Mathematica Societatis János Bolyai. Qualitative Theory of Differential Equations, vol. 30, pp. 363–379 (1979)
30. Heitzler, D., et al.: Competing G protein-coupled receptor kinases balance G protein and β-arrestin signaling. Mol. Syst. Biol. **8**(590) (2012)
31. Hemery, M., François, P.: In silico evolution of biochemical log-response. J. Phys. Chem. B **19**, 2235–2243 (2019)
32. Hong, C.I., Zámborszky, J., Csikasz-Nagy, A.: Minimum criteria for DNA damage-induced phase advances in circadian rhythms. PLoS Comput. Biol. **5**(5), e1000384 (2009)

33. Hsiao, Y.T., Lee, W.P.: Reverse engineering gene regulatory networks: coupling an optimization algorithm with a parameter identification technique. BMC Bioinform. **15**(Suppl. 15), S8 (2014)

34. Huang, C.Y., Ferrell, J.E.: Ultrasensitivity in the mitogen-activated protein kinase cascade. PNAS **93**(19), 10078–10083 (1996)

35. Kageyama, H., Nishiwaki, T., Nakajima, M., Iwasaki, H., Oyama, T., Kondo, T.: Cyanobacterial circadian pacemaker: Kai protein complex dynamics in the KaiC phosphorylation cycle in vitro. Mol. Cell **23**(2), 161–171 (2006)

36. Markevich, N.I., Hoek, J.B., Kholodenko, B.N.: Signaling switches and bistability arising from multisite phosphorylation in protein kinase cascades. J. Cell Biol. **164**(3), 353–359 (2004)

37. Oishi, K., Klavins, E.: Biomolecular implementation of linear I/O systems. IET Syst. Biol. **5**(4), 252–260 (2011)

38. Qian, L., Soloveichik, D., Winfree, E.: Efficient turing-universal computation with DNA polymers. In: Sakakibara, Y., Mi, Y. (eds.) DNA 2010. LNCS, vol. 6518, pp. 123–140. Springer, Heidelberg (2011). https://doi.org/10.1007/978-3-642-18305-8_12

39. Rizk, A., Batt, G., Fages, F., Soliman, S.: A general computational method for robustness analysis with applications to synthetic gene networks. Bioinformatics **12**(25), il69-il78 (2009)

40. Rizk, A., Batt, G., Fages, F., Soliman, S.: Continuous valuations of temporal logic specifications with applications to parameter optimization and robustness measures. Theor. Comput. Sci. **412**(26), 2827–2839 (2011)

41. Rubens, J.R., Selvaggio, G., Lu, T.K.: Synthetic mixed-signal computation in living cells. Nat. Commun. **7**, 11658 (2016)

42. Sarpeshkar, R.: Analog synthetic biology. Philos. Trans. R. Soc. Lond. A: Math. Phys. Eng. Sci. **372**(2012), 20130110 (2014)

43. Sauro, H.M., Kim, K.: Synthetic biology: it's an analog world. Nature **497**(7451), 572–573 (2013)

44. Uhlendorf, J., et al.: Long-term model predictive control of gene expression at the population and single-cell levels. Proc. Natl. Acad. Sci. USA **109**(35), 14271–14276 (2012)

45. Valiant, L.: Probably Approximately Correct. Basic Books (2013)

46. Vasic, M., Soloveichik, D., Khurshid, S.: *CRN++*: Molecular Programming Language. In: Doty, D., Dietz, H. (eds.) DNA 2018. LNCS, vol. 11145, pp. 1–18. Springer, Cham (2018). https://doi.org/10.1007/978-3-030-00030-1_1

47. Vecchio, D.D., Abdallah, H., Qian, Y., Collins, J.J.: A blueprint for a synthetic genetic feedback controller to reprogram cell fate. Cell Syst. **4**, 109–120 (2017)

Designing Distributed Cell Classifier Circuits Using a Genetic Algorithm

Melania Nowicka[1,2]([⊠]) [iD] and Heike Siebert[1]

[1] Freie Universitaet, 14195 Berlin, Germany
m.nowicka@fu-berlin.de
[2] Max Planck Institute for Molecular Genetics, 14195 Berlin, Germany

Abstract. Cell classifiers are decision-making synthetic circuits that allow in vivo cell-type classification. Their design is based on finding a relationship between differential expression of miRNAs and the cell condition. Such biological devices have shown potential to become a valuable tool in cancer treatment as a new type-specific cell targeting approach. So far, only single-circuit classifiers were designed in this context. However, reliable designs come with high complexity, making them difficult to assemble in the lab. Here, we apply so-called Distributed Classifiers (DC) consisting of simple single circuits, that decide collectively according to a threshold function. Such architecture potentially simplifies the assembly process and provides design flexibility. We present a genetic algorithm that allows the design and optimization of DCs. Breast cancer case studies show that DCs perform with high accuracy on real-world data. Optimized classifiers capture biologically relevant miRNAs that are cancer-type specific. The comparison to a single-circuit classifier design approach shows that DCs perform with significantly higher accuracy than individual circuits. The algorithm is implemented as an open source tool.

Keywords: Synthetic biology · Boolean modeling · Genetic algorithms · miRNA profiling · Cell classifiers · Cancer

1 Introduction

Synthetic biology has shown its immense potential in recent years in a wide array of applications. This is particularly true for the medical field, where synthetic biological systems are developed for versatile employment from diagnostics to treatment [24,28]. Research in design and construction of cell classifier circuits touches on both these areas. Cell classifiers are molecular constructs capable of sensing certain markers in the environment, processing the input and reacting with a signal-specific output. A prime example for this are miRNA-based classifiers that distinguish cell states, e.g., as healthy or diseased, based on their miRNA expression profiles applying boolean logic (Fig. 1A) [15,26]. These circuits can be delivered to cells on plasmids or viral vectors and trigger the production of a desired output, e.g., a toxic compound causing cell apoptosis in diseased cells (Fig. 1B).

© Springer Nature Switzerland AG 2019
L. Bortolussi and G. Sanguinetti (Eds.): CMSB 2019, LNBI 11773, pp. 96–119, 2019.
https://doi.org/10.1007/978-3-030-31304-3_6

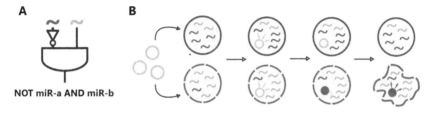

Fig. 1. (A) An exemplary boolean design of a two miRNA-input cell classifier. (B) A schema showing two types of cells, healthy (solid line) and diseased (dashed line). The classifiers are delivered to the cells, sense the internal input levels and respond with respect to a given cell condition.

A variety of different approaches to designing synthetic circuits is available [12,17,25]. However, to confront many application-derived limitations, circuit designs must be often tailored to rigorous specifications. Since cell classifiers must be feasible to implement in the lab, many constraints are posed on the building blocks of these circuits that need to be encoded in the design problem. So far, only a few different methods for designing single-circuit classifiers were described [1,15]. Mohammadi et al. proposed two different heuristic approaches [15]. The procedure performing with the highest accuracy in terms of sample classification allows to optimize a classifier's topology using a mechanistic model of the circuit and a predefined set of biochemical parameters. Another approach was presented by Becker et al. [1]. The authors propose a method for finding globally optimal classifiers represented by boolean functions based on binarized miRNA expression data. To search through the entire space of solutions in a short time frame the authors apply logic solvers. Becker et al. compare their results to the previously mentioned state-of-the-art method demonstrating significant improvement in binary classification of presented classifiers [1].

While this research shows that theoretically single-circuit classifiers can perform such classification tasks [1,15], there is a number of challenges for the approach in application. Depending on the heterogeneity of the data, to obtain a clear-cut classification often a circuit of high complexity is needed. Generally, the cost both in time and money for classifier circuit construction in the lab goes up the larger and more complex the circuit architecture gets, quickly becoming not feasible at all [15]. A further problem is the robustness needed for reliable performance when faced with uncertainty and noise in signals and wide ranging possibilities for perturbations of the classifier functionality in natural environments. To address these issues the principles of distributed classification, as inherent in many natural systems such as the immune system and shown to be an effective strategy, e.g., in machine learning, can be exploited [19,22]. Here, the idea is to design a set of different classifier circuits, also called distributed classifier, that perform classification in an integrated manner. Such a set can consist of rather simple classifiers that still perform more accurately than a complex single circuit classifier, since the individual classification results are aggregated which compensates for individual mistakes. A theoretical design of such a distributed classifier based on synthetic gene circuits was presented by Didovyk et al. [3]. The classifier is optimized by training a starting population of simple circuits on the available data similarly to machine learning algorithms, i.e., by presenting learning examples and successively removing low-performance circuits. While this work considers only a quite specific scenario being designed for bacterial cell cultures, it highlights the potential of the underlying idea of using distributed classifiers.

Here, we adapt the distributed classifier approach proposed by Didovyk et al. [3] to the problem of cell classifier design. We define a *Distributed Classifier* (DC) as a set of single-circuit classifiers that decide collectively based on a threshold function. Biologically, the threshold may correspond to a certain concentration of the drug that allows to treat the cells or fluorescent marker allowing to classify the cell type [3,14,15]. According to Mohammadi et al. [15] such threshold manipulation may be achieved by changing the biochemical parameters of a circuit model. Due to the high complexity of the problem, we apply a heuristic approach to design and optimize DCs, namely, a genetic algorithm (GA). GAs are evolution-inspired metaheuristics that allow to optimize populations of individuals [13]. Such evolutionary approaches were successfully applied to various biological questions [11], e.g., design of synthetic networks and, in particular, design of single-circuit classifiers [15,25]. Due to the high flexibility of GAs in terms of design and parameters, the algorithm may be efficiently adapted to the distributed classifier problem.

In this article, we illustrate the potential of distributed classifiers in application, in particular, in cancer cell classification. The following section contains preliminaries including the definition of a single-circuit and distributed classifier. Section 3 describes the architecture of the proposed genetic algorithm for the design and optimization of DCs. In Sect. 4 we present case studies performed on real-world breast cancer data and compare the results with a single-circuit design method proposed by Becker et al. [1]. Finally, we discuss the distributed classifier performance and comment on potential future work.

2   Preliminaries

In this section we describe the data we employ to designing classifiers, introduce single-circuit and distributed classifiers and propose binary classification measures that allow to evaluate their performance.

2.1   miRNA Expression Data

The proposed method is a boolean approach and utilizes binarized and annotated data. While our focus is on miRNA expression profiles, the approach can naturally be applied to any data set of the format introduced below.

In cancer research, differentially expressed miRNAs provide a valuable source of information about tumor development, progression and response to a therapy [8,9]. Thus, dysregulated miRNAs have been considered as potential biomarkers for cancer diagnosis and treatment. One of the approaches allowing to distinguish up- and down-regulated miRNAs is discretization of the expression data into a finite number of states. Discretization provides clear and interpretable information about the miRNA behaviour and makes the learning process from the data more efficient [6]. However, the procedure is also related to a potential information loss. We comment on this issue in Sect. 5.

Table 1. A miRNA expression data set.

ID	Annots	miR-a	miR-b	miR-c
1	0	0	1	0
2	0	0	1	0
3	1	1	0	0
4	1	0	0	0
5	1	1	0	0

We define a data set $D = (S, A)$ as a finite set of samples $S \subseteq \{0,1\}^m$, where $m \in \mathbb{N}$ is the number of miRNAs and $A : S \longrightarrow \{0,1\}$ is sample annotation. Presented as a table, the first column includes unique sample IDs and the second the annotation of samples (Annots), where 0 is a label assigned to negative class samples (healthy) and 1 to positive (cancerous). The following columns are miRNA expression profiles describing the miRNA regulation among the samples. miRNAs are binarized into two states: up- (1/positive) and down-regulated (0/negative), according to a given threshold. An example of a data set is presented in Table 1. A miRNA is non-regulated if for every sample its state is either 0 or 1 (e.g., Table 1, miR-c). Some miRNAs can perfectly separate the samples into the two categories implied by the annotation (e.g., Table 1, miR-b).

2.2 Single-Circuit Classifier

A single-circuit cell classifier may be represented by a boolean function $f : S \longrightarrow \{0,1\}$. To make a classifier feasible to construct in the lab additional constraints must be imposed on the function. We adopt here the constraints introduced by Mohammadi et al. [15]. Accordingly, the function should be given in *Conjunctive Normal Form* (CNF), i.e., a conjunction of clauses where each clause is a disjunction of negated (negative) or non-negated (positive) literals. Here, the literals correspond to the miRNAs and clauses to the gates. It may consist of: (i) negative literals only in 1-element clauses (NOT gates) (ii) at most 3 positive literals per clause (OR gate) (iii) up to 10 literals (miRNAs) and up to 6 clauses (gates) in total (iv) including at most 4 NOT gates and 2 OR gates. A circuit topology presented as a CNF satisfying the above-mentioned constraints directly corresponds to the biological model of the circuit employed by Mohammadi et al. [15]. An example of a classifier is presented below.

$$\neg miR\text{-}a \wedge (miR\text{-}b \vee miR\text{-}c) \tag{1}$$

The function should output 1/True in case of cancerous and 0/False in case of healthy cells. The example function presented in Eq. 1 classifies a cell as positive/1 if $miR\text{-}a$ is down-regulated and at least one of the other miRNAs ($miR\text{-}b$ or $miR\text{-}c$) is up-regulated.

2.3   Distributed Classifier

Here, we introduce a concept of *Distributed Classifier* (*DC*) for the cell classification problem. A DC is a finite set $DC = \{f_1, ..., f_c\}$, where f_i is a boolean function $f_i : S \longrightarrow \{0, 1\}$, to which we will refer from now on as a *Rule*, $c \leq c_{max}$, $c \in \mathbb{N}$, is the *DC* size and $c_{max} \in \mathbb{N}$ is an upper bound for the DC size. Motivated by Sect. 2.2 a *Rule* must be a boolean function in a *Conjunctive Normal Form* consisting of at most two single-literal clauses. An example of a *DC* is presented below.

$$\{miR\text{-}a,\ miR\text{-}b \wedge \neg miR\text{-}c,\ miR\text{-}a \wedge miR\text{-}d\}.$$

We assume that each *Rule* in the set must be unique, i.e,, we do not allow copies of *Rules* in the *DC*. Also, two identical miRNA IDs cannot occur in one *Rule*, i.e., a trivial false function is not allowed ($a \wedge \neg a$). Thus, the functions are in a minimised form ($a \wedge a = a$). The DC categorizes cells according to a threshold function $F_{DC} : S \longrightarrow \{0, 1\}$ with

$$F_{DC}(s) = \begin{cases} 0, & \sum_{i=1}^{c} f_i(s) < \theta \\ 1, & \sum_{i=1}^{c} f_i(s) \geq \theta, \end{cases} \tag{2}$$

where $s \in S$ is a sample and $\theta \in [0, c]$ is a threshold. Here, we use $\theta = \lfloor \alpha \cdot c \rceil$ as the threshold, where α is a ratio that allows to calculate the decision threshold based on the classifier size. The threshold is then rounded half up. F_{DC} returns 1/True if a certain number of *Rules* (θ) outputs 1/True, e.g., $\alpha = 0.5$ for $c = 5$ indicates that at least 3 *Rules* must output 1/True to classify a cell as positive.

Depending on α one may receive different results. In case of a very low threshold, e.g., if only one *Rule* outputing 1/True results in *DC* outputing 1/True, the DC becomes simply a disjunction of *Rules*. Note, that the function may then classify in favor of the positive class, as the decision to classify a sample as positive is in fact made by only one rule. This effect is already reduced by not considering 2-literal OR gates as rules. Otherwise, if the threshold is c ($\alpha = 1$), i.e., all the rules must output 1 for the DC to output 1, the function takes a form of a conjunction of clauses staying close to the single-circuit classifier. Unlike the disjunction, a conjunction may classify in favor of the negative class which may decrease the sensitivity of the method. Applying intermediate thresholds results in different combinations of those functions, therefore, different classification performance. In terms of cell classifiers applied as a cancer treatment, one may consider the following problem: in case of high α, the classifier may misclassify the diseased cells resulting in false negatives. Thus, the treatment may be less effective. However, low α may result in misclassification of healthy cells which makes the treatment more toxic as the drug is released in those cells (false positives). Here, one should consider what type of errors is less desirable and apply a suitable threshold. We discuss this issue further in Sect. 4.

2.4   Evaluation

Here, we introduce the measures we employ to evaluate DCs in terms of binary classification. Many metrics that may be applied are available [20]. However, real-world expression data is often heavily imbalanced, i.e., the samples are not equally represented in the two classes. Data imbalance may significantly influence the classification results [27]. Balanced Accuracy (BACC) is an intuitive and easily interpretable metric that allows to balance the importance of samples in both classes (Eq. 3) [20]. Thus, as a main measure of classifier's performance we apply BACC.

$$BACC(DC, D) = \frac{\frac{TP}{P} + \frac{TN}{N}}{2} \tag{3}$$

where D is a given data set, P and N are the numbers of positive and negative samples in D, TP is the number of samples correctly classified as positive and TN is the number of samples correctly classified as negative. TP and TN are threshold-dependent values, i.e., they may change while applying different threshold values for a given classifier. To evaluate other aspects of classifier's performance we employ additional common metrics such as sensitivity ($TP/(TP + FN)$), specificity ($TN/(TN + FP)$) and accuracy ($(TP + TN)/(P + N)$). Sensitivity represents the ability of the method to correctly distinguish samples belonging to the positive class, while specificity shows the ability to correctly distinguish those belonging to the negative class. Accuracy gives information about the proximity of results to the true values, but does not take data imbalance into account.

3   Genetic Algorithm

In this section we present the architecture of a GA applied to design and optimization of DCs. In the following sections we describe the core structure of the algorithm as well as the used parameters and operators.

3.1   General Description

The input miRNA expression data must be formatted as described in Sect. 2.1. To optimize the DCs, seven parameters must be specified: $iter$ - number of GA's iterations, ps - population size, i.e., the number of DCs allowed in the population, cp - crossover probability, mp - mutation probability, ts - tournament size, c_{max} - maximal size of a classifier, i.e., the number of single-circuit classifiers in a DC, α - the decision threshold ratio. As an output, the algorithm returns a list of all best solutions found over the GA's iterations according to their balanced accuracy (DC_{best}). In case of single-circuit classifiers, besides the accuracy, the complexity of a solution is also taken into account [1,15]. Thus, we choose the solution consisting of the lowest number of rules as the optimal one. The algorithm starts with a random generation of an initial population (Algorithm 1, line 1). Next, the population is evaluated and a list of best solutions DC_{best} is created (Algorithm 1, 2). Having the initial population generated, the algorithm starts with a first generation. At the beginning, ps individuals are selected in so-called tournaments as potential parents to be recombined, i.e., randomly exchange genes. (Algorithm 1, 4–7). Many selection operators are described in the literature. Tournament selection allows to increase the chance of very good solutions to be selected as parents while maintaining the diversity in the population and can be efficiently implemented [23]. Next, the crossover occurs with the probability cp (Algorithm 1, 8–13). Crossover allows to generate new solutions (children), based on previously selected individuals (parents). Here, a child classifier may be created by copying rules from parent classifiers by randomly choosing which parent the next rule is duplicated from. As classifier sizes may differ, we propose two recombination strategies described further in Sect. 3.4. Next, individuals in the new population may mutate with the probability mp (Algorithm 1, 14). At the end of each iteration the list of best solutions (DC_{best}) is updated (Algorithm 1, 15). All the described steps in a generation are repeated $iter$ times (Algorithm 1, 3–16). Below we explain the details of the algorithm design.

Algorithm 1. A genetic algorithm for designing DCs.

Data: dataset D

Parameters : number of iterations $iter$, population size ps, crossover probability cp, mutation probability mp, tournament size ts, maximal size of DC c_{max}, threshold ratio α

Output: DC_{best}

1 $Population \longleftarrow$ InitializePopulation(D, ps, c_{max})
2 $DC_{best} \longleftarrow$ Evaluate($Population$, D, α)
3 **for** $i = 0$ to $iter$ **do**
4   **for** $i = 0$ to $ps/2$ **do**
5          $Parent_1$, $Parent_2 \longleftarrow$ SelectParents($Population$)
6          $Parents \longleftarrow$ Add($Parent_1$, $Parent_2$)
7   **end**
8   **for** $i = 0$ to $ps/2$ **do**
9          $Parent_1$, $Parent_2 \longleftarrow$ RandomlyChooseParents(Parents)
10         $Child_1$, $Child_2 \longleftarrow$ Crossover($Parent_1$, $Parent_2$, cp, c_{max})
11         $NewPopulation \longleftarrow$ Add($Child_1$, $Child_2$)
12   RemoveUsedParents($Parent_1$, $Parent_2$, $Parents$)
13   **end**
14     $Population \longleftarrow$ Mutate($NewPopulation$, D, mp, c_{max})
15     $DC_{best} \longleftarrow$ Evaluate($Population$, D, α)
16 **end**

3.2   Population

Individual Encoding. An individual (i.e., a DC) is encoded as a vector of single rules (genes). A unique ID and a fitness score is assigned to each individual. Both, the distributed classifier and single rules must satisfy the previously described constraints (see Sect. 2.3). Note, rules must consist of unique miRNA-inputs and DCs must consist of unique rules.

Initial Population. An initial population of a given size (ps) is generated randomly, i.e., each classifier and each single rule in the classifier is randomly initialized. Individuals in the population may be of a different size c and maximally of a size c_{max}. Thus, to generate a new individual, c must first be defined. Then, each single rule is generated in a few steps. First, the rule size $(RuleSize)$ and $RuleSize$ miRNA IDs are randomly chosen. Then, for each miRNA the sign (positive/negative) is randomly assigned. This procedure (Algorithm 1, 1) is described in details in the appendix (Algorithm 2).

3.3   Fitness Function and Evaluation

As described in Sect. 2.4, to evaluate the classification performance of a distributed classifier we apply balanced accuracy as the fitness function. To count TPs and TNs we iterate over samples and evaluate the performance of a DC according to the threshold function described in Sect. 2.3. The fitness score is calculated separately for each DC in the population (Algorithm 1, 2, 15). As mentioned before, each iteration of the GA is completed by the update of the list of the best found solutions (DC_{best}). If the newly generated DCs perform with higher BACC than the solutions currently stored in DC_{best}, the new best DCs replace the previously found classifiers. If the new DCs have identical scores as the solutions in DC_{best} they are added to the list of the best solutions (Algorithm 1, 15). The classification threshold is a parameter specified by the user. In Sect. 4 we discuss the influence of different thresholds on the results.

3.4   Operators

Selection. Parents, to be potential candidates for recombination, are chosen in a process of *tournament selection* (Algorithm 1, 4–7). In each selection iteration two parents are chosen in separate tournaments. To select one parent, a number of *ts* individuals is randomly chosen from the current population to participate in a tournament. The winning candidate is an individual with the best fitness score. The first and the second parent must be different individuals. Thus, in each iteration, after choosing the first parent, its ID is temporarily blocked to be re-selected. The steps are repeated to form a population of selected individuals of the size *ps*. For more details see Appendix (Algorithm 3).

Crossover. In each crossover iteration two parents are randomly chosen from a population of selected individuals to recombine and generate two new individuals. Crossover (Algorithm 1, 8–13) occurs with the probability cp. To decide whether parents exchange information a random number p is chosen. If $p \leq cp$ then the two randomly chosen parents recombine. Otherwise, parents are copied to a new population. If chosen parents are of the same size we perform uniform crossover (Fig. 2A). To create two new individuals, rules from the first and second parent are paired off. Then, the first rule in each pair is assigned with equal probability to either the first or second child, while the second rule is assigned to the other child. The step is repeated until all the rules from parents were utilized and the children consist of the same number of rules as the parents. Otherwise, if the sizes of parents differ, to preserve a chance for each rule to be exchanged, we apply an index-based crossover (Fig. 2B). Here, the rules from the first and second parent are paired off according to a randomly chosen index specifying the position of a shorter parent in relation to the other one (see example in Fig. 2B). Paired rules are crossovered uniformly. Rules that cannot be paired (due to different sizes) may be copied to a randomly chosen child. Note, the index-based crossover may shorten the size of an individual as additional rules cannot be copied to the larger classifier. Details on the implementation of the index-based crossover may be found in the Appendix (Algorithms 4 and 5).

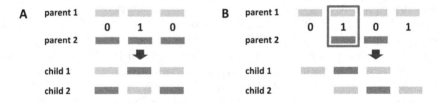

Fig. 2. Two crossover strategies applied in the presented GA. Yellow (light) and green (dark) boxes represent rules in different DCs (parents or children). (A) Uniform crossover. (B) Index-based crossover. The crossover index is marked by a red frame. (Color figure online)

Mutation. Mutation (Algorithm 1, 14) may occur on two levels: both, rules and inputs may mutate. A rule may (i) be removed from a classifier, (ii) be added to a classifier and (iii) be copied from one classifier to another. As mentioned before, index-based crossover may shorten the classifier. Here, two possibilities to extend the size of a classifier are available: a new rule may be initialized and added to a classifier or copied from another classifier. These two options balance the influence of crossover on the size of classifiers. An input may (i) be removed from a rule, (ii) be added to a rule, (iii) may change the sign (i.e., become a negative or positive input respecting the constraints described in Sect. 2.3). Rules, being larger components affecting the classifier size, mutate with a lower probability than inputs (0.2). Note, the maximal size of a classifier (c_{max}) must be preserved. For more details see Appendix (Algorithm 6).

4   Case Study

In this section, we illustrate the potential of DCs in application by performing case studies on real-world breast cancer data. We first describe the data sets used to evaluate DC performance. Then, we present results of parameter tuning and cross-validation. We analyze the classifier performances, as well as the relevance of chosen miRNAs. Finally, we compare DCs with a single-circuit classifier design approach.

4.1   Breast Cancer Data

To evaluate the performance of our approach we use Breast Cancer data sets previously applied by Becker et al. [1] and Mohammadi et al. [15] to the design of single-circuit classifiers. Originally the data was described by Farazi et al. [5] and pre-processed by Mohammadi et al. [15]. The details about the samples and miRNAs may be found in Table 2. The data set *All* includes samples of different breast cancer subtypes. This allows to compare breast cancer samples with the control samples. The following data sets are subsets representing different breast cancer subtypes containing information about the differences between particular subtypes and the control. Note, the data sets are significantly imbalanced as the negative class is heavily underrepresented. The data is formatted according to the description presented in Sect. 2.1. In terms of cell classifiers, non-regulated miRNAs do not carry any information. Thus, we remove them from the data sets before optimizing the classifiers. The last two columns of Table 2 include numbers of miRNAs before and after the filtering procedure.

Table 2. Breast Cancer data description.

Dataset	Samples	Positive	Negative	miRNAs	filtered miRNAs
All	178	167	11	478	57
Triple-	82	71	11	456	52
Her2+	86	75	11	438	19
ER+ Her-	32	21	11	392	18
Cell Line	17	6	11	375	59

4.2   Parameter Tuning

To tune the parameters of the genetic algorithm we applied a random search approach. The random search method allows to obtain results similar to the grid search approach, while decreasing the computational cost [2]. This provides an opportunity to extend the range of tested parameters. To tune the parameters we used the Breast Cancer All data set. We performed 3-fold cross-validation and repeated each GA run 10 times to obtain the average balanced accuracy on the test data. We have randomly chosen 300 combinations of 5 parameters in following ranges: $iter$: 25–100, step 25; ps: 50–300, step 50; cp: 0.1–1.0, step 0.1; mp: 0.1–1.0, step 0.1; ts: 10–50%, step 10% (of ps). We tuned the parameters for $\alpha = 0.50$ as in intermediate threshold ratio and $c_{max} = 5$ and chose a following set of parameters based on average scores: $iter = 75$, $ps = 200$, $cp = 1.0$, $mp = 0.3$, $ts = 10\%$ (20 individuals). We applied those parameters to all case studies presented in the following sections. One may expect that the parameters optimized for a given decision threshold may further improve the performance of classifiers. We comment on it briefly in Sect. 5.

4.3 Cross-Validation

To evaluate the classifiers accuracy we performed 3-fold cross-validation for the breast cancer data sets presented in Sect. 4.1. We partition the data sets into 3 folds nearly equal in terms of the number of samples representing each class per fold. For each fold we run the algorithm once. For all tests we apply $c_{max} = 5$. The classifier size $c_{max} = 5$ allows to preserve the maximal number of miRNA inputs as proposed for single-circuit classifiers [1,15]. Maintaining similar complexity of classifiers allows to compare the DC-based method to another approach.

We test eight different values of α: 0.25, 0.35, 0.40, 0.50, 0.60, 0.65, 0.75, 0.85, to evaluate the influence of the threshold function on the classification accuracy. As mentioned before, the results might be influenced by the parameter tuning being done for $\alpha = 0.5$. The best results are presented in Table 3 (complete results for different α values may be found in the Appendix, Table 5). The DCs presented in the results are the first best shortest classifiers found by the algorithm. If identical BACC values for the testing data were obtained for more than one α, we present results for a DC with the highest BACC value on the training data. In case of equal training BACC values, we present an exemplary result for a chosen threshold. Table 3 includes the α-s and performance scores. All scores except of $BACC_{train}$ were calculated on the testing data.

Table 3. Results of 3-fold cross-validation. For the Breast Cancer All data set we found DCs performing with identical score values for two α values (0.50, 0.60) and for ER+Her- for 6 different α values (0.35, 0.50, 0.60, 0.65, 0.75, 0.85)

Dataset	α	Sensitivity	Specificity	ACC	BACC	$BACC_{train}$
All	0.50	0.92	0.92	0.92	0.92	0.98
Triple-	0.85	0.92	0.75	0.89	0.83	0.98
Her2+	0.75	0.99	0.61	0.94	0.80	0.96
ER+ Her-	0.50	0.90	0.64	0.82	0.77	0.93
Cell Line	0.25	1.00	1.00	1.00	1.00	1.00

High BACC values obtained for the training data sets, as well as the average final population BACC values (0.91), show that the populations converge over the iterations resulting in high-performing DCs. The BACC values measured for the testing data sets are significantly higher for the largest and the smallest data sets than for the intermediate-size ones. Note, that for the data sets Her2+ and ER+ Her- the number of relevant miRNAs is significantly lower than for the other data sets. Thus, the space of available solutions is also substantially decreased. Cell Line data set includes 6 different miRNAs that perfectly separate samples [1]. Thus, excellent performance was expected for this particular data set. The accuracy is higher than BACC for all data sets as the metric is not sensitive to data imbalance.

The sensitivity is high for all data sets meaning that the method success-fully classifies samples belonging to a positive class. However, the specificity is substantially decreased for Her2+ and ER+ Her- data sets. Note, the data sets are significantly imbalanced, i.e., the negative class is strongly underrepresented. Thus, even small number of errors results in substantially decreased specificity.

The best α values differ among the data sets. For the largest one, α is equal or not much higher than 50%. The data sets of intermediate sizes (Triple- and Her2+) favoured two more extreme α values. For the ER+Her- several α values returned identical results (Appendix, Table 5). For the smallest data set the lowest α value resulted in the highest BACC. Thus, the threshold seems to be data-specific and should be adjusted to the data set for the DC to perform well.

Applying a certain threshold caused a shift in the rates of certain types of errors. Here, we analyze false positive rates (FP_{rate}) and false negative rates (FN_{rate}) observed among all data sets for two extreme applied α values. In case of a low threshold (0.25, $FP_{rate} = 0.34$) the shift is displayed towards mis-classification of the negative samples in comparison to a very high threshold (0.85, $FP_{rate} = 0.27$). The high threshold (0.85, $FN_{rate} = 0.13$) causes more frequent misclassification of positive samples in comparison to the low one (0.25, $FN_{rate} = 0.04$). The influence of a certain threshold on the shift should be fur-ther investigated. Complete information about FP_{rate}s and FN_{rate}s for different thresholds may be found in Appendix, Table 6.

The tests were performed using Allegro CPU Cluster provided by Freie Uni-versitaet Berlin[1]. An average run-time is 45 min for one cross-validation fold of the largest data set employed in the case studies. The tests may be performed on a personal computer. However, the breast cancer data sets consist of up to 180 samples and up to 60 relevant miRNAs. Thus, one should consider performing extended scalability tests to estimate the run-time limits of the method.

4.4   Analysis of Input Viability

In this section we analyze miRNA inputs that occur in two exemplary classifiers. We chose the best performing classifiers for the largest data set (All) representing all subtypes and the smallest Cell Line data set.

---

[1] https://www.allegro.imp.fu-berlin.de/Cluster.

For breast cancer All two different α values resulted in the highest BACC. We found that classifiers for each cross-validation fold in the data set are identical for both α values. Also, all the classifiers are of the same size $c = 4$. In this case the applied α does not change the threshold function between both values (0.50 and 0.60), i.e., for all data sets at least 2 *Rules* must output 1 to classify a cell as positive. Below we present a DC found for the third cross-validation fold of the All data set. The classifier consists of 4 different 1-input rules. We analyzed the miRNAs and found that all of them may be relevant for cancer sample classification. The classifier is presented below.

$$\{\neg miR\text{-}378, \ miR\text{-}200c, \ \neg miR\text{-}145, \ \neg miR\text{-}451\text{-}DICER1\}$$

miR-378, miR-145, miR-451-$DICER1$ are described as down-regulated in breast cancer [4,5], e.g., the study by Ding et al. [4] has shown that underexpression of miR-145 is related to increased proliferation of breast cancer cells. Also, miR-378 occurred as down-regulated in the best 1-input single-circuit classifier presented previously by Becker et al. for the same data set [1]. miR-200c is marked as up-regulated in breast cancer in [21].

Another classifier we present is a DC for the third cross-validation fold for the Cell Line data set:

$$\{\neg miR\text{-}146a, \ \neg miR\text{-}143\}$$

For most of the α values the performance of found DCs was very low for this particular fold in the Cell Line data set (BACC = 0.50). A perfect classifier of size 2 performing with BACC = 1.00 on both training and testing data was found with $\alpha = 0.25$, i.e., one of 2 rules must output 1 to classify the cell as positive. We found that both, miR-146a and miR-143, are described as down-regulated in breast cancer [10,16].

4.5   Comparison to Other Methods

We optimized single-circuit classifiers with the ASP-based method proposed by Becker et al. [1] by performing 3-fold cross-validation using the same data sets and identical division into folds. The objective function of the ASP algorithm is based on the minimization of the total number of classification errors. Note that the ASP method may return several optimal classifiers. Different combinations of FPs and FNs influence the balanced accuracy. Thus, to increase the chance of ASP to perform well, we have chosen the best classifiers according to their BACC. Here, we do not compare our results to Mohammadi et al. [15] as the approach did not perform better than the ASP-based approach as described by Becker et al. in terms of binary classification [1] (Table 4).

Table 4. Comparison of results of 3-fold cross-validation for the ASP-based approach proposed by Becker et al. [1] and for the GA (as in Table 3).

Dataset	Method	Sensitivity	Specificity	ACC	BACC	$BACC_{train}$
All	GA	0.92	0.92	0.92	**0.92**	0.98
	ASP	0.96	0.47	0.93	0.72	0.92
Triple-	GA	0.92	0.75	0.89	**0.83**	0.98
	ASP	0.89	0.44	0.83	0.67	0.96
Her2+	GA	0.99	0.61	0.94	0.80	0.96
	ASP	1.00	0.61	0.95	**0.81**	0.96
ER+ Her-	GA	0.90	0.64	0.82	**0.77**	0.93
	ASP	0.90	0.64	0.82	**0.77**	0.93
Cell Line	GA	1.00	1.00	1.00	**1.00**	1.00
	ASP	0.83	1.00	0.93	0.92	1.00

The DC-based method outperformed the single-circuit approach in 3 of 5 case studies. For two other data sets the resulting BACC (test) values are either identical (ER+Her-) or very similar (Her2+). This may imply that further improvement of classifier performance for those data sets is not possible with the currently applied techniques. The training BACC values are also significantly higher for the DC-based approach. Note, the DC-based design method explores a different search space than the single circuit approach. Although single circuits are also allowed as 1-rule classifiers, their complexity is substantially lower in comparison to single circuits. Additionally, ASP returns globally optimal solutions, i.e., it adjusts the classifier perfectly to the training data, which may cause overfitting. Although, the classifiers obtain high BACC on the training data (average for all data sets: 0.95), the classifiers may be too specific to perform well on the testing data.

The scalability of the ASP-based approach was previously shown by Becker et al. [1]. As mentioned before, the ASP-based approach optimizes much simpler classifiers than the GA-based method. Thus, the run-times of both approaches are not easily comparable.

5   Discussion

In this article, we introduced a new approach to cell classifier design. The concept of DCs proposed by Didovyk et al. [3] was re-formalized in the context of miRNA-based cell classification. We designed and implemented a genetic algorithm that allows design and optimization of *DC*s. We performed case studies on real-world data and compared our results to a single-circuit design method obtaining significantly higher or similar accuracy.

DCs show immense potential as an alternative to single-circuit designs. Presented case studies demonstrate the DC's ability to perform classification on real-world cancer data. The results obtained on the training data show that the proposed genetic algorithm allows to optimize classifiers that achieve high accuracy. The cross-validation demonstrates that the optimized DCs classify unknown data with high accuracy. The data sets for which the algorithm returns the worst results (Her2+, ER+Her-) are ones with the lowest number of relevant miRNAs. Thus, the number of possible solutions is significantly decreased in contrast to other data sets. The best performing decision thresholds differ significantly among data sets. However, higher α values seem to be more efficient. Testing a wide range of thresholds while optimizing the classifiers is strongly recommended. The comparison to a single-circuit design method shows that DCs outperformed single-circuit classifiers on most of the presented data sets according to balanced accuracy. Although the GA performs better on the largest and the smallest data sets than on the intermediate-size ones, the results obtained for both compared methods for Her2+ and ER+Her- are very similar which may suggest that for those data sets significant improvement is not possible. The improvements in binary classification may be a result of applying a different strategy to cell classifier design. Here, single-circuit decision is complemented by a collective classification based on a threshold function. Thus, the DCs may be more resistant to data noise than single-circuit classifiers.

Generally, the problem of designing reliable and efficient DCs begins with the initial data processing. As mentioned before, the data sets employed for the case studies are significantly imbalanced. Although we apply an objective function that allows to partially overcome this issue, one may consider applying data balancing methods such as weighted schemes that balance the sample importance [7]. Furthermore, our approach to the design of DCs is based on binarized data sets. As mentioned before, data discretization allows obtaining clear-cut information about miRNA regulation and efficient exploration of the search space. One advantage of this data processing procedure is absorption of noise coming from, e.g., lab artifacts. However, simultaneously some information that may be valuable for the classification is lost. Considering binarization according to a given threshold, miRNAs having their concentrations significantly higher (or lower, respectively) than the threshold may be more informative. Thus, one may introduce a multi-objective function that allows to optimize both, the accuracy and the use of particular miRNAs according to, e.g., a weighted scheme favoring more reliable miRNAs.

Adapting the ASP approach to classifier design, one could apply ASP to the optimization of DCs, obtain globally optimal solutions and compare with the heuristic approach. However, ASP searches through the entire solution space; thus, the run-time may be significantly increased with the rising number of possible combinations. As we expect that this may significantly limit the ASP-based optimization, one may explore other possibilities. In the proposed GA the initial population is generated randomly, i.e., there is no preference in choosing particular miRNAs or gate signs that built rules. One may optimize the initial population by creating rules taking such preferences into account. The ASP allows to optimize single short classifiers with relaxed constraints in a short time, e.g., allowing up to a certain number of errors. This may generate a pool of rules that are pre-optimized resulting in a better starting point for the algorithm.

Although the results demonstrate that classifiers perform with high accuracy, the possibilities to further develop the presented method should be explored. Certainly, the approach must be tested using more data representing variety of cancer types. Although the proposed genetic algorithm performed well on the presented case studies, particular parts of the algorithm may be improved. In this work we do not tune parameters for different applied thresholds due to time-consuming calculations. It should be further investigated whether the parameters may be optimized for certain thresholds to improve the performance of classifiers. Additionally, different selection operators may be tested to evaluate the influence of a chosen operator on the results [23]. Although tournament selection is described in the literature as a well-performing operator, some other operators may be more accurate for particular problems than the commonly recommended ones.

Although DCs are not yet applied in terms of cancer cell classification, the approach should be further investigated. DCs are designed based on available building blocks that are in fact single-circuit classifiers. Mohammadi et al. [15] presented a biochemical model of a single-circuit classifier that allows to manipulate the output compound concentration. Thus, the biological output threshold for a given classifier may be adjusted to perform the classification in living cells. As the on-off single-circuit response may be regulated on the biological level, the sum of their outputs should also be adaptable for a given DC. This needs to be investigated through further work in the lab.

Data and Software Availability. The algorithm is implemented in Python 3. The scripts, as well as the data used to tune the parameters and test the algorithm's performance including the results, are available at GitHub [18].

Acknowledgements. We would like to thank to P. Mohammadi, Y. Benenson and N. Beerenwinkel (ETH Zurich) for sharing the breast cancer data with us. MN would like to thank to J. Bartoszewicz (RKI, Berlin) for his valuable comments and support with cluster handling.

Appendix

Algorithm 2. The algorithm describes the generation of an initial population of size ps (Sect. 3.2). ps individuals are created randomly, i.e., the number of rules is randomly chosen and the rules are randomly generated. To create an individual its size must be first specified (line 2). Then, c rules must be generated in a

few steps. First, the size of a rule ($RuleSize$) and miRNA IDs ($miRNAs$) must
be randomly chosen (lines 4–7). Then, the sign (positive/negative) is randomly
assigned to the miRNAs (line 8). Note that in case of $RuleSize = 2$, the miRNAs
are connected with an AND. c rules generated as described above create an
individual which may be added to a population (line 12). The steps are repeated
until the population consists od ps individuals (lines 1–11).

Algorithm 2. Initialization of a first population.

Data: dataset D
Parameters: population size ps, maximal size of a DC c_{max}
Output: *Population*

1 **for** $i = 1$ **to** ps **do**
  /* randomly choose the size of a new classifier */
2    $c \longleftarrow$ RandomlyChooseInRange(1, c_{max})
3   **for** $i = 1$ **to** c **do**
    /* randomly choose the size of a new rule */
4       $RuleSize \longleftarrow$ RandomlyChooseInRange(1, 2)
    /* randomly choose miRNA IDs */
5       $miRNAs \longleftarrow$ RandomlyChooseIDs(D, $RuleSize$)
    /* randomly assign miRNA signs */
6       $miRNAs \longleftarrow$ RandomlyAssignSigns()
    /* create a new rule */
7       $Rule \longleftarrow$ CreateARule($miRNAs$)
    /* add a new rule to a classifier */
8       $Individual \longleftarrow$ Add($Rule$)
9   **end**
  /* add a new classifier to a population */
10    $Population \longleftarrow$ Add($Individual$)
11 **end**

Algorithm 3. The algorithm describes the selection of parents that are potential candidates to recombine (Sect. 3.4). The parents are chosen in tournaments
of size ts, i.e., ts candidates are randomly chosen from the population to participate in a tournament (lines 1–5, 7–11). In each round 2 parents are selected from
the population. The winning candidates are individuals with the highest BACC
(lines 5, 11). After the first parent is selected its ID is temporarily blocked to
be re-selected (line 6). This allows to diverse the population of selected parents.
The new population of selected parents is then utilized to perform crossover.

Algorithm 3. Selection of parents

Input: *Population*
Parameters: population size *ps*, tournament size *ts*
Output: $Parent_1$, $Parent_2$
/* repeat adding to a tournament *ts* times */
1 **for** $i = 1$ **to** *ts* **do**
   /* randomly choose an individual's ID */
2     $Candidate \longleftarrow$ RandomlyChooseInRange(1, ps)
   /* add a candidate ID to a tournament */
3     $Candidates \longleftarrow$ Add($Candidate$)
4 **end**
   /* choose the best parent in a tournament */
5 $Parent_1 \longleftarrow$ SelectBest($Candidates$)
   /* block a chosen ID to be re-selected */
6 $ps \longleftarrow$ BlockID($Parent_1$, ps)
7 **for** $i = 0$ **to** *ts* **do**
   /* randomly choose an individual's ID */
8     $Candidate \longleftarrow$ RandomlyChooseInRange(1, ps)
   /* add a candidate ID to a tournament */
9     $Candidates \longleftarrow$ Add($Candidate$)
10 **end**
   /* choose the best parent in a tournament */
11 $Parent_2 \longleftarrow$ SelectBest($Candidates$)

Algorithm 4. The algorithm describes the crossover procedure performed on the population of selected parents (Sect. 3.4). Each couple of parents chosen randomly from the population of selected parents exchange genes with the probability *cp*. If the randomly chosen *probability* is lower than *cp* the parents undergo the crossover (lines 2–13). Otherwise, the parents are copied directly to a new population (line 15). If parents are of the same size, uniform crossover is performed (line 11–12). Otherwise, index-based crossover is applied (lines 7–9). Both procedures are described in details in Sect. 3.4.

Algorithm 5. The algorithm describes the index-based crossover that we apply if the sizes of parents differ to preserve a chance for each rule to be exchanged. Here, the rules from the first and second parent are paired off according to a randomly chosen index specifying the position of a shorter parent in relation to the other one. The index is chosen randomly and is in range between 1 and $ParentSize_1$-$ParentSize_2$ (Algorithm 4, line 7). Paired rules are crossovered uniformly. Rules that cannot be paired (due to different sizes) may be copied to a randomly chosen child. As a result, the number of rules in each child is between the minimum and the maximum size of the two parents. Note, the index-based crossover may shorten the size of an individual as additional rules cannot be copied to the larger classifier. This procedure is described in details in Sect. 3.4.

Algorithm 4. Crossover

Input: $Parent_1$, $Parent_2$, crossover probability cp
Output: $Child_1$, $Child_2$
   /* randomly choose the probability of crossover */
1 $probability \longleftarrow$ DrawProbability(0,1)
   /* if $probability \leq cp$ perform crossover */
2 **if** $probability \leq cp$ **then**
      /* assign a longer parent to $Parent_1$ */
3     $Parent_1$, $Parent_2 \longleftarrow$ AssignParentsBySize($Parent_1$, $Parent_2$)
      /* assign sizes of parents */
4     $ParentSize_1 \longleftarrow$ Size($Parent_1$)
5     $ParentSize_2 \longleftarrow$ Size($Parent_2$)
      /* if parents sizes differ perform index-based crossover */
6   **if** $ParentSize_1 \neq ParentSize_2$ **then**
        /* randomly choose the crossover index */
7        $CrossoverIndex \longleftarrow$ RandomlyChooseInRange(1, $ParentSize_1$ - $ParentSize_2$)
        /* perform index based crossover */
8        $Child_1$, $Child_2 \longleftarrow$ IndexCrossover($Parent_1$, $Parent_2$, $ParentSize_1$, $ParentSize_2$, $CrossoverIndex$)
9        $Population \longleftarrow$ Add($Child_1$, $Child_2$)
10   **else**
        /* if parents have identical sizes perform uniform crossover */
11       $Child_1$, $Child_2 \longleftarrow$ UniformCrossover($Parent_1$, $Parent_2$)
        /* add children to a new population */
12       $Population \longleftarrow$ Add($Child_1$, $Child_2$)
13   **end**
14 **else**
      /* if $probability > cp$ copy parents to a new population */
15     $Population \longleftarrow$ Add($Parent_1$, $Parent_2$)
16 **end**

Algorithm 6. The algorithm describes mutation (Sect. 3.4). Mutation may occur on two levels: both, rules and inputs may mutate. A rule may (i) be removed from a classifier, (ii) be added to a classifier and (iii) be copied from one classifier to another (lines 5–17). An input may (i) be removed from a rule, (ii) be added to a rule, (iii) may change the sign i.e., become a negative or positive input (lines 19–32). Rules, being larger components affecting the classifier size, mutate with a lower probability than inputs (0.2). Note, the maximal size of a classifier (c_{max}) must be preserved.

Algorithm 5. Index-based crossover

Input: $Parent_1$, $Parent_2$, $ParentSize_1$, $ParentSize_2$, $CrossoverIndex$
Output: $Child_1$, $Child_2$

1 **for** $i = 1$ **to** $ParentSize_1$ **do**
   /* decide whether the rule will be exchanged */
   /* $SwapMask$=1 corresponds to rule exchange */
   /* $SwapMask$=0 corresponds to copying without exchanging */
2    $SwapMask \longleftarrow$ RandomlyChooseInRange(0, 1)
   /* 1 - rule is exchanged */
3   **if** $SwapMask = 1$ **then**
   /* if the rules do not pair off */
   /* i.e., there is no possibility to exchange rules */
4   **if** $i < CrossoverIndex$ OR $i \geq CrossoverIndex + ParentSize_2$
   then
   /* copy a rule from $Parent_1$ to $Child_2$ */
5          $Child_2 \longleftarrow$ CopyRule($Parent_1$, i)
6   **else**
   /* if the rules pair off exchange rules */
   /* copy a rule from $Parent_2$ to $Child_1$ */
7          $Child_1 \longleftarrow$ CopyRule($Parent_2$, i)
   /* copy a rule from $Parent_1$ to $Child_2$ */
8          $Child_2 \longleftarrow$ CopyRule($Parent_1$, i)
9   **end**
10   **else**
   /* 0 - rule is not exchanged */
11   **if** $i < CrossoverIndex$ OR $i \geq CrossoverIndex + ParentSize_2$
   then
   /* copy a rule from $Parent_1$ to $Child_1$ */
12          $Child_1 \longleftarrow$ CopyRule($Parent_1$, i)
13   **else**
   /* else copy rules to the parents without exchanging */
   /* copy a rule from $Parent_1$ to $Child_1$ */
14          $Child_1 \longleftarrow$ CopyRule($Parent_1$, i)
   /* copy a rule from $Parent_2$ to $Child_2$ */
15          $Child_2 \longleftarrow$ CopyRule($Parent_2$, i)
16   **end**
17   **end**
18 **end**

Algorithm 6. Mutation

Input: *Population*, maximal size of a DC c_{max}
Output: *Population*

1   **for** $i = 1$ **to** *ps* **do**
    /* randomly choose the probability of mutation */
2   *probability* ⟵ DrawProbability(0,1)
    /* if *probability* ≤ *mp* perform mutation */
3   **if** *probability* ≤ *mp* **then**
       /* choose the mutation level */
       /* 1 corresponds to mutation of a rule */
       /* 2-4 corresponds to mutation of an input */
4   *MutationLevel* ⟵ RandomlyChooseInRange(1, 5)
5   **if** *MutationLevel* = *1* **then**
          /* choose the mutation type */
6   *MutationType* ⟵ DrawItem(*add, remove, copy*)
7   **switch** *MutationType* **do**
8   **case** *add*
                /* add rule */
9   AddRule(*Population, i, c_{max}*)
10   **end**
11   **case** *copy*
                /* copy rule */
12   CopyRule(*Population, i, c_{max}*)
13   **end**
14   **case** *remove*
                /* remove rule */
15   RemoveRule(*Population, i*)
16   **end**
17   **endsw**
18   **else**
          /* choose the mutation type */
19   *MutationType* ⟵ DrawItem(*add, remove, sign*)
20   **switch** *MutationType* **do**
21   **case** *add*
                /* add input */
22   *Rule* ⟵ DrawRule(1, *ps*)
23   AddInput(*Population, i, Rule*)
24   **end**
25   **case** *remove*
                /* remove input */
26   RemoveInput(*Population, i, c_{max}, Rule*)
27   **end**
28   **case** *sign*
                /* change sign of an input */
29   ChangeInputSign(*Population, i, Rule*)
30   **end**
31   **endsw**
32   **end**
33   **end**
34   **end**

Table 5. Results of 3-fold cross-validation.

Dataset	α	Sensitivity	Specificity	ACC	BACC	BACC$_{train}$
All	0.85	0.89	0.83	0.88	0.86	0.96
	0.75	0.94	0.81	0.93	0.87	0.98
	0.65	0.95	0.72	0.93	0.83	0.98
	0.60	0.92	0.92	0.92	**0.92**	0.98
	0.50	0.92	0.92	0.92	**0.92**	0.98
	0.40	0.94	0.64	0.92	0.79	0.99
	0.35	0.97	0.72	0.96	0.85	0.99
	0.25	0.96	0.72	0.94	0.84	1.00
Triple-	0.85	0.92	0.75	0.89	**0.83**	0.98
	0.75	0.96	0.67	0.92	0.81	0.99
	0.65	0.94	0.58	0.89	0.76	1.00
	0.60	0.93	0.64	0.89	0.78	1.00
	0.50	0.93	0.64	0.89	0.78	1.00
	0.40	0.94	0.56	0.89	0.75	1.00
	0.35	0.94	0.53	0.89	0.74	0.99
	0.25	0.94	0.53	0.89	0.74	1.00
Her2+	0.85	0.99	0.44	0.92	0.72	0.96
	0.75	0.99	0.61	0.94	**0.80**	0.96
	0.65	0.99	0.53	0.93	0.76	0.96
	0.60	1.00	0.53	0.94	0.76	0.96
	0.50	1.00	0.53	0.94	0.76	0.96
	0.40	0.99	0.53	0.93	0.76	0.96
	0.35	1.00	0.53	0.94	0.76	0.96
	0.25	1.00	0.53	0.94	0.76	0.93
ER+ Her-	0.85	0.90	0.64	0.82	**0.77**	0.93
	0.75	0.90	0.64	0.82	**0.77**	0.93
	0.65	0.90	0.64	0.82	**0.77**	0.93
	0.60	0.90	0.64	0.82	**0.77**	0.93
	0.50	0.90	0.64	0.82	**0.77**	0.93
	0.40	0.90	0.53	0.78	0.72	0.93
	0.35	0.90	0.64	0.82	**0.77**	0.93
	0.25	0.90	0.53	0.78	0.72	0.91
Cell Line	0.85	0.67	1.00	0.87	0.83	1.00
	0.75	0.67	1.00	0.87	0.83	1.00
	0.65	0.67	1.00	0.87	0.83	1.00
	0.60	0.67	1.00	0.87	0.83	1.00
	0.50	0.67	1.00	0.87	0.83	1.00
	0.40	0.67	1.00	0.87	0.83	1.00
	0.35	1.00	0.89	0.93	0.94	1.00
	0.25	1.00	1.00	1.00	**1.00**	1.00

Table 6. FP_{rate} and FN_{rate} values for different thresholds (for all datasets).

Threshold	FP_{rate}	FN_{rate}
0.85	0.27	0.13
0.75	0.23	0.11
0.65	0.31	0.11
0.60	0.26	0.12
0.50	0.26	0.12
0.40	0.35	0.11
0.35	0.34	0.04
0.25	0.34	0.04

References

1. Becker, K., Klarner, H., Nowicka, M., Siebert, H.: Designing miRNA-based synthetic cell classifier circuits using answer set programming. Front. Bioeng. Biotechnol. **6**, 70 (2018). https://doi.org/10.3389/fbioe.2018.00070
2. Bergstra, J., Bengio, Y.: Random search for hyper-parameter optimization. J. Mach. Learn. Res. **13**, 281–305 (2012)
3. Didovyk, A., Kanakov, O.I., Ivanchenko, M.V., Hasty, J., Huerta, R., Tsimring, L.: Distributed classifier based on genetically engineered bacterial cell cultures. ACS Synth. Biol. **4**(1), 27–82 (2015). https://doi.org/10.1021/sb500235p
4. Ding, Y., et al.: miR-145 inhibits proliferation and migration of breast cancer cells by directly or indirectly regulating TGF-β1 expression. Int. J. Oncol. **50**(5), 1701–1710 (2017). https://doi.org/10.3892/ijo.2017.3945
5. Farazi, T.A., et al.: MicroRNA sequence and expression analysis in breast tumors by deep sequencing. Cancer Res. **71**(13), 4443–4453 (2011). https://doi.org/10.1158/0008-5472.CAN-11-0608
6. Gallo, C.A., Cecchini, R.L., Carballido, J.A., Micheletto, S., Ponzoni, I.: Discretization of gene expression data revised. Brief. Bioinform. **17**(5), 758–770 (2016). https://doi.org/10.1093/bib/bbv074
7. Haixiang, G., Yijing, L., Shang, J., Mingyun, G., Yuanyue, H., Bing, G.: Learning from class-imbalanced data: review of methods and applications. Expert Syst. Appl. **73**, 220–239 (2017). https://doi.org/10.1016/J.ESWA.2016.12.035
8. Iorio, M.V., Croce, C.M.: MicroRNA dysregulation in cancer: diagnostics, monitoring and therapeutics. A comprehensive review. EMBO Mol. Med. **4**(3), 143–59 (2012). https://doi.org/10.1002/emmm.201100209
9. Lan, H., Lu, H., Wang, X., Jin, H.: MicroRNAs as potential biomarkers in cancer: opportunities and challenges. BioMed Res. Int. **2015**, 125094 (2015). https://doi.org/10.1155/2015/125094
10. Li, Y., Xu, Y., Yu, C., Zuo, W.: Associations of miR-146a and miR-146b expression and breast cancer in very young women. Cancer Biomarkers **15**(6), 881–887 (2015). https://doi.org/10.3233/CBM-150532
11. Manning, T., Sleator, R.D., Walsh, P.: Naturally selecting solutions: the use of genetic algorithms in bioinformatics. Bioengineered **4**(5), 266–78 (2013). https://doi.org/10.4161/bioe.23041

12. Marchisio, M., Stelling, J.: Computational design of synthetic gene circuits with composable parts. Bioinformatics **24**(17), 1903–1910 (2008). https://doi.org/10.1093/bioinformatics/btn330
13. McCall, J.: Genetic algorithms for modelling and optimisation. J. Comput. Appl. Math. **184**(1), 205–222 (2005). https://doi.org/10.1016/J.CAM.2004.07.034
14. Miki, K., et al.: Efficient detection and purification of cell populations using synthetic MicroRNA switches. Cell Stem Cell **16**(6), 699–711 (2015). https://doi.org/10.1016/j.stem.2015.04.005
15. Mohammadi, P., Beerenwinkel, N., Benenson, Y.: Automated design of synthetic cell classifier circuits using a two-step optimization strategy. Cell Syst. **4**(2), 207–218.e14 (2017). https://doi.org/10.1016/j.cels.2017.01.003
16. Ng, E.K.O., Li, R., Shin, V.Y., Siu, J.M., Ma, E.S.K., Kwong, A.: MicroRNA-143is downregulated in breast cancer and regulates DNA methyltransferases 3A inbreast cancer cells. Tumor Biol. **35**(3), 2591–2598 (2014). https://doi.org/10.1007/s13277-013-1341-7
17. Nielsen, A.A., et al.: Genetic circuit design automation. Science **352**, 6281 (2016). https://doi.org/10.1126/science.aac7341
18. Nowicka, M.: A genetic algorithm to designing distributed cell classifier circuits (2019). https://github.com/MelaniaNowicka/RAccoon
19. Palmer, E.: The T-Cell antigen receptor: a logical response to an unknown ligand. J. Recept. Signal Transduct. **26**(5–6), 367–378 (2006). https://doi.org/10.1080/10799890600919094
20. Ramola, R., Jain, S., Radivojac, P.: Estimating classification accuracy in positive-unlabeled learning: characterization and correction strategies. In: Pacific Symposium on Biocomputing, vol. 24, pp. 124–135 (2019)
21. Sánchez-Cid, L., et al.: MicroRNA-200, associated with metastatic breast cancer, promotes traits of mammary luminal progenitor cells. Oncotarget **8**(48), 83384–83406 (2017). https://doi.org/10.18632/oncotarget.20698
22. Schapire, R.E.: The strength of weak learnability. Mach. Learn. **5**(2), 197–227 (1990). https://doi.org/10.1007/BF00116037
23. Shukla, A., Pandey, H.M., Mehrotra, D.: Comparative review of selection techniques in genetic algorithm. In: 2015 International Conference on Futuristic Trends on Computational Analysis and Knowledge Management (ABLAZE), pp. 515–519. IEEE, February 2015. https://doi.org/10.1109/ABLAZE.2015.7154916
24. Slomovic, S., Pardee, K., Collins, J.J.: Synthetic biology devices for in vitro and in vivo diagnostics. Proc. Natl. Acad. Sci. **112**(47), 14429–14435 (2015). https://doi.org/10.1073/pnas.1508521112
25. Smith, R.W., van Sluijs, B., Fleck, C.: Designing synthetic networks in silico: a generalised evolutionary algorithm approach. BMC Syst. Biol. **11**(1), 118 (2017). https://doi.org/10.1186/s12918-017-0499-9
26. Xie, Z., Wroblewska, L., Prochazka, L., Weiss, R., Benenson, Y.: Multi-input RNAi-based logic circuit for identification of specific cancer cells. Science **333**(6047), 1307–1311 (2011). https://doi.org/10.1126/science.1205527
27. Yang, K., Li, J., Gao, H.: The impact of sample imbalance on identifying differentially expressed genes. BMC Bioinform. **7**(Suppl. 4), S8 (2006). https://doi.org/10.1186/1471-2105-7-S4-S8
28. Ye, H., Fussenegger, M.: Synthetic therapeutic gene circuits in mammalian cells. FEBS Lett. **588**(15), 2537–2544 (2014). https://doi.org/10.1016/j.febslet.2014.05.003

Extending a Hodgkin-Huxley Model for Larval *Drosophila* Muscle Excitability via Particle Swarm Fitting

Paul Piho[1](✉), Filip Margetiny[2], Ezio Bartocci[4], Richard R. Ribchester[2,3], and Jane Hillston[1]

[1] School of Informatics, University of Edinburgh, Edinburgh, UK
paul.piho@ed.ac.uk
[2] Centre for Discovery Brain Sciences, University of Edinburgh, Edinburgh, UK
[3] Euan MacDonald Centre for Motor Neurone Disease Research,
University of Edinburgh, Edinburgh, UK
[4] Faculty of Informatics, TU Wien, Vienna, Austria

Abstract. We present a model of excitability in larval *Drosophila* muscles. Our model was initially based on modified Hodgkin-Huxley equations, adapted to represent variable, regenerative depolarisations (action potentials) we have occasionally observed in intracellular recordings and that can be triggered by excitatory junction potentials at neuromuscular synapses. We modified several kinetic equations describing voltage sensitive Ca^{2+} and K^+ ionic currents, previously used to predict excitability in muscle cells of the mammalian cardiac atrioventricular node. The resulting nonlinear differential equations had multiple unknown parameters. Thus, to fit the model to experimental observations of variable excitability, we developed a new implementation of particle swarm optimisation. This GPU-based implementation allows us to adopt an ensemble model approach in which each experimental observation is used to find a plausible parameterisation, resulting in a set of models accounting for cell-to-cell variability of muscle excitability in *Drosophila* larvae, and with potential applications to population-based modeling of other excitable cell types.

1 Introduction

The control of muscle contraction is fundamental to behaviour. In mammals and other vertebrates, muscle contraction is the end result of a signalling cascade that begins with excitation of motor neurones in the brain stem or spinal cord, triggering waves of regenerative depolarization and repolarization (action potentials) that are propagated by saltatory conduction along each myelinated nerve axon. Activation of muscle fibres takes place at neuromuscular junctions (NMJs). When a muscle action potential is propagated along the muscle surface membrane it leads to conformational change in muscle proteins, enabling contraction.

© Springer Nature Switzerland AG 2019
L. Bortolussi and G. Sanguinetti (Eds.): CMSB 2019, LNBI 11773, pp. 120–139, 2019.
https://doi.org/10.1007/978-3-030-31304-3_7

In this paper we investigated the mechanisms of muscle excitability in the abdominal muscles of larval fruit flies (*Drosophila melanogaster*). These invertebrate muscles do not express the voltage-sensitive sodium channels that are essential for generation of action potentials in vertebrate muscle fibres. Larval muscles do, however, contain voltage-sensitive calcium (Ca^{2+}) and potassium (K^+) channels: a similar situation to some non-skeletal muscle fibres in vertebrates, for example in the atrioventricular (AV) node of the mammalian heart [3,8,12]. However, unlike mammalian cardiac AV node cells, the functional significance of voltage-sensitive Ca^{2+} and K^+ channels in larval *Drosophila* muscle fibres is unclear, since activation of these channels is not necessary for muscle contraction and the summative effects of a short burst of excitatory NMJ potentials (EJPs) is sufficient for evoking contractile responses [14,22]. Nevertheless, larval *Drosophila* muscle fibres are capable of regenerative, Ca^{2+}-dependent depolarization and we seek to establish the physiological role of these phenomena. As part of that ongoing investigation, we have developed a computational model of excitability at larval *Drosophila* NMJs. Our approach was initially based on a modified Hodgkin-Huxley model of the firing properties of muscle cells in the mammalian cardiac AV node, which also depends mainly on a combination of interacting voltage-sensitive Ca^{2+} currents and K^+ currents [8]. In addition, in order to parameterise the model we have applied a new GPU-based implementation of the particle swarm optimisation method, following a similar procedure to others [15]. Initial investigations of this model suggest that the quantitative characteristics of the regenerative responses that have been observed experimentally can be accounted for by a computational model of this type.

The rest of the paper is structured as follows. In Sect. 2 we present an overview of the relevant biological background and a brief summary of Hodgkin-Huxley models of action potentials. In Sect. 3 we introduce the computational model of excitability at larval *Drosophila* neuromuscular junctions. In Sect. 4 we set up the parameter study for the presented model in the context of particle swarm optimisation and describe the preprocessing steps taken for isolating individual examples of actions potentials from the experimental data. In Sect. 5 we discuss the results of the parameter fitting for the considered model and finally, in Sect. 6 we give conclusions and some further research directions.

2   Background

2.1   Action Potentials

Action potentials are primary mechanisms of cell-to-cell communication in nervous and neuromuscular systems and they occur when transmembrane voltage undergoes rapid depolarisation then repolarisation. In neurones, this is often referred to as spiking or firing. The changes in membrane potential are caused by the flow of charged ions along their extracellular-intracellullar concentration gradients through voltage-gated ion channels in the cell membrane, proteins that typically incorporate selectivity filters for Na^+, Ca^{2+} or K^+ ions. In the resting state, the inside voltage is often more negative than $-70\,\mathrm{mV}$ with respect to

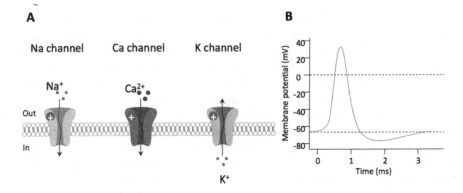

Fig. 1. A. Voltage sensitive ion channels in excitable cell membranes with the normal direction of ionic flux when open indicated. Several subtypes of Na, Ca and K channels are expressed in different cell types and species, differing in protein structure and activation/inactivation kinetics. B. Schematic illustration of regenerative depolarization (action potential) and recovery of the transmembrane resting potential following activation of ion channels like those shown in A. The magnitude and time course of these phenomena vary between cell types, depending on the number, density and types of voltage-sensitive ion channels.

the outside, due to the open states of voltage-insensitive channels. The opening of voltage-sensitive channels becomes regenerative when membrane potential is depolarised beyond critical threshold values that are unique for each type of channel protein. Opening of Na^+ or Ca^{2+} channels admits positively charged ions into the cell, further depolarising the membrane and causing more channels to open. This positive feedback is frequently sufficient to bring about a rapid reversal in the membrane potential (inside becoming positive, rather than negative). The polarity of the membrane is then restored by a combination of delayed voltage-dependent inactivation and delayed activation of other channels, typically K^+-channels, that enable flux of positive ions along concentration gradients from inside to out. This interplay between ion channels of different types is the basis of the depolarization/repolarization that is used to propagate signals along axons and between cells (Fig. 1).

In vertebrate skeletal muscle, action potentials are trigged by axonal contacts at neuromuscular junctions (NMJs). Each presynaptic motor nerve terminal contains neurotransmitter molecules (acetylcholine) packaged into 30 nm spheres (synaptic vesicles), some of which are tethered to the intracellular surface nerve membrane at "active zones". An incoming nerve action potential triggers fusion of about 50 vesicles with the nerve terminal membrane, releasing their contents into the synaptic cleft. This process of exocytosis is executed following influx of Ca^{2+} ions through Ca^{2+}-selective, voltage sensitive ion channels in the nerve terminal membrane. These ions then bind to signaling proteins integrated into the active zone molecular complex [23,24]. Molecules of neurotransmitter released by exocytosis diffuse rapidly across the narrow (50 nm) synaptic cleft between motor nerve terminal and muscle fibre, where they bind to specific protein receptors located in high density ($> 10^5$ μm^{-2}) at the crests of membrane folds of the motor endplate, the muscle surface opposed to the sites of presynaptic neurotransmitter release. Activation of these receptors generates an inward postsynaptic ionic current, which depolarizes the motor endplate membrane. When the membrane potential at the motor endplate reaches around $-65\,mV$, voltage-sensitive Na-channels (NaV channels) located in the crypts of the junctional folds are activated, leading to a regenerative depolarization that is similar in character to the neuronal action potential [11,29,30]. The muscle action potential is propagated along the muscle surface membrane and into a network of invaginations known as t-tubules. Here, proteins are coupled to those controlling the release of Ca^{2+} from the sarcoplasmic reticulum, an intracellular membrane-bound storage depot [4]. Binding of released Ca^{2+} brings about an energy-dependent conformational change in other muscle proteins, enabling force generation or muscle shortening via recycling of molecular cross-bridges between an orderly array of cytoskeletal filaments comprising the protein molecules actin and myosin [26]. Neuromuscular function is similarly initiated and executed in invertebrate muscles, including those of *Drosophila* larvae. The most distinctive chemical and structural differences are that larval NMJs utilise glutamate as a neurotransmitter and the postsynaptic membrane folds, rich in glutamate receptors, are more extensive than in vertebrates and is normally referred to as the sub-synaptic reticulum.

Action potentials in vertebrate muscle fibres are obligatory for excitation-contraction coupling: if NaV channels in muscle are selectively blocked pharmacologically, then synaptically-evoked endplate potentials (EPPs) at neuromuscular junctions, though tens of millivolts in amplitude, fail to trigger muscle contraction [17,29]. By contrast, muscle fibres in the abdominal muscles of larval *Drosophila*, do not express NaV channels. Instead, they contain voltage-sensitive Ca^{2+} and K^+ channels. But as noted above, the functional significance of voltage-sensitive Ca^{2+} and K^+ channels in larval *Drosophila* muscle fibres is unclear. It is generally regarded that they are of little physiological significance since they are only reliably observed in recordings from muscles in which extracellular Ca^{2+} concentration is increased beyond normal physiological maxima, or when membrane K^+ permeability is reduced by adding selective channel blocking drugs [6,10,21,22].

However, action potentials are also occasionally observed in larval muscle fibres under more normal physiological recording conditions [25,31]. Figure 2(A) shows an intracellular microelectrode recording obtained from a filleted preparation of a 3rd instar larval *Drosophila*, which clearly shows a train of spikes: regenerative depolarising action potentials. Larval fillet preparations and intracellular recordings were made using standard techniques [27]. The preparation was bathed in a normal HL3.1 physiological saline (containing 1.5 mM Ca^{2+}; 4 mM Mg^{2+}) without any ion channel blockers. The muscles were impaled with glass microelectrodes, resistance 10–40 MΩ. Segmental nerves were aspirated into a fire-polished 10 μm diameter suction pipette/electrode and stimulated

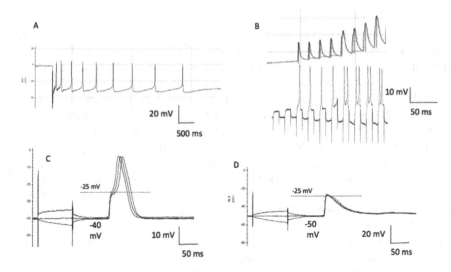

Fig. 2. Trace obtained during impalement of Muscle 4, in normal HL3.1 bathing medium and in the absence of any ion channel blockers. B. Combined optical recording of twitch contractions (arbitrary units) of Muscle 12 in a 3rd instar larval fillet preparation (upper trace) and simultaneous intracellular recording of membrane potential in response to progressive 2 nA increments in the strength of current pulses injected through the recording microelectrode (lower trace) in normal HL3.1 medium. Baseline drift was due to slight movement of the preparation during recording. C: Intracellular recording of a nerve-evoked excitatory synaptic (junctional) potential (EJP) in Muscle 6, sufficient to activate a regenerative action potential (three successive sweeps at 2 s intervals). D: Spontaneous hyperpolarisation of the resting potential by about 10 mV, in the same muscle fibre as C, abolished the regenerative depolarisation, leaving only a large EJP in response to nerve stimulation. The prepulses in C, D are responses to ±1 nA rectangular pulses injected through the recording microelectrode, used to check membrane integrity (resistance and capacitance).

with 0.2 ms pulses 1−10 V in amplitude. Trigger and current pulses were delivered and recordings were captured via a Digidata 1550B interface using pClamp-10 software (Molecular Devices, San Jose, USA). Images were captured using a QImaging Optimos camera (Teledyne Photometrics, Tucson, USA) driven by public domain Micromanager software (micro-manager.org). Images were post-processed and muscle contractions recording in FiJi (imagej.net/Fiji) using the Muscle Motion plugin (github.com/l-sala/MUSCLEMOTION).

A brief discharge of action potentials diminished in frequency as the resting membrane potential spontaneously hyperpolarised. The identified muscle in this case was Muscle 4 but we have observed similar phenomena in intracellular recordings from muscles 5, 6, 7, 12, and 13. Figure 2(B) shows combined optical recording of twitch contractions of Muscle 12 in 3rd instar larval fillet preparation (upper trace) and simultaneous intracellular recording of membrane potential in response to progressive 2 nA increments in the strength of current pulses injected through the glass recording microelectrode (lower trace), in normal HL3.1 medium. Contractile responses were only elicited when membrane depolarisation exceeded the firing threshold for regenerative responses. Summative contractile responses were evoked when membrane depolarisation was sufficient to evoke action potential doublets.

The experience from the Ribchester Lab is that about 10% of freshly-dissected larval preparations bathed in normal (or even reduced) Ca^{2+} containing media show action potentials and these are associated with brisk muscle contractions (c.f. Fig. 2(B)). The mechanism of these regenerative responses, which activate at a much higher threshold than vertebrate muscle action potentials, is wholly consistent with published data on the voltage-dependence of Ca^{2+} channels and K^+-channels expressed in larval muscle: specifically, a form of L-type Ca^{2+} channel with an activation threshold of about -25 mV, as well as several types of K^+-channels [6,21]. Sixteen of these recordings were from muscle fibres that showed sufficient membrane integrity and stability to warrant further analysis and simulation.

2.2   Hodgkin-Huxley Type Models

In 1952 Hodgkin and Huxley proposed and tested a model to account for the propagation of action potentials in the squid giant axon, the most favourable preparation at that time for comparing empirical data with computational analysis [7]. The Hodgkin-Huxley formulation was based on the notion that membrane ionic permeability is voltage- and time-dependent and that permeabilities to ions, specifically Na^+ and K^+, are associated with distinct activation and inactivation kinetics. In their model, the cell membrane is represented as a dielectric separating conducting ionic media, thus conferring transmembrane capacitance, in parallel with batteries representing transmembrane voltages. Selective ionic permeabilities were represented by separate variable conductances. Based on this abstraction they applied and numerically solved a set of nonlinear ordinary differential equations (ODEs) to describe the flow of membrane current and to predict the change in transmembrane voltage during the action potential [7].

The voltage-sensitive ionic permeabilities envisaged by Hodgkin and Huxley were subsequently shown to be mediated by protein molecules embedded in membranes and that functioned as ion channels in their open state [18]. Subsequently such models were adapted to other excitable cell types, including cardiac and skeletal muscle, and are now widely used in membrane biophysics due to their computational efficiency and relative mathematical simplicity.

2.3   Particle Swarm Optimisation

PSO is a stochastic optimisation technique for continuous non-linear functions introduced by Eberhart and Kennedy in [5] and is inspired by social behaviour of bird flocking or fish schooling. The algorithm initialises and maintains a swarm of particles where each particles represents a random solution with a velocity in the search space. Each particle moves through the search space based on its own the best solution, and the best global solution, obtained thus far. It was demonstrated in [32] that a GPU based implementation can result in performance improvements for large swarm sizes and many dimensional problems.

3   Model

The structure of the model presented here is based on a Hodgkin-Huxley type model [7] of myocyte action potentials in the AV node of the mammalian heart published by Inada *et al.* [9]. We based the model on nodal cells of myocardium, as we hypothesised the same ionic properties ($Ca^{2+} : K^+$ gating) underlie the generation of action potentials in larval *Drosophila* muscle. Our model assumed one cellular compartment (inside-outside) and was modified to accommodate different kinetics appropriate to the larval muscle fibres. Functional homologues for channels known to occur in *Drosophila* muscle but which are absent from the cardiac muscle were added.

The model represents the change in voltage across the cell membrane based on the temperature, membrane capacitance and atmospheric pressure (all treated as constants and specified in Appendix A) and a sum of ionic currents flowing through open ion channels. The total voltage change was determined as a function of ionic current based on the following equation

$$\frac{d}{dt}V = \frac{-I_{total}}{C_m} + \frac{d}{dt}V_{init}, \qquad I_{total} = I_{Cv} + I_{Kv1} + I_{Kv2} + I_{Kv3} + I_b + I_f$$

giving the change in membrane voltage as a function of time and ionic currents in *Drosophila* muscle cells. V_{init} is a function representing the magnitude and time course of initial depolarisation that results from activation of ligand-gated glutamate channels by neurotransmitter at the NMJs and which then triggers activation of the voltage-gated currents. This function aims to account for the dataset consisting of evoked responses as the synaptic signal which it represents is not integral to action potential occurrence but is present in our dataset.

Ion channels in the model are characterised using sets of ODEs, which are used to determine the expected proportion of channels which are in open (conducting) state, as opposed to closed (non-conducting) states. The proportion of channels in the open state is dependent on their activation and inactivation rates as a function of membrane voltage, values of which are dependent on a set of equations expressing sensitivity of the channel to voltage and time.

The total current passing through the ion channels is dependent on the conductance of ion channels which represents the population of channels present on the cell surface. In this paper, we explore different parametrisations of the conductance values to identify the channels which contribute the most to the characteristics of the *Drosophila* muscle action potential.

The model under consideration consists of 6 ion channels and the initialisation current – one channel (Cv2) modelling the inward currents, three (Kv1, Kv2, Kv3) modelling the outward current and finally two pacemaking currents

(I_b, HCN) modelling channels which conduct inwards at highly negative membrane voltages and outwards in more positive voltages. The ODE formulations of the channels, along with their empirically found parametrisations of the activation and inactivation rates, are from papers [9] and [1]. In the following we give a brief description of the channels and their functions in the model. The ODE formulations of the channels are given in Appendix A.

Cv2 Current. The channel gives the inward Ca^{2+} current underlying muscle activation in *Drosophila* embryos. The formulation of Cv2 model was taken from the model of rabbit atrioventricular cell by Inada *et al.* [9], due to their functional resemblance to mammalian L-type channels.

Kv1 Current (Shaw). Kv1 channel in *Drosophila* larvae conducts a transient outward potassium current. It is a voltage dependent, fast inactivating potassium channel, which controls (and prevents) repetitive firing of the cell by prolonging and enhancing hyperpolarisation of the cell in response to depolarisation. The formulation for Kv1 current used in this paper taken from its mammalian homologue in rat Purkinje cell neurons [1].

Kv2 Current (Shab). Kv2 carries a delayed-rectifier potassium current. The channel slowly opens and closes in response to depolarising voltage. The delayed activation kinetics are important to control the duration of action potential in 3rd instar *Drosophila* and mammalian neurons. The formulation of Kv2 current considered here was taken from [9] formulation for IKr.

Kv3 Current (Shaker). Kv3 channels are low conductance ion channels activated at depolarised voltages which generate atypical, delayed voltage-dependent slowly activating and non-inactivating currents. These contribute to maintaining of the resting membrane potential but have little effect on action potential parameters. In traditional Hodgkin-Huxley type models these channels could be considered K^+ leak channels. As for Kv1 current the formulation for Kv3 is modelled after its mammalian homologue Kv3.3 from rat Purkinje cells [1].

Background Current. I_b As there is only one type of functional voltage-gated excitatory ion channel, with relatively high activating threshold (around $-25\,\mathrm{mV}$), in order to observe spontaneous action potentials a depolarising driving force (pacemaker current) is necessary. The formulation for fast pacemaking background current was taken from rabbit heart background pacemaker as in [9].

HCN Current. I_f In previous versions (Margetiny, unpublished), the fits of the model to experimental data were seen to improve when an HCN channel (or channel with HCN-like kinetics) was added. However, whether or not such a channel occurs in *Drosophila* muscle tissue is unknown. We consider the original model from rabbit cardiomyocyte [9] in the context of regenerative responses in *Drosophila* muscles. Similarly to background current, HCN is hypothesised to be a multiple-ion permeable channel, albeit with much slower kinetics.

Initialisation Current. The initialisation current is modelled through its time-dependent effect on the voltage by

$$V_i(t) = \beta_i \left(\frac{t}{\alpha_i}\right) \exp\left(1 - \frac{t}{\alpha_i}\right)$$

4   Parameter Estimation Problem

The general parameter estimation problem we are aiming to solve is the following: what are the parameters $\boldsymbol{\theta}$ such that the deterministic model $f(t, \boldsymbol{\theta})$ serves as a good predictor to the voltage response during an action potential as observed in the experimental time-series data. In the following we describe the available experimental data as well as the preprocessing steps.

4.1   Data Preprocessing

The available data is in the form of time series measurements of voltage response to stimulus provided by current injections. Figure 3a and b give two examples of available time series. Figure 3a shows the voltage response to a single induced synaptic stimulus triggering an action potential while Fig. 3b shows five consecutive stimuli. Note that in this case only one of the current injections has triggered the action potential behaviour.

The measurements from the experiments started with a depolarising current intended to test the membrane resistance and is not relevant to the fitting of the model (c.f. Fig. 2(C)). The timings for testing the membrane resistance are consistent throughout the dataset and thus we have simply dropped measurements before 0.2 s. For hence forward we are considering the time series with the prepulse removed.

From there, we need to identify the parts of the time series corresponding to the action potential behaviour. The method for identifying parts of the time series is done in the following way. For each time series

- we identify the indices $\{i_1, \cdots, i_n\}$ corresponding to peaks in the time series using standard implementations of peak finding algorithms.
- we identify the indices $\{j_1, \cdots, j_n\}$ corresponding to where the peaks start. For that we first use Savgol-Goyal high pass filter [19] to smooth the time series resulting in a series for which we can numerically calculate derivatives. Working backwards from a peak we find where the derivatives change sign.
- we split the time series into n parts corresponding to single instances of action potentials in the following way. The k-th series corresponds to the values of the original time series between the indices j_k and j_{k+1}. We normalise the time by taking the initial time to be 0.0. For each point between j_k and j_{k+1} we consider the time passed since the measurement at index j_k. Secondly, we normalise voltages by considering the voltage differences between the base of the peak at index j_k each point in the new time series.

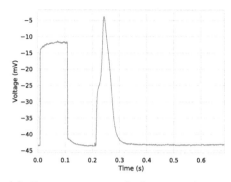
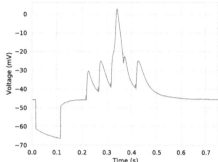

(a) Response to a single synaptic (NMJ) stimulus, triggering a regenerative response (same recording as Figure 2C). Prepulse is a membrane resistance test pulse.

(b) Five consecutive synaptic stimuli, only the third of which triggered as regenerative response

Fig. 3. Time series data for voltage response during action potential.

- Note that not all such generated time series correspond to action potentials. We pick a threshold voltage of $-10\,\mathrm{mV}$ for the peaks that are likely to correspond to the action potential phenomenon.
- In order to reduce the computational load we are going to consider a subsample of the generated time-series.

Figure 4 gives examples of the results of the process. In particular, the dots represent an 18 point sub-sample of the experimental time series showing the action potential. The number of sampling points is an arbitrary choice and can be easily changed as long as the sub-sample sufficiently captures the shape characteristics of the traces. The resulting dataset from the available recordings consists of 16 instances of action potentials.

4.2 PSO Fitting

For this preliminary study of fitting the Hodgkin-Huxley type action potential model we used a standard particle swarm optimisation (PSO) algorithm over a given search space. The fitness calculations are given by the following. Given a single action potential time series consisting of points $(t_0, V_0), \cdots, (t_m, V_m)$ and a model $f(t, \boldsymbol{\theta})$ parametrised by $\boldsymbol{\theta}$ we consider the simple distance measure

$$K(\boldsymbol{\theta}) = \sum_{i=0}^{m} (f(t_i, \boldsymbol{\theta}) - f(0, \boldsymbol{\theta}) - V_i)^2$$

where $f(t_i, \boldsymbol{\theta}) - f(0, \boldsymbol{\theta})$ gives the difference of the voltages predicted by the model at time 0 and t_i and V_i gives the same quantity for the experimental time series. Similarly to the multi-swarm method presented in [15] we perform the

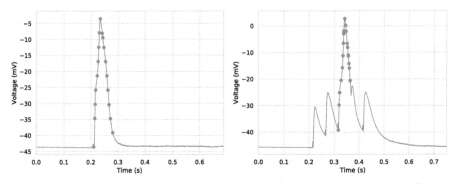

(a) Single synaptic response same as Fig. 3a)

(b) Consecutive synaptic responses (same recording as Fig. 3b)

Fig. 4. Generated time-series sub-sampled at 18 points.

optimisation algorithm for multiple initialisations of the swarm. In our case the initialisations are provided by the distinct time-series of action potentials.

4.3   Implementation

The standard PSO algorithm adapted for the described problems proceeds through the following steps:

1. Particles are initialised with uniformly sampled values from the search space and velocities. An alternative initialisation of particles through Latin hypercube sampling, as done in [20] can be considered in further work.
2. For each particle the set of ODEs giving the corresponding system dynamics is solved.
3. Based on the ODE solutions each particle gets a reward value.
4. Global maximum reward is found.
5. The location of a particle, $\boldsymbol{\theta}$, in the search space is updated based on the global maximum $\boldsymbol{\theta}_{\mathrm{gbest}}$ and individual best previous location of the particle $\boldsymbol{\theta}_{\mathrm{ibest}}$. In particular, the update between the i-th and $(i+1)$-th iteration for an individual particle takes the following form

$$\boldsymbol{\theta}_{i+1} = \boldsymbol{\theta}_i + w\boldsymbol{v}_i + c_1 r_{i_1}(\boldsymbol{\theta}_{\mathrm{ibest}_i} - \boldsymbol{\theta}_i) + c_2 r_{i_2}(\boldsymbol{\theta}_{\mathrm{gbest}_i} - \boldsymbol{\theta}_i)$$

   where r_{i_1}, r_{i_2} are random numbers in the interval $[0, 1]$ and w (weight given to previous velocities), c_1 (called cognitive weight) and c_2 (called social weight) are parameters of the optimisation algorithm.
6. Go back to Step 2.

In the implementation we made use of the fact that PSO is easily parallelisable on graphics processing units (GPUs) [15, 32] so that each particle is assigned a single GPU thread. In particular, for the standard PSO given above all steps other than

Step 2 are easily parallelisable. Reward evaluations are slightly more complex consisting of two steps: the integration step for solving the system of ODEs and the actual reward calculation. However, storing the trajectories resulting from numerical integration in memory would severely limit the scalability of the algorithm to large numbers of particles. Instead, we can update the value of the reward function on the fly after each step of the numerical integration. This way only the point necessary for the next iteration of numerical integration is stored in memory. The integration for reward calculations in this paper was performed by the simple Euler forward method. For the model presented in this paper this was found to be sufficient but other fixed time-step methods, like Runge-Kutta fourth-order method, can be easily considered. Finally, the boundary conditions are enforce in the following way: if the particle is about to violate the boundary for a given parameter its position in the component of this parameter is set to the boundary value while reversing the relevant component of its velocity.

5   Results

The parameters under investigation are the conductance values for each of the channels $(g_{Cv}, g_{Kv1}, g_{Kv2}, g_{Kv3}, g_{I_b}, g_f)$, the reversal potential E_b for the background pacemaker current and the shape parameters α_i, β_i for the voltage change due to the initial current injection. The bounds for each of the parameter values are set to encompass a range of physiologically plausible values. For conductances this was the interval between $0.0\,\mu S$ and $0.016\,\mu S$. The viable shape parameters for the initialisation current were set so that the induced voltage would reach its peak between values $10\,mV$ and $25\,mV$ before $0.5\,ms$ in order to feasibly set up the action potential. We conducted two sets of experiments: (a) parameter g_f was held at 0.0 ,disabling the channel in the model and (b) the HCN channel corresponding to g_f conductance values was enabled. For both sets of experiments we ran the PSO on the action potential traces 6 times – each with different random seeding and a varying weight parameter w from $\{0.7, 0.72, 0.74, 0.76, 0.78, 0.80\}$ to further perturb the behaviour of the particle swarms between different experiments in order to find as many different optima as possible. We set the values $c_1 = c_2 = 2.0$ as in [5]. The effect of varying the social and cognitive weights for this problem was not explored and is left for further work. We ran a fixed number (2000) of iterations. Little can be said about the convergence properties of the PSO for the given fitness function, but in experiments we saw that 2000 iterations generally allowed the swarms to settle to some local optimal values.

In the multi-swarm implementation we associated each instance of observed action potential with its own fitness function. Not averaging over the collected AP samples and running the PSO algorithm on each sample multiple times allows us to effectively find a set of plausible parametrisations of the proposed model. This gives an alternative way to generate a population of models aiming to take into account the cell-to-cell variability similarly to [2,13]. Figure 5 shows the model fitted to sub-samples of the time series shown in Fig. 4. The results

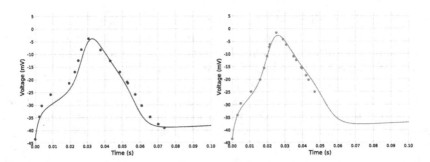

Fig. 5. Blue and red trajectories correspond to best model fittings for the time-series presented in Fig. 4(a) and (b) respectively. (Color figure online)

Table 1. Mean and standard deviation of fitted conductance values for the model with the HCN channel disabled.

	mean	std.		mean	std.
g_{Cv2}	0.0106	0.0031	g_{Kv3}	0.0083	0.0053
g_{Kv1}	0.0093	0.0043	g_b	0.0102	0.0038
g_{Kv2}	0.0091	0.0041	E_b	2.8	23.4

of the fitting are summarised by combined box and violin plots in Figs. 6 and 7, describing the shape of the distributions of found parameter values. Table 1 gives mean and standard deviation summary statistics for the model fitting with the HCN channel disabled. We have discarded parameters which give rise to voltage responses that do not recover to the interval -50 mV to -30 mV after the occurrence of an action potential or result in overly low fitness values.

From the results of parameter fitting with HCN channel disabled we first note that the summary plots indicate that conductance values of g_{Cv2} and g_b close to 0 are unlikely to fit the experimental traces well. This seems to confirm the necessity for involvement of both Cv2 channels and a pacemaker current I_b to facilitate action potentials in *Drosophila*. In addition, slightly tighter interquartile range for the Kv2 channel conductance compared to the other two K^+ channels (Kv1, Kv3) points towards its more significant effect in shaping of the action potentials while Kv1 and Kv3 are permitted to vary more. This is consistent with previously understood physiology of *Drosophila* muscle ionic activity as Kv1 and Kv3 are expected to be important in regulation of repeated firing and unlikely to influence the parameters of a single action potential. Surprisingly, the modelling experiments might indicate a higher reversal potential for I_b than originally expected (median -1.5 mV with mean 2.8 mV and standard deviation 23.4, as opposed to -22 mV). A higher reversal potential for I_b may suggest a more complex current consisting of several ion channel conductances, or a single channel which is more biased towards inward current than previously estimated. This inspires further modelling and experimental enquiry into the nature of channel or channels responsible for the generation of background

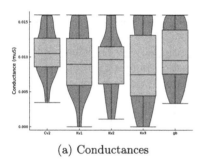

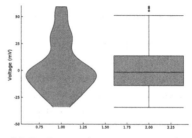

(a) Conductances   (b) Background current reversal potential.

Fig. 6. Violin and box plots of fitted parameter values for the model the HCN channel disabled.

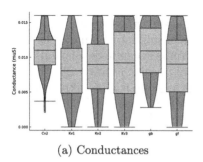

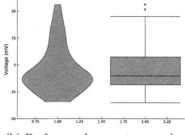

(a) Conductances   (b) Background current reversal potential.

Fig. 7. Violin and box plots of fitted parameter values for the model including the HCN channel.

pacemaking current. Further, we experimented with addition of HCN channels. The resulting dispersion of fitted conductance values g_f is similar to Kv1 and Kv3 indicating that the presence of the HCN channel in the model is of little importance for fitting *Drosophila* action potentials. Moreover the fitness calculations do not show quantitatively better fits being achieved with the addition of the HCN channel.

All optimisation runs were conducted with 2000 iterations where a swarm of 64 particles was assigned to each of the 16 trajectories, thus simulating 1024 particles in parallel. Each run took approximately 4 minutes to complete on a machine equipped with Nvidia Titan X GPU with the equivalent single-threaded execution on a laptop CPU taking around 19 min. Additionally, the GPU implementation provided better scalability for the same optimisation problem, with the total of 2048 and 4096 particles taking 7 and 12 min in the case of the GPU implementation and 38 and 78 min in the case of the CPU implementations.

6   Conclusion

We have presented a new model of excitability of the abdominal muscles of larval *Drosophila*, observed experimentally following excitatory depolarisation at a minority of NMJs, and used this model to explore techniques for modulation of its parameters via a novel GPU-based implementation of the particle swarm optimisation method. This approach was computationally very efficient and supported an ensemble model view, allowing each action potential recording to be used to obtain a plausible set of parameters that might be used, for example, to account for cell-to-cell variability in the incidence, magnitude and time course of regenerative action potentials in larval muscle recordings.

From a functional standpoint, our unpublished preliminary data suggest that when freely moving larvae undergo rapid peristaltic locomotion, for example to escape a potential predator there is insufficient time for more than one brisk and powerful twitch contraction per abdominal segment. Combined measurements of synaptic potentials and muscle shortening indicate that single EJPs are not sufficient to account for this escape behaviour (M.Fjeldstad and R.R.Ribchester, *unpublished*). Thus, we hypothesise that rapid contractile responses of larval muscle fibres are enabled by an endogenous mechanism that modulates the muscle fibre resting membrane potential and this permits synaptic depolarization to trigger a regenerative response (Fig. 2(C, D)). This results, as in vertebrate muscle, in brisk muscle contraction (Fig. 2(B)). This hypothesis implies that further analysis of the characteristics and mechanism of regenerative membrane depolarization in larval *Drosophila* muscle will yield deeper insight into their function. Computational modelling of these events and exploring the scope and causes of their variability from moment to moment will facilitate the analysis.

Due to the nature of the model, distinct channels expressed in the model are capable of compensating for each other resulting in widely dispersed viable parameter values. Thus analysis of correlations between the parameter values as well as refinements and alterations of the model would be of interest for further work. In many such scenarios the small size of the currently available dataset would be a limiting factor. On a practical level, dynamic modelling (including parameter fitting) in real time would be a valuable tool that could complement experimental approaches, such as the dynamic clamp technique: an experimental procedure that enables electrophysiologists to explore the consequences and potential functional significance of varying the specificity and kinetics of different ionic currents and determining their transitory effects on membrane potential [16, 28]. Computational speed is an essential consideration for real time feedback between dynamic modelling and dynamic clamp to be feasible and practical.

Acknowledgements. We thank Mr Keiran Brown for assistance with intracellular recordings and Professor Mark Boyett, Manchester University, for sharing source code for action potential modelling in myocytes of the rabbit AV node. FM is supported by an MRC PhD Studentship in the University of Edinburgh Doctoral Training Programme in Precision Medicine. PP is supported by EPSRC grant EP/L01503X/1 (CDT in Pervasive Parallelism) and STSM Grant from COST Action IC1406 High-Performance Modelling and Simulation for Big Data Applications.

Appendix A Model

The following appendix gives the ODE formulations of the channels considered in this paper. The ODEs are characterised by general Hodgkin-Huxley scheme, with a transitions between channels in open and shut states modelled as a first order chemical reaction

$$\text{open} \underset{\beta}{\overset{\alpha}{\rightleftharpoons}} \text{shut}$$

Transition rate expressions α and β provided for the channels in this model correspond to number of openings or closures of channel per second.

The functions for d_{L_∞}, m_∞, n_∞, y_∞ determine the proportion of channels in a particular (generally open) state under equilibrium conditions. This value changes as a function of the membrane voltage via a change in voltage dependent rates α and β. While change in equilibrium proportion happens instantly, change in real proportion of channels in open state does not: the rate at which the proportion of d_L, m, n, d, y changes towards its equilibrium value is given by a differential equation. The time constant τ is an expression of how fast this equilibrium is achieved and is dependent on the innate properties of channel and its sensitivity to voltage. Channels Cv1 and Kv2 have more than one inactivation mode – one happening at a slower time-scale than the other. These are denoted $f_{L,fast}$, $f_{L,slow}$ for the Cv1 channel and $p_{a,fast}$ and $p_{a,slow}$ for the Kv2 channel.

Reversal potential E_K for potassium and calcium channels at temperature 294.15 K (laboratory conditions) were calculated using the Nernst equation based on expected intracellular and extracellular concentrations of potassium and calcium

$$E = \frac{RT}{zF} \ln \frac{[\text{ion concentration ouside}]}{[\text{ion concentration inside}]}$$

where R is the universal gas constant, T the temperature in Kelvins, F the Faraday constant and z valency of the ion. For *Drosophila* the intracellular concentrations are estimated to be 0.05 mmol and 140 mmol for calcium and potassium respectively. Extracellular solution for the experiments used 5.0 mmol potassium and 1.5 mmol calcium concentration. Finally, Q_{10} is experimentally [1] determined time change constant for the Kv1 potassium channel.

Background Current I_b

$$I_b(V) = g_b(V - E_b)$$

HCN Current I_f

$$y_\infty(V) = \frac{1.0}{1.0 + \exp(\frac{V + 83.19}{13.56})}$$

$$\frac{d}{dt}y = \frac{y_\infty - y}{\tau_y}$$

$$\tau_y(V) = 0.250 + 2.0\exp(-\frac{(V + 70.0)^2}{500.0})$$

$$I_f(V) = g_f y(V - E_f)$$

Cv1 Current

$$d_{L\infty}(V) = \frac{1}{1+\exp\left(\frac{V+18.2}{-5}\right)}$$

$$\alpha_{d_L}(V) = \frac{-26.12(V+35.0)}{\exp\left(\frac{V+35.0}{-2.5}\right)-1}$$

$$+\frac{-78.11V}{\exp\left(-0.208V\right)-1}$$

$$\beta_{d_L}(V) = \frac{10.52(V-5.0)}{\exp\left(0.4\times(V-5.0)\right)-1.0}$$

$$\frac{d}{dt}d_L = \frac{d_{L\infty}-d_L}{\tau_{d_L}}$$

$$\frac{d}{dt}f_{L,fast} = \frac{f_{L,fast\infty}-f_{L,fast}}{\tau_{L,fast}}$$

$$\frac{d}{dt}f_{L,slow} = \frac{f_{L,slow\infty}-f_{L,fast}}{\tau_{L,slow}}$$

$$\tau_{d_L} = \frac{1}{\alpha_{d_L}+\beta_{d_L}}$$

$$I_{Cv2}(V) = g_{Cv}d_L(0.675f_{L,fast}+0.325f_{L,slow})(V-E_{CaL})$$

Kv2 Current

$$p_{a,fast\infty}(V) =$$

$$p_{a,slow\infty}(V) = \frac{1}{1+\exp\left(\frac{V+10.22}{-8.5}\right)}$$

$$\tau_{p_a,fast}(V) = \frac{1}{17\exp\left(0.0398V\right)+0.221\exp\left(-0.051V\right)}$$

$$\tau_{p_a,slow}(V) = 0.33581$$

$$+0.90673\exp\left(\frac{-(V+10.0)^2}{988.05}\right)$$

$$\frac{d}{dt}p_{a,fast} = \frac{p_{a,fast\infty}-p_{a,fast}}{\tau_{p_a,fast}}$$

$$p_{i\infty}(V) = \frac{1}{1+\exp\left(\frac{V+4.9}{15.14}\right)}\times$$

$$\left(1-0.3\exp\left(\frac{-V^2}{500}\right)\right)$$

$$\alpha_{p_i}(V) = 92.01\exp\left(-0.0183V\right)$$

$$\beta_{p_i}(V) = 603.6\exp\left(0.00942\right)$$

$$\tau_{p_i} = \frac{1}{\alpha_{p_i}+\beta_{p_i}}$$

$$\frac{d}{dt}p_i = \frac{p_{i\infty}-p_i}{\tau_{p_i}}$$

$$I_{Kv2}(V) = g_{Kv2}\left(0.9p_{a,fast}+0.1p_{a,slow}\right)p_i(V-E_K)$$

Kv1 Current

$$\alpha_n(V) = 0.12889\exp\left(\frac{-V+45.0}{-33.90877}\right)$$

$$\beta_n(V) = 0.12889\exp\left(\frac{-V+45.0}{12.42101}\right)$$

$$n_\infty(V) = \frac{\alpha_n(V)}{\alpha_n(V)+\beta_n(V)}$$

$$\tau_n = Q_{10}\frac{1}{\alpha_n+\beta_n}$$

$$\frac{d}{dt}n = \frac{n_\infty-n}{\tau_n}$$

$$I_{Kv1} = g_{Kv1}n^4(V-E_K)$$

Kv3 Current

$$\alpha_{mShaw}(V) = 0.22 \exp\left(\frac{V+16}{26.5}\right)$$

$$\tau_{mShaw} = \frac{1}{\alpha_{mShaw} + \beta_{mShaw}}$$

$$\beta_{mShaw}(V) = 0.22 \exp\left(\frac{-V+16}{26.5}\right)$$

$$\frac{d}{dt}m = \frac{m_\infty - m}{\tau_{mShaw}}$$

$$m_\infty(V) = \frac{\alpha_n}{\alpha_n + \beta_n}$$

$$I_{Kv3} = g_{Kv3}m^4(V - E_K)$$

References

1. Akemann, W., Knöpfel, T.: Interaction of Kv3 potassium channels and resurgent sodium current influences the rate of spontaneous firing of Purkinje neurons. J. Neurosci. **26**(17), 4602–4612 (2006). https://doi.org/10.1523/JNEUROSCI.5204-05.2006
2. Britton, O.J., et al.: Experimentally calibrated population of models predicts and explains intersubject variability in cardiac cellular electrophysiology. Proc. Natl. Acad. Sci. **110**(23), 2098–2105 (2013). https://doi.org/10.1073/pnas.1304382110
3. Choisy, S., Cheng, H., Orchard, C., James, A., Hancox, J.: Electrophysiological properties of myocytes isolated from the mouse antrioventricular node: L-type ICA, IKr, If, and Na-Ca exchange. Physiol. Rep. **3**, e12633 (2015)
4. Dulhunty, A.: Excitation-contraction coupling from the 1950s to the new millenium. Clin. Exp. Pharmacol. Physiol. **33**, 763–772 (2006)
5. Eberhart, R., Kennedy, J.: A new optimizer using particle swarm theory. In: Proceedings of the Sixth International Symposium on Micro Machine and Human Science, MHS 1995, pp. 39–43, October 1995. https://doi.org/10.1109/MHS.1995.494215
6. Gielow, M., Gu, G., Singh, S.: Resolution and pharmacological analysis of the voltage-dependent clacium channels of Drosophila larval muscles. J. Neurosci. **15**, 6085–6093 (1995)
7. Hodgkin, A., Huxley, A.: A quantitative description of membrane current and its application to conductance and excitation in nerve. J. Physiol. **11**(4), 500–544 (1952)
8. Inada, S., Hancox, J., Zhang, H., Boyett, M.: One-dimensional mathematical model of the atrioventricalur node including atrio-nodal, nodal and nodal-his cells. Biophys. J. **97**, 2117–2127 (2009)
9. Inada, S., Hancox, J., Zhang, H., Boyett, M.: One-dimensional mathematical model of the atrioventricular node including atrio-nodal, nodal, and nodal-his cells. Biophys. J. **97**(8), 2117–2127 (2009). https://doi.org/10.1016/j.bpj.2009.06.056
10. Lee, J., Ueda, A., Wu, C.: Distinct roles of Drosophila cacophany and DmcalD Ca(2+) channels in synaptic homeostasis: genetic interactions with slowpoke Ca(2+)-activated BK channels in presynaptic excitability and postsynaptic response. Dev. Neurobiol. **74**, 1–15 (2014)
11. Martin, A.: Amplification of neurotransmission by postjunctional folds. Proc. Biol. Sci. **258**, 321–326 (1994)

12. Munk, A., Adjemian, R., Zhao, J., Ogbabhebriel, A., Shrier, A.: Electrophysiological prperties of morphologically distinct cells isolated from the rabbit atrioventricular node. J. Physiol. (Lond.) **493**(Pt 3), 801–818 (1996)

13. Muszkiewicz, A., et al.: Variability in cardiac electrophysiology: using experimentally-calibrated populations of models to move beyond the single virtual physiological human paradigm. Prog. Biophys. Mol. Biol. **120**(1–3), 115–127 (2016). https://doi.org/10.1016/j.pbiomolbio.2015.12.002

14. Newman, Z., et al.: Input-specific plasticity and homeostasis at the Drosophila larval neuromuscular junction. Neuron **93**, 1388–1404 (2017)

15. Nobile, M.S., Besozzi, D., Cazzaniga, P., Mauri, G., Pescini, D.: A GPU-based multi-swarm PSO method for parameter estimation in stochastic biological systems exploiting discrete-time target series. In: Giacobini, M., Vanneschi, L., Bush, W.S. (eds.) EvoBIO 2012. LNCS, vol. 7246, pp. 74–85. Springer, Heidelberg (2012). https://doi.org/10.1007/978-3-642-29066-4_7

16. Ortega, F., Butera, R., Christini, D., White, J., Dorval, A.: Dynamic clamp in cardiac and neuronal systems using RTXI. Methods Mol. Biol. **1183**, 327–354 (2014)

17. Ribchester, R., et al.: Progressive abnormalities in skeletal muscle and neuromuscular junctions of trangenic mice expressing the Huntingdon's disease mutation. Eur. J. Neurosci. **20**, 3092–3114 (2004)

18. Sakmann, B., Neher, E.: Patch clamp techniques for studying ionic channels in excitable membranes. Annu. Rev. Physiol. **46**, 455–472 (1984)

19. Savitzky, A., Golay, M.J.E.: Smoothing and differentiation of data by simplified least squares procedures. Anal. Chem. **36**, 1627–1639 (1964)

20. Schutte, J.F., Reinbolt, J.A., Fregly, B.J., Haftka, R.T., George, A.D.: Parallel global optimization with the particle swarm algorithm. Int. J. Numer. Methods Eng. **61**(13), 2296–2315 (2004). https://doi.org/10.1002/nme.1149

21. Singh, S., Wu, C.: Properties of potassium currents and their role in membrane excitability in Drosophila larval muscle fibers. J. Exp. Biol. **152**, 59–76 (1990)

22. Singh, S., Wu, C.: Ionic currents in larval muscles of Drosophila. Int. Rev. Neurobiol. **43**, 191–220 (1999)

23. Slater, C.: The functional organization of motor nerve terminals. Prog. Neurobiol. **134**, 55–103 (2015)

24. Südof, T.: Neurotransmitter release: the last millisecond in the life of a synaptic vesicle. Neuron **80**, 675–690 (2013)

25. Suzuki, N., Kano, M.: Development of action potential in larval muscle fibers in Drosophila melanogaster. J. Cell. Physiol. **93**, 383–388 (1977)

26. Sweeney, H., Hammers, D.: Muscle contraction. Cold Spring Harbour Perspect. Biol. **10**, a023200 (2018)

27. West, R.J.H., Briggs, L., Perona Fjeldstad, M., Ribchester, R.R., Sweeney, S.T.: Sphingolipids regulate neuromuscular synapse structure and function in Drosophila. J. Comp. Neurol. **526**(13), 1995–2009 (2018). https://doi.org/10.1002/cne.24466

28. Wilders, R.: Dynamic clamp: a powerful tool in cardiac electrophysiology. J. Physiol. (Lond.) **576**, 349–359 (2006)

29. Wood, S., Slater, C.: The contribution of postsynaptic folds to the safety factor for neurotrnsmission in rat fast- and slow-twitch muscles. J. Physiol. (Lond.) **500**(Part 1), 165–176 (1997)

30. Wood, S., Slater, C.: Safety factor at the neuromuscular junction. Prog. Neurobiol. **64**, 393–429 (2001)
31. Yamaoka, K., Ikeda, K.: Electrogenic responses elicited by transmembrane depolarizing current in aerated body wall muscles of Drosophila melanogaster larvae. J. Comp. Physiol. **163**, 705–714 (1988)
32. Zhou, Y., Tan, Y.: GPU-based parallel particle swarm optimization, In: Proceedings of the IEEE Congress on Evolutionary Computation, CEC 2009, Trondheim, Norway, pp. 1493–1500 (2009). https://doi.org/10.1109/CEC.2009.4983119

Cell Volume Distributions in Exponentially Growing Populations

Pavol Bokes[1]([✉]) and Abhyudai Singh[2]

[1] Department of Applied Mathematics and Statistics, Comenius University,
84248 Bratislava, Slovakia
pavol.bokes@fmph.uniba.sk
[2] Department of Electrical and Computer Engineering, University of Delaware,
Newark, DE 19716, USA
absingh@udel.edu

Abstract. Stochastic effects in cell growth and division drive variability in cellular volumes both at the single-cell level and at the level of growing cell populations. Here we consider a simple and tractable model in which cell volumes grow exponentially, cell division is symmetric, and its rate is volume-dependent. Consistently with previous observations, the model is shown to sustain oscillatory behaviour with alternating phases of slow and fast growth. Exact simulation algorithms and large-time asymptotics are developed and cross-validated for the single-cell and whole-population formulations of the model. The two formulations are shown to provide similar results during the phases of slow growth, but differ during the fast-growth phases. Specifically, the single-cell formulation systematically underestimates the proportion of small cells. More generally, our results suggest that measurable characteristics of cells may follow different distributions depending on whether a single-cell lineage or an entire population is considered.

Keywords: Cell growth · Cell division · Cell size

1 Introduction

Each living cell is an individual entity occupying a given volume enclosed by the cell membrane [1]. Homeostatis of cell volume is due to balance between cell growth and division. Growth in cell volume is understood to occur continuously in time and is often assumed to be exponential. Cell division is typically represented as a discrete event at which the volume of a mother cell abruptly changes into the volume of either daughter cell [19]. Specifically, symmetric division means that each daughter obtains exactly one half of their mother's volume. The contents of a mother cell, including its transcriptome and proteome, are also divided between its daughters. Fluctuations in cell volume due to cell growth and division can therefore correlate with gene-expression noise [3].

There are (at least) two alternative approaches to the modelling of cell-growth dynamics. In the first approach, one follows a single cell line, discarding the other

© Springer Nature Switzerland AG 2019
L. Bortolussi and G. Sanguinetti (Eds.): CMSB 2019, LNBI 11773, pp. 140–154, 2019.
https://doi.org/10.1007/978-3-030-31304-3_8

daughter cell at each division. In the single-cell approach, the time-dependent cell volume can be represented by a piecewise deterministic [6] or drift-jump Markovian process [15]. One is interested in the probability distribution of the random process, in particular at steady state. In the second approach, one follows both daughter cells, and is interested in the dynamics of the population size as well as the distribution of cell volumes among the population, in particular in the large-time limit. A question of interest is whether the probability distribution obtained from the single-cell approach and the population distribution obtained from the population approach are the same or different. The difference between single-cell and population approaches can be relevant in a number of applications, e.g. in cancer biology, which allow for experimental setups in which the reproductive history of a cell can be traced [10]. We will examine this problem for a particular type of volume-growth model.

The maintenance of homeostasis requires that cells actively control their proliferation [14,16,18]. The necessary feedback can be exerted through e.g. the cell's age [9], its current size [13], the size at its inception [2], or by a combination of these mechanisms [11,17]. In this manuscript we specifically focus on size-based regulation. Within the framework of a relatively simple model, we will proceed towards the following goals: (i) develop exact and efficient stochastic algorithms to simulate the volume growth process; (ii) characterise the large-time asymptotic behaviour of the process by formulating and solving a master equation; and (iii) draw conclusions about the similarities and differences between the single-cell and the whole-population approach to modelling cell growth.

The outline of the paper is as follows: in Sect. 2 we introduce and logarithmically transform the model. In Sect. 3 we present an iterative algorithm for simulating the single-cell version of the model and a recursive algorithm for the simulation of the whole-population version. In Sect. 4 we introduce the concept of periodicity in the context of the current model. In Sects. 5 and 6 we formulate the master equation and provide tractable closed-form formulae for large-time solutions in the single-cell and whole-population cases. In Sect. 7 we present the main implications of the current work on the dynamics of cell growth and cell volume distributions. In Sect. 8 we extract conclusions from the presented analysis.

2   Model Fundamentals

Our model for cell-volume growth is based on the following fundamentals:

1. Cell volume grows exponentially in time. Specifically,

$$V(t) = V(0)2^t, \tag{1}$$

where t is the time since birth. Time is measured in units of the volume doubling time.

2. A mother cell can divide into two daughter cells. Either daughter cells obtains one half of the mother's volume. From a single-cell viewpoint, at the time of division the volume changes abruptly according to the mapping

$$V \to \frac{V}{2}, \tag{2}$$

where the volume on the left-hand side represents the volume of the mother right before the division and a daughter's volume is on the right-hand side.

3. Cells divide with a volume dependent stochastic rate $\gamma(V)$. We specifically focus on the case of

$$\gamma(V) = \begin{cases} 0 & \text{if } V < V_c, \\ \alpha & \text{if } V \geq V_c, \end{cases} \tag{3}$$

where α is a constant rate and V_c is a critical volume threshold.

It turns out that it is much more convenient to use a logarithmic transformation of cell volume defined by

$$u = 1 + \log_2 \frac{V}{V_c}. \tag{4}$$

By (3), cells cannot divide before they reach the critical volume V_c. Hence the cell volume is always greater than $\frac{V_c}{2}$, and the log-volume, as defined by (4), is always positive.

The model fundamentals (1)–(3), when expressed in the language of log-volume, read as follows:

1. Dividing (1) by the critical volume V_c and taking the binary logarithm gives

$$u(t) = u(0) + t, \tag{5}$$

   i.e between divisions a cell's log-volume grows linearly with unit rate.

2. Taking the binary logarithm of (2) divided by V_c gives

$$u \to u - 1, \tag{6}$$

   meaning that, upon division, a cell's log-volume decreases by one.

3. Since requiring $V \geq V_c$ is equivalent to $u \geq 1$, the dependence (3) of the stochastic rate on volume translates into

$$\gamma(u) = \begin{cases} 0 & \text{if } u < 1, \\ \alpha & \text{if } u \geq 1, \end{cases} \tag{7}$$

in terms of the log-volume.

Table 1 summarises the model fundamentals in the linear and logarithmic volume scales.

Table 1. Model formulation in terms of Volume V (left column) and Log-Volume $u = 1 + \log_2 \frac{V}{V_c}$ (right column).

	Volume V	Log-Volume ($U = 1 + \log_2 \frac{V}{V_c}$)
Growth	$V(t) = V(0)2^t$	$u(t) = u(0) + t$
Division Map	$V \rightarrow \frac{V}{2}$	$u \rightarrow u - 1$
Division Rate	$\gamma(V) = \begin{cases} 0 & \text{if } V < V_c, \\ \alpha & \text{if } V \geq V_c, \end{cases}$	$\gamma(u) = \begin{cases} 0 & \text{if } u < 1, \\ \alpha & \text{if } u \geq 1, \end{cases}$

3   Stochastic Simulation

In this section we present an algorithmic approach the model of cell-volume growth based on the fundamentals presented in Sect. 2. Hereby we distinguish two versions of the model.

Single-Cell Version. Upon cell division, one of the daughter cells is followed, the other discarded. We are interested in the probabilistic description of the cell volume along an arbitrarily chosen lineage.

Population Version. Both daughter cells are followed upon cell division. We are interested in the growth of the number of offspring and in the distribution of volume across the population.

Throughout this section we will operate with log-volumes of cells rather than their volumes (see Sect. 2 for explanation).

A common building block in both versions of the algorithm is to sample the waiting time τ for division of a cell which currently has log-volume u. The waiting time consists of two parts: the deterministic time required to reach the critical log-volume of one; the stochastic time that it takes to divide once the critical threshold has been passed. Since log-volume grows with unit rate, the deterministic waiting time is $1 - u$ if $u < 1$ and zero if it is greater than one. After crossing the threshold division has a constant propensity α to occur; it follows that the stochastic waiting time will be exponentially distributed with mean $1/\alpha$. Putting the deterministic and stochastic parts of the waiting time τ together, we find that it can be sampled as

$$\tau = \max\{1 - u, 0\} - \frac{\ln\theta}{\alpha}, \tag{8}$$

where θ is drawn from the uniform distribution on the unit interval. We have thereby used the well known fact that $-\ln\theta$ is exponentially distributed with unit mean.

We are now ready to go through the individual steps of the single-cell version of the simulation algorithm (Algorithm 1). The algorithm requires the following inputs: the model parameter α which gives the (post-threshold) division rate; the initial log-volume u_0; the time points $t_0, t_1, \ldots, t_m$ at which we wish the log-volume to be recorded. We assume that these time points are ordered from the

---

Algorithm 1. Single-cell version

Require: Timepoints $t_0, \ldots, t_m$; division rate α; initial log-volume u_0
Ensure: Sampled values $u_1, \ldots, u_m$ of log-volumes at the given timepoints

1: Initialise current time and log-volume: $t \leftarrow t_0$; $u \leftarrow u_0$
2: **while** $t < t_m$ **do**
3:     Draw θ from the unit-interval uniform distribution
4:     Set $\tau \leftarrow \max\{1 - u, 0\} - \frac{\ln\theta}{\alpha}$
5:     **for all** i such that $t \leq t_i < t + \tau$ **do**
6:         Set $u_i \leftarrow u + t_i - t$.
7:     **end for**
8:     Update time and log-volume: $t \leftarrow t + \tau$; $u \leftarrow u + \tau - 1$
9: **end while**

---

lowest to the largest. Time t_0 is understood to be the initial time at which the log-volume is given by the initial value u_0. The algorithm returns the sampled value u_1, ..., u_m of log-volumes at the given time points.

The algorithm starts by initialising the current time t and log-volume u with the initial values t_0 and u_0 (Line 1 in Algorithm 1). The next steps are repeated while the current time t is less that the largest time point t_m at which a recording of the log-volume is sought (Lines 2–9): first, the waiting time until the next division is sampled in Lines 3–4 using the formula (8); second, log-volumes are recorded at all recording times t_i which fall between the current time t and the time $t + \tau$ of next division (Lines 5–7); third, the current time and log-volume are updated to the post division values (Line 8).

We are now well positioned to proceed to the population version of the simulation algorithm (Algorithm 2). The population version is only marginally more elaborate than the single-cell version thanks to the use of recursion. Algorithm 2 requires the same input as Algorithm 1, but provides a different output, returning for each given time point a list of log-volumes across the whole population. By a list we understand a collection of elements (here log-volumes), some of which may be present in the list multiple times. We may append a number to a list; we may query how many times a given element is present in the list; we may query the total number of elements in the list—here the population size.

The algorithm proceeds as follows. First, we make sure that the lists are empty initially (Lines 1–3). Then we make a call to the procedure CELL (Line 15). The CELL procedure is defined recursively in Lines 4–14. The procedure calculates the contribution made by a cell that is introduced into the population at time t_0 with log-volume u_0, and by the entire offspring of that cell, to the lists of log-volume recordings. The cell's individual contribution is calculated in Lines 5–9. Comparing Lines 5–9 in Algorithm 2 to the corresponding passage in Algorithm 1 (Lines 3–7), we note that the two sections of code differ only in that the cell's log-volume is either stored as a single value (Algorithm 1, Line 6) or added to a list potentially containing multiple values (Algorithm 2, Line 8). The contribution of the cell's offspring to the log-volume recordings is calculated in

Algorithm 2. Population version

Require: Timepoints $t_0, \ldots, t_m$; division rate α; mother cell's log-volume u_0
Ensure: Lists $\mathcal{U}_0, \ldots, \mathcal{U}_m$ of log-volumes at the given timepoints

  1: **for** $i = 1, \ldots m$ **do**
  2:     Initialise $\mathcal{U}_i$ to an empty list
  3: **end for**

  4: **procedure** CELL(t, u)
  5:     Draw θ from the unit-interval uniform distribution
  6:     Set $\tau \leftarrow \max\{1 - u, 0\} - \frac{\ln\theta}{\alpha}$
  7:     **for all** i such that $t \leq t_i < t + \tau$ **do**
  8:        Append the value $u + t_i - t$ to the list $\mathcal{U}_i$
  9:     **end for**
10:     **if** $t + \tau < t_m$ **then**
11:        CELL$(t + \tau, u_0 + \tau - 1)$
12:        CELL$(t + \tau, u_0 + \tau - 1)$
13:     **end if**
14: **end procedure**

15: CELL(t_0, u_0)

Lines 11–12 by making a recursive call to the CELL procedure for either of its daughter cells. Cells that are born after the last recording time point t_m cannot make contribution to the log-volumes recordings. For this reason, the recursive calls are made only if the mother cell divides before the time t_m of last recording (Line 10). In this manner, we make sure that the recursion does not continue *ad infinitum*.

4   Periodicity

Regardless of the particulars of the growth-control mechanism, a minimalistic model based on exponential growth and symmetric division, which we shall consider here, exhibits a type of periodic behaviour [4,5,7]. Specifically, the volume, measured in units of its doubling time, of a daughter cell at time $t > 0$ is equal to the volume of the mother cell at time $t = 0$ multiplied by 2^{t-n}, where n is the daughter cell's generation. A couple of important observation follow immediately from this. First, possible cell volumes are restricted to a discrete set of values at any given time. Second, cell-volume measurements taken at different times cannot be equal unless the times of measurement differ by an integer multiple of the volume doubling time. We will see in what follows that this periodicity with respect to the volume doubling time has important consequences for the cell growth process that persist even in the asymptotic limit of large times.

    Let t_0 be the initial time and u_0 be the log-volume of the mother cell at the initial time. These two are input values in stochastic simulation. At time $t > t_0$, the log-volume $u(t)$ of a daughter cell is constrained to the discrete set

$$u(t) \in \{\varphi + n, n \in \mathbb{Z}\}, \tag{9}$$

where $0 \leq \varphi < 1$ is a phase defined by

$$\varphi = u_0 + t - t_0 - \lfloor u_0 + t - t_0 \rfloor, \tag{10}$$

where $\lfloor a \rfloor$ denotes the floor of a (the nearest integer lower than the real number a). The choice of the value of n within the discrete set (9) depends inversely on the number of divisions of the first mother cell up to the daughter.

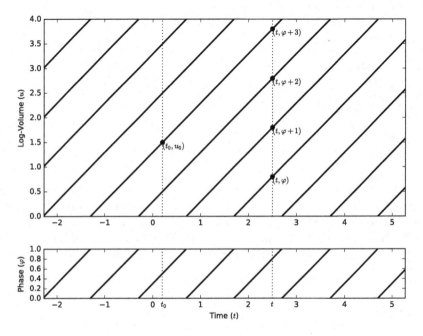

Fig. 1. Periodicity of the cell-volume process. At any given time, the log-volume belongs to a discrete constraint set (top panel). Which of the constraint sets applies is determined by the phase φ, which is a one-periodic function of time t (bottom panel).

The constraint (9) holds regardless whether the single-cell or the population approach to modelling cell-volume dynamics is taken. Additionally, it holds regardless of the choice of the log-volume-dependent division rate $\gamma(u)$. Specifically, for the threshold-like dependence (7) we know that the log-volume has to be positive, implying that the n in the constraint (9) has to be non-negative.

The presence of the constraint (9) is best explained graphically (Fig. 1, top panel). While the cell does not divide, its log-volume increases along a straight line with unit slope. When it divides, the log-volume transfers to a parallel line with intercept one unit lower. Regardless of the timing of cell divisions, the log-volume trajectories are constrained to a discrete union of parallel lines whose intercepts differ by an integer. The constraint (9) is obtained by taking a cross-section at time t of these parallel lines.

The constraint sets (9) are parametrised by the phase φ, which is a one-periodic function of time t (Fig. 1, bottom panel). The log-volume $u(t)$ visits each constraint set periodically with unit period. Different phases give disjoint constraint sets. The union of all constraint sets over phases $0 \le \varphi < 1$ gives the entire state space of real log-volumes.

Discrete Markov chains whose state space is partitioned into disjoint classes which are periodically (with discrete period) visited by the chain are called periodic Markov chains [12]. By analogy, we refer to the cell-volume process also as periodic. Periodicity has consequences for the large-time behaviour of a process. Large-time behaviour of aperiodic processes is typically given by a steady-state distribution. Contrastingly, periodic processes retain the dependence of the phase even in the large-time limit.

In the next two sections, we will characterise the large-time behaviour of the periodic cell-volume process using first the single-cell and then the population approach.

5   Large-Time Single-Cell Behaviour

In the previous section we showed that the log-volume $u(t)$ of a cell at time t is constrained to the set of values $n + \varphi(t)$, where n is an integer the phase $\varphi(t)$ is a function of time t (and of initial data). The probabilities

$$p_n(t) = \text{Prob}[\, u(t) = n + \varphi(t)\,] \tag{11}$$

have a discontinuity at any time t for which at which $\varphi(t)$ has a discontinuity (cf. Fig. 1, bottom panel). We then have a consistency condition

$$p_n(t^-) = \text{Prob}[\, u(t) = n + 1\,] = p_{n+1}(t^+), \quad \text{whenever } \varphi(t) = 0. \tag{12}$$

Away from the discontinuities, the probabilities (11) satisfy a system of balance equations

$$\frac{\mathrm{d}p_n(t)}{\mathrm{d}t} = \gamma(n+1+\varphi(t))p_{n+1}(t) - \gamma(n+\varphi(t))p_n(t), \quad 0 < \varphi(t) < 1. \tag{13}$$

Integrating the system (13) forward in time, one obtains the probabilities $p_n(t)$ until a discontinuity in $\varphi(t)$ is encountered. The consistency condition (12) needs to be applied at times of discontinuity to calculate from the probabilities right before the discontinuity their values right after the discontinuity. The integration of the system (13) can then be restarted with the post-discontinuity values.

The system (13) comprises an infinite number of coupled linear ordinary differential equations with non-constant coefficients. In general, the system (13) can be solved by truncating to a finite number of equations and using a numerical solver. In the specific case of a threshold dependence (7) of the division rate on the log-volume, we will be able to find an explicit large-time solution to (13) subject to (12).

As time progresses, the log-volume distribution becomes independent of the specifics of the initial condition and depend on time only via the phase φ. Let us denote this distribution by $\pi_n(\varphi)$. It satisfies a system of balance equations

$$\frac{d\pi_n(\varphi)}{d\varphi} = \gamma(n+1+\varphi)\pi_{n+1}(\varphi) - \gamma(n+\varphi)\pi_n(\varphi), \tag{14}$$

which looks similar to (13), differing in that the independent variable is now the phase φ, which is restricted to the range $0 \leq \varphi \leq 1$. The consistency condition (12) translates into

$$\pi_n(1) = \text{Prob}[u(t) = n+1] = \pi_{n+1}(0), \tag{15}$$

which provide a set of boundary conditions for the system (14). For threshold-type dependence (7) of division rate on log-volume, we have $\gamma(n+\phi) = \alpha(1 - \delta_{n,0})$, so that the system (14) simplifies to

$$\frac{d\pi_0(\varphi)}{d\varphi} = \alpha\pi_1(\varphi), \quad \frac{d\pi_n(\varphi)}{d\varphi} = \alpha(\pi_{n+1}(\varphi) - \pi_n(\varphi)), \quad n = 1, 2 \ldots \tag{16}$$

We look for a solution to (16) subject to the boundary conditions (15) in the form of an exponential

$$\pi_n(\varphi) = c_n e^{-\mu\varphi}, \quad n \geq 1, \tag{17}$$

where μ is an eigenvalue and c_n are eigenvector components. Inserting the ansatz (17) into (16) we find

$$-\mu c_n = \alpha(c_{n+1} - c_n), \quad n \geq 1. \tag{18}$$

Substituting the ansatz (17) into (15) we find that

$$c_{n+1} = c_n e^{-\mu}, \quad n \geq 1. \tag{19}$$

Substituting (19) into (18) and simplifying yields the characteristic equation

$$\mu = \alpha(1 - e^{-\mu}). \tag{20}$$

Elementary analysis shows that the characteristic Eq. (20) has a positive solution μ only if $\alpha > 1$. Furthermore, it is a unique positive solution and lies in the interval $\alpha - 1 < \mu < \alpha$. From now on we require that $\alpha > 1$ holds and we take for μ the unique positive solution to the characteristic Eq. (20). The condition $\alpha > 1$ guarantees the (post-threshold) dominance of division over growth, which is critical for the maintenance of cell volume homeostasis.

The recursive relation (19) implies that $c_n = c_0 e^{-\mu n}$, which, if substituted into (17), yields

$$\pi_n(\varphi) = c_0 e^{-\mu(n+\varphi)}, \quad n \geq 1. \tag{21}$$

Positivity of μ guarantees that $\pi_n(\varphi)$ has a finite ℓ^1 norm and can be normalised into a probability distribution. Integrating the first equation in (16) subject to $\pi_0(0) = 0$ leads to

$$\pi_0(\varphi) = c_0 \frac{ae^{-\mu}(1 - e^{-\mu\varphi})}{\mu} = c_0 \frac{1 - e^{-\mu\varphi}}{e^{\mu} - 1}, \tag{22}$$

in which the second equality is due to (20). The normalisation constant c_0 can be determined from the relation

$$1 = \sum_{n=0}^{\infty} \pi_n(\varphi) = c_0 \left(\frac{1 - e^{-\mu\varphi}}{e^{\mu} - 1} + \sum_{n=1}^{\infty} e^{-\mu(n+\varphi)} \right)$$

$$= c_0 \left(\frac{1 - e^{-\mu\varphi}}{e^{\mu} - 1} + \frac{e^{-\mu\varphi}e^{-\mu}}{1 - e^{-\mu}} \right) = \frac{c_0}{e^{\mu} - 1},$$

from which

$$c_0 = e^{\mu} - 1 \tag{23}$$

follows. Inserting (23) into (21) and (22) finalises our analysis.

In summary, we approximate the probability $p_n(t)$ that the cell's log-volume is equal to $n + \varphi(t)$ in the large-time regime by a phase-dependent distribution

$$p_n(t) \sim \pi_n(\varphi(t)), \quad t \gg 1, \tag{24}$$

where $\pi_n(\varphi)$ is given explicitly by

$$\pi_0(\varphi) = 1 - e^{-\mu\varphi}, \quad \pi_n(\varphi) = (e^{\mu} - 1)e^{-\mu(n+\varphi)}, \quad n = 1, 2, \ldots, \tag{25}$$

and μ is the unique positive solution to the transcendental characteristic equation (20), which exists provided that $a > 1$.

6   Large-Time Population Behaviour

Assume that at the initial time t_0 the population consisted of a single mother cell with log-volume u_0. Algorithm 2 ensures that at $t > t_0$ the log-volumes of its progeny are contained in a list $\mathcal{U}(t)$. Lists differ from sets in that they can contain the same element multiple times. Due to periodicity of the cell-volume process, $\mathcal{U}(t)$ can only contain elements with values $n + \varphi(t)$, where n is an integer and $\varphi(t)$ is the phase as defined by (10). Define by $f_n(t)$ the number of times a particular value $n + \varphi(t)$ is present in the list. The consistency condition

$$f_n(t^-) = f_{n+1}(t^+), \quad \text{whenever } \varphi(t) = 0, \tag{26}$$

holds at times of discontinuity of $\varphi(t)$. Provided that the numbers $f_n(t)$ are sufficiently large, we can treat them as continuous quantities that satisfy a population balance equation

$$\frac{\mathrm{d}f_n(t)}{\mathrm{d}t} = 2\gamma(n + 1 + \varphi(t))f_{n+1}(t) - \gamma(n + \varphi(t))f_n(t), \quad 0 < \varphi(t) < 1, \tag{27}$$

away from the times of discontinuity of $\varphi(t)$. The population balance Eq. (27) differs from the probability balance Eq. (13) in the factor 2 multiplying the first term on the right-hand side of (27). It is easy to verify that $f_n(t) = 2^{-n}p_n(t)$, where $p_n(t)$ is a solution to the probability balance Eq. (13), is in fact a solution to the population balance Eq. (27). It does not however satisfy the consistency condition (26). In order to satisfy the consistency condition, we modify the solution to

$$f_n(t) = c\, 2^{-n+t-\varphi(t)} p_n(t), \qquad (28)$$

where c is a tunable constant. Since $t - \varphi(t)$ is constant in any interval in which $\varphi(t)$ is continuous, the function (28) is a constant multiple of $2^{-n}p_n(t)$ in any such interval, and as such satisfies the population balance Eq. (27). Further, it is easy to verify that the consistency condition (26) is met by (28). In the regime of large times, we can approximate $p_n(t)$ by $\pi_n(\varphi(t))$ to obtain

$$f_n(t) \sim c\, 2^{-n+t-\varphi(t)} \pi_n(\varphi(t)), \quad t \gg 1. \qquad (29)$$

The total number of progeny at time t is given by

$$f(t) = \sum_{n=0}^{\infty} f_n(t) \sim c\, 2^{t-\varphi(t)} \left(1 - e^{-\mu\varphi(t)} + \sum_{n=1}^{\infty} (e^{\mu} - 1)2^{-n} e^{-\mu(n+\varphi(t))} \right)$$

$$= c\, 2^{t-\varphi(t)} \left(1 - e^{-\mu\varphi(t)} + (e^{\mu} - 1)e^{-\mu\varphi(t)} \frac{e^{-\mu}}{2 - e^{-\mu}} \right)$$

$$= c\, 2^{t-\varphi(t)} \frac{2 - e^{-\mu} - e^{-\mu\varphi(t)}}{2 - e^{-\mu}}. \qquad (30)$$

Finally,

$$\frac{f_n(t)}{f(t)} \sim \frac{2 - e^{-\mu}}{2 - e^{-\mu} - e^{-\mu\varphi(t)}} 2^{-n} \pi_n(\varphi(t)), \quad t \gg 1, \quad n = 0, 1, \ldots, \qquad (31)$$

gives the proportion of cells which have log-volume $n + \varphi(t)$ at a large time t.

7   Results

In this Section we use the simulation and analytic methods presented in the previous Sections to examine the dynamical behaviour of our model for cell-volume growth.

Figure 2 shows the cell count, on a logarithmic scale, as function of time measured in units of volume doubling time. Although the overall trend is characterised by an exponential increase, the cell count is nevertheless subject to periodically recurring cycles of fast growth alternating with slow growth. The cyclic behaviour is sustained even at large times. The simulation-based cell count (blue lines) exhibits low-copy-number noise at earlier times. The analysis-based cell count (orange lines) faithfully reproduces the large-time cyclic behaviour of

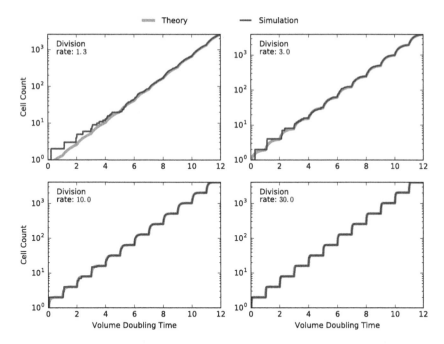

Fig. 2. Size of the cell population derived from an individual mother cell as function of time measured in units of volume doubling time. Panels differ in the choice of the division rate α that applies after the critical cell volume has been reached by a growing cell. Results of stochastic simulation by Algorithm 2 (blue lines) cross-validate the theoretical prediction (30) (orange lines). The log-volume of the mother cell at initial time (here $t_0 = 0$) is set to the critical value of $u_0 = 1$ in all examples. The undetermined constant c in the theoretical result (30) has been chosen so as to perfectly fit the simulation result at the last timepoint (here $t_m = 12$). The cell count was obtained by counting the total number of elements in the lists $\mathcal{U}_i$ returned by Algorithm 2. (Color figure online)

the results of simulation. The four panels of Fig. 2 differ in the choice of volume-dependent division rate $\gamma(V)$. The value specified within the figure panels gives the division rate α that applies once a volume threshold has been crossed (cf. Eq. (3)). For large values of the division rate α (bottom right panel in particular), the control of cell volume becomes near deterministic: division initiates almost immediately (i.e. with a very high rate) after the critical volume threshold is reached. We observe that in the near-deterministic regime of cell-volume control, the growth dynamics assumes a step-like pattern. Away from the deterministic regime, i.e. for lower values of the post-threshold division rate, the cycles of fast and slow growth are less pronounced.

In the following computational experiment, we let a colony of cells, derived from a single progenitor, grow until it counts in thousand individuals, and then study in detail a single ensuing period of cyclic growth. The top panel in Fig. 3 shows the dependence of the cell count on phase, which is consistent with the

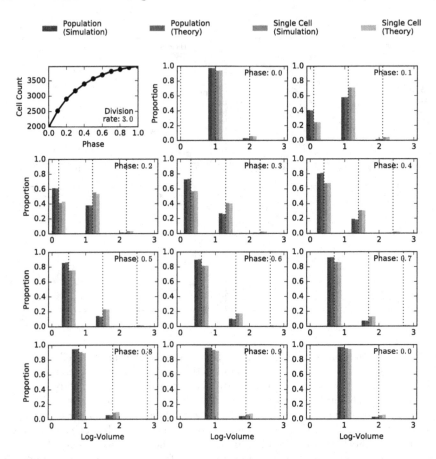

Fig. 3. Discrete log-volume distributions at different phases of the growth cycle for post-threshold division rate set to $\alpha = 3$. (Color figure online)

time dependence of cell count reported in Fig. 2. In the remaining panels of Fig. 3, blue-coloured bars represent the distributions of log-volume in the cell population at different phases of the period. At any phase of the cycle, the log-volume distribution is discrete (cf. Sect. 4). The support of the distribution, which is indicated by vertical dotted lines in the panels of Fig. 3, travels to the right as phase increases. At the end of the period, the support of the distribution, as well as the distribution itself, returns to where it started at the beginning of the cycle. The theoretical proportions (31) (bars of lighter shade of blue) are in a good agreement with the results of simulation by Algorithm 2 (bars of darker shade of blue).

In order to compare the population and single-cell versions of the model, we juxtapose the proportion of cell population with a particular log-volume (Fig. 3, blue bars) to the probability of observing the log-volume within a single-cell lineage (Fig. 3, orange bars). The single-cell probabilities were estimated from an ensemble of 4096 independent sample paths generated by Algorithm 1 (Fig. 3, bars of darker shade of orange). In each single-cell simulation, we used the same initial condition as in the population simulation; we also skipped the same amount of time, before analysing a single period of cyclic behaviour, as in the population simulation. The simulation-based estimates, both single-cell and population-wide versions, are cross-validated with the theoretical probability values (25) (Fig. 3, bars of lighter shade of orange). Comparing the results of the population and single-cell versions of the model, we observe that the single-cell approach tends to underestimate the proportion of small cells and overestimate the proportion of big cells in the phases of fast growth (Fig. 3, phases 0.1 to 0.7, say). In the later phases of slow growth, the two approaches lead to the same values of log-volume.

8   Conclusions

We considered a model for the growth of cellular volume in which cells grow exponentially and divide with a volume-dependent rate, resulting in two daughter cells each inheriting half of their mother's volume. Two versions of the model are systematically compared: in the first version, a single cell lineage is described by a stochastic, Markovian, model; in the second, an exponentially growing cell population is followed. We constructed simulation algorithms for both model versions which are exact in the sense that they require no numerical discretisation technique to sample the underlying stochastic process [8].

The model, in either of the two formulations, is shown to sustain cyclic behaviour with alternating phases of slow and fast growth. The phases of fast growth occur when most cells are large enough to divide, whereas the phases of slow growth take place when most cells are too small to divide. The cyclic behaviour, resulting from the periodicity of the stochastic process, is intriguing as it suggests possible biases from specific experimental designs (e.g. choice of measurement times). Notably, periodicity is a consequence of a fully symmetric division, and even small amounts of asymmetry in a more general model situations are expected to eventually break periodicity.

Our computational analysis suggests that the population-based approach leads to greater proportions of small cells and smaller proportions/probabilities of large cells in the fast-growth phases than the single-cell approach. This observation is consistent with the fact that the population approach includes twice as many daughter cells than mother cells do in comparison with the single-cell approach. Additionally, our results provide a quantitative evaluation of this effect and tractable analytic solutions valid in the large time asymptotic regime.

Thus, our results provide insights into the dynamics of the process of cell growth and suggest commonalities as well as differences between single-cell Markovian modelling and whole-population simulation. We expect that the methods explored in this paper can be applicable in related and more complex descriptions of cell growth and division.

References

1. Alberts, B., Johnson, A., Lewis, J., Raff, M., Roberts, K., Walter, P.: Molecular Biology of the Cell. Garland Science, New York (2002)
2. Amir, A.: Cell size regulation in bacteria. Phys. Rev. Lett. **112**(20), 208102 (2014)
3. Antunes, D., Singh, A.: Quantifying gene expression variability arising from randomness in cell division times. J. Math. Biol. **71**(2), 437–463 (2015)
4. Bell, G.I., Anderson, E.C.: Cell growth and division: I. a mathematical model with applications to cell volume distributions in mammalian suspension cultures. Biophys. J. **7**(4), 329 (1967)
5. Bernard, E., Doumic, M., Gabriel, P.: Cyclic asymptotic behaviour of a population reproducing by fission into two equal parts. arXiv preprint arXiv:1609.03846 (2018)
6. Davis, M.: Piecewise-deterministic markov processes: a general class of non-diffusion stochastic models. J. R. Stat. Soc. B **46**, 353–388 (1984)
7. Diekmann, O., Heijmans, H.J., Thieme, H.R.: On the stability of the cell size distribution. J. Math. Biol. **19**(2), 227–248 (1984)
8. Gillespie, D.: Exact stochastic simulation of coupled chemical reactions. J. Phys. Chem. **81**, 2340–61 (1977)
9. Hannsgen, K.B., Tyson, J.J., Watson, L.T.: Steady-state size distributions in probabilistic models of the cell division cycle. SIAM J. Appl. Math. **45**(4), 523–540 (1985)
10. Kretzschmar, K., Watt, F.M.: Lineage tracing. Cell **148**, 33–45 (2012)
11. Modi, S., Vargas-Garcia, C.A., Ghusinga, K.R., Singh, A.: Analysis of noise mechanisms in cell-size control. Biophys. J. **112**(11), 2408–2418 (2017)
12. Norris, J.R.: Markov Chains. Cambridge Univ Press, Cambridge (1998)
13. Perthame, B.: Transport Equations in Biology. Springer, Berlin (2006)
14. Robert, L., Hoffmann, M., Krell, N., Aymerich, S., Robert, J., Doumic, M.: Division in escherichia coli is triggered by a size-sensing rather than a timing mechanism. BMC Biol. **12**(1), 17 (2014)
15. Schuss, Z.: Theory and Applications of Stochastic Processes: An Analytical Approach. Springer, Berlin (2009)
16. Taheri-Araghi, S., et al.: Cell-size control and homeostasis in bacteria. Curr. Biol. **25**, 385–391 (2015)
17. Thomas, P.: Analysis of cell size homeostasis at the single-cell and population level. Front. Phys. **6**, 64 (2018)
18. Vargas-Garcia, C.A., Ghusinga, K.R., Singh, A.: Cell size control and gene expression homeostasis in single-cells. Curr. Opin. Syst. Biol. **8**, 109–116 (2018)
19. Vargas-Garcia, C.A., Soltani, M., Singh, A.: Conditions for cell size homeostasis: a stochastic hybrid system approach. IEEE Life Sci. Lett. **2**(4), 47–50 (2016)

Transient Memory in Gene Regulation

Calin Guet[2], Thomas A. Henzinger[2], Claudia Igler[2], Tatjana Petrov[1(✉)],
and Ali Sezgin[3]

[1] Department of Computer and Information Sciences/Centre for the Advanced Study
of Collective Behaviour, University of Konstanz, 78464 Konstanz, Germany
tatjana.petrov@gmail.com
[2] IST Austria, Am Campus 1, 34000 Klosterneuburg, Austria
[3] Aselsan, Ankara, Turkey

Abstract. The expression of a gene is characterised by its transcription
factors and the function processing them. If the transcription factors are
not affected by gene products, the regulating function is often represented as a combinational logic circuit, where the outputs (product) are
determined by current input values (transcription factors) only, and are
hence independent on their relative arrival times. However, the simultaneous arrival of transcription factors (TFs) in genetic circuits is a strong
assumption, given that the processes of transcription and translation of
a gene into a protein introduce intrinsic time delays and that there is
no global synchronisation among the arrival times of different molecular
species at molecular targets.

In this paper, we construct an experimentally implementable genetic
circuit with two inputs and a single output, such that, in presence of
small delays in input arrival, the circuit exhibits qualitatively distinct
observable phenotypes. In particular, these phenotypes are long lived
transients: they all converge to a single value, but so slowly, that they
seem stable for an extended time period, longer than typical experiment
duration. We used rule-based language to prototype our circuit, and we
implemented a search for finding the parameter combinations raising the
phenotypes of interest.

The behaviour of our prototype circuit has wide implications. First, it
suggests that GRNs can exploit event timing to create phenotypes. Second, it opens the possibility that GRNs are using event timing to react
to stimuli and memorise events, without explicit feedback in regulation.
From the modelling perspective, our prototype circuit demonstrates the
critical importance of analysing the transient dynamics at the promoter
binding sites of the DNA, before applying rapid equilibrium assumptions.

Tatjana Petrov's research was supported by SNSF Advanced Postdoc. Mobility Fellowship grant number $P300P2\_161067$, the Ministry of Science, Research and the Arts
of the state of Baden-Württemberg, and the DFG Centre of Excellence 2117 'Centre
for the Advanced Study of Collective Behaviour' (ID: 422037984). Claudia Igler is the
recipient of a DOC Fellowship of the Austrian Academy of Sciences. Thomas A. Henzinger's research was supported in part by the Austrian Science Fund (FWF) under
grant Z211-N23 (Wittgenstein Award).

© Springer Nature Switzerland AG 2019
L. Bortolussi and G. Sanguinetti (Eds.): CMSB 2019, LNBI 11773, pp. 155–187, 2019.
https://doi.org/10.1007/978-3-030-31304-3_9

Keywords: Gene regulation · Stochastic modelling ·
Long lived transients · DNA looping

1   Introduction

The fundamental conceptual breakthroughs related to how a gene is turned on
and off, have inspired a large body of theoretical and experimental work on
gene regulation, including the explanation of stochastic switching between lysis
and lysogeny of phage [25], all the way to more complex logic gate formalisms
that attempt to abstract more complex biological behaviour. Synthetic biology
enthusiasts often use analogies with how electronic circuits are manipulated by
computers [13,24], and have demonstrated success in engineering simple genetic
circuits that are encoded in DNA and perform their function *in vivo*. However,
such digital (in the sense that the expression states are encoded through Boolean
values) and combinational design (in the sense that the output is a pure function
of present input only, different to the *sequential* design) quickly becomes infeasible in experiment, because the cellular environment is resource-limited and
highly crosstalk-prone. The effective engineering of biological systems needs to
take into account the intrinsic properties of the biological medium, so as not to
fight against the principles of tinkering that characterise biology [16], but rather
to make use of them. Significant conceptual challenges remain related to the still
unsatisfactory quantitative but also qualitative understanding of the underlying
processes [20]. Understanding time-dependent phenomena is fundamental in this
complex picture of the cell that unravels itself at the molecular scale, especially
since cells do not have computer-like clocking mechanisms, beyond circadian and
cell cycle ones. A major question emerges as to what are the macroscopic effects
of small delays in the arrival times of different molecules at molecular targets.

Gene expression in a single cell is modelled by a stochastic process which
captures the stochastic switching among possible configurations at the DNA
(the architectural configuration of which is often termed *promoter logic*, e.g.
shown in Fig. 4), and their effect on the copy number of other species involved
in regulation, such as mRNA, proteins and transcription factors (TFs). The
switching mechanism depends on the binding affinities of the TFs and RNA
polymerase to their respective binding sites as well as the concentrations of those
proteins in the cell. Stochastic dynamics of such gene regulatory process typically
has a single equilibrium, as a consequence of reversibility of reactions occurring
at the DNA binding sites. Sometimes, the transient regime of the distribution
among DNA configurations is rapid and robust to possible delays in arrival of
TFs. In such cases, it is satisfactory to use the statistical thermodynamics model,
which has shown unquestionable success (e.g. [1,37]). It estimates the probability
of being in any of the possible DNA binding configurations from their relative
binding energies (Boltzmann weights) and the protein concentrations, both of
which can often be experimentally accessed. While this model takes into account
the stochasticity inherent to the DNA binding configurations (unlike the also
widely used deterministic limit [18]), it neglects the transient probabilities in

the promoter logic before the equilibrium is reached. The question arises: In which ways does the transient regime at the DNA (promoter) affect the shape and duration of observable protein dynamics? Can it happen that the observable transients move towards the unique equilibrium so slowly, that they are mistaken for steady state dynamics?

In this paper, as a proof-of-concept, we construct a prototype genetic circuit based on two different transcription factors that regulate the same gene, without feedback. Our circuit demonstrates that, for gene regulation, qualitatively distinct transients may take extraordinarily long times to disappear, and the observable phenotype in the transient can be highly sensitive to the order of arrival of TFs in the system. In particular, the transient phenotype may appear to be stable even though it is not, creating an effect of *long lived transients*[7]. Our prototype circuit is realistic, experimentally implementable in the sense that the mechanism can be implemented by the current technology and kinetic rate values are in realistic ranges. The behaviour of this circuit suggests that the genetic circuit can memorise the order of arrival of TFs, although there is no explicit feedback at the gene regulatory level.

2   Preliminaries and Background

A gene is expressed at a basal rate, whenever the RNA polymerase (RNAP) is bound to its promoter region at the DNA. *Activators* are transcription factors (TFs) that bind to specific locations on the DNA, or to other TFs, and enhance the expression of gene g by promoting the binding of RNAP. *Repressors* reduce the expression of gene g, by directly blocking the binding of RNAP, or indirectly, by inhibiting the activators, or promoting direct repressors. The mechanism of how and at which rates the molecular species are interacting is transparently written in a list of reactions. Reactions are equipped with the stochastic semantics which is valid under mild assumptions [12].

Definition 1. A reaction system; is a pair (S, R), such that $S = \{S_1, \ldots, S_n\}$ is a finite set of species, and $R = \{r_1, \ldots, r_r\}$ is a finite set of reactions. The state of a system can be represented as a multi-set of species, denoted by $\mathbf{x} = (x_1, ..., x_n) \in \mathbf{N}^n$. Each reaction is a triple $r_j \equiv (\mathbf{a}_j, \boldsymbol{\nu}_j, c_j) \in \mathbf{N}^n \times \mathbf{N}^n \times \mathbb{R}_{\geq 0}$, written down in the following form:

$$a_{1j}S_1, \ldots, a_{nj}S_n \xrightarrow{c_j} a'_{1j}S_1, \ldots, a'_{nj}S_n, \text{ such that } \forall i.a'_{ij} = a_{ij} + \nu_{ij}.$$

The vectors $\mathbf{a}_j$ and $\mathbf{a}'_j$ are often called respectively the *consumption* and *production* vectors due to jth reaction, and c_j is the respective *kinetic rate*. If the jth reaction occurs, after being in state $\mathbf{x}$, the next state will be $\mathbf{x}' = \mathbf{x} + \boldsymbol{\nu}_j$. This will be possible only if $x_i \geq a_{ij}$ for $i = 1, \ldots, n$.

Stochastic Semantics. The species multiplicities follow a continuous-time Markov chain (CTMC) $\{X(t)\}_{t \geq 0}$, defined over the state space $S = \{\mathbf{x} \mid \mathbf{x}$

is reachable from $\mathbf{x}_0$ by a finite sequence of reactions from $\{r_1, \ldots, r_r\}\}$. In other words, the probability of moving to the state $\mathbf{x} + \boldsymbol{\nu}_j$ from $\mathbf{x}$ after time Δ is

$$\mathsf{P}(X(t + \Delta) = \mathbf{x} + \nu_j \mid X(t) = \mathbf{x}) = \lambda_j(\mathbf{x})\Delta + o(\Delta),$$

with λ_j the propensity of jth reaction, assumed to follow the principle of mass-action: $\lambda_j(\mathbf{x}) = c_j \prod_{i=1}^{n} \binom{x_i}{a_{ij}}$. The binomial coefficient $\binom{x_i}{a_{ij}}$ reflects the probability of choosing a_{ij} molecules of species S_i out of x_i available ones.

Example 1 (basal gene expression). Basal gene expression with RNAP binding can be modelled with four reactions, where the first reversible reaction models binding between the promoter site at the DNA and the polymerase, and the second two reactions model the protein production and degradation, respectively:

$$\mathsf{DNA, RNAP} \leftrightarrow \mathsf{DNA.RNAP} \text{ at rates } k, k^-$$
$$\mathsf{DNA.RNAP} \rightarrow \mathsf{DNA.RNAP + P} \text{ at rate } \alpha$$
$$\mathsf{P} \rightarrow \emptyset \text{ at rate } \beta.$$

The state space of the underlying CTMC $S \cong \{0, 1\} \times \{0, 1, 2, \ldots\}$, such that $s_{(1,x)} \in S$ denotes an active configuration (where the RNAP is bound to the DNA) with $x \in \mathbf{N}$ protein copy number, as depicted in Fig. 2.

Example 2 (adding repression). Repressor blocking the polymerase binding can be modelled by adding a reaction

$$\mathsf{DNA}, R \leftrightarrow \mathsf{DNA}.R$$

In this case, there are three possible promoter configurations, that is, $S \cong \{\mathsf{DNA}, \mathsf{DNA.RNAP}, \mathsf{DNA}.R\} \times \{0, 1, 2, \ldots\}$, where $D_0 = \{\mathsf{DNA}, \mathsf{DNA}.R\}$ are inactive promoter states.

Computing the Transient. Using the vector notation $\mathbf{X}(t) \in \mathbf{N}^n$ for the marginal of process $\{X(t)\}_{t \geq 0}$ at time t, we can compute this transient distribution by integrating the *chemical master equation* (CME). Denoting by $p^{(t)}(\mathbf{x}) = \mathsf{P}(\mathbf{X}(t) = \mathbf{x})$, the CME for state $\mathbf{x} \in \mathbf{N}^n$ reads

$$\frac{d}{dt} p^{(t)}(\mathbf{x}) = \sum_{j=1, \mathbf{x} - \nu_j \in S}^{r} \lambda_j(\mathbf{x} - \boldsymbol{\nu}_j) p^{(t)}(\mathbf{x} - \boldsymbol{\nu}_j) - \sum_{j=1}^{r} \lambda_j(\mathbf{x}) p^{(t)}(\mathbf{x}). \tag{1}$$

The solution may be obtained by solving the system of differential equations, but, due to its high (possibly infinite) dimensionality, it is often statistically estimated by simulating the traces of $\{X_t\}$, known as the stochastic simulation algorithm (SSA) in chemical literature [12]. As the statistical estimation often remains computationally expensive for desired accuracy, for the case when the deterministic model is unsatisfactory due to the low multiplicities of many molecular species [19], different further approximation methods have been proposed, major challenge to which remains the quantification of approximation accuracy (see [36] and references therein for a thorough review on the subject).

2.1   Transients in Gene Expression Without Feedback

We will further focus on regulation of single gene without feedback. This allows a circuit-view, where activators and repressors are inputs, and the average transient protein expression is the output. Since there is a single DNA molecule per cell, each state counts one copy of the current DNA configuration, and zero copies of all other DNA binding configurations. Hence, the expression state for a single gene of interest consists of two layers: the proteins that we see, and the regulatory configuration of the DNA (for example, two activators and polymerase are bound) (Fig. 3). Such two-layer model allows us to study the transient of coupled promoter state and protein count. In order to focus our analysis on the effect of input timing perturbations, yet to keep our model simple, we chose not to involve further mechanistic details, such as the steps involving mRNA.

The following observation states that in general, when there is no feedback, computing the output does not require integrating the Master equation for the entire CTMC, but only for a CTMC controlling the switching among the DNA configurations (depicted left in Fig. 3).

Lemma 1. Let $\{X(t)\}_{t \geq 0}$ be the CTMC for a model of single gene regulation without feedback, over the state space $S = S_0 \uplus S_1 = (D_0 \uplus D_1) \times \{0, 1, \ldots\}$, where $D_0 = \{D_{01}, D_{02}, \ldots\}$ are inactive DNA configurations, and $D_1 = \{D_{11}, D_{12}, \ldots\}$ are active DNA configurations (RNAP bound). Let the reaction system;(S, R) be such that all reactions are of one of the following types (for some $i \geq 0$ and $j \geq 0$):

$$\text{(de)activation: } D_{0i} \leftrightarrow D_{1i} \text{ at rates } k_i, k_i^-$$

$$\text{switching: } D_{0i} \leftrightarrow D_{0j} \text{ at rates } k_{0ij}, k_{0ij}^-$$

$$\text{switching: } D_{1i} \leftrightarrow D_{1j} \text{ at rates } k_{1ij}, k_{1ij}^-$$

$$\text{protein syhnthesis: } D_{1i} \to D_{1i} + P \text{ at rate } \alpha$$

$$\text{protein degradation: } P \to \emptyset \text{ at rate } \beta.$$

Then, the average amount of protein in a population follows the differential equation

$$\frac{\mathrm{d}}{\mathrm{d}t} \langle x_p(t) \rangle = P(\mathbf{X}(t)|_D \in D_1)\alpha - \beta \langle x_p(t) \rangle, \tag{2}$$

where $\langle x_p(t) \rangle$ denotes the average amount of the protein molecules at time t, and process $\mathbf{X}(t)|_D$ is the projection of process $\mathbf{X}(t)$ to states at the promoter, that is $\mathbf{X}(t)|_D = d$ if and only if $(\mathbf{X}(t) \in \cup_{i \geq 0}(d, i))$. In other words, $P(\mathbf{X}(t)|_D \in D_1)$ denotes the marginal probability that the promoter is in active state (bound RNAP) at time t.

The proof is discussed in Appendix, Sect. 1.C.

Corollary 1. Let $\pi_1 = \lim_{t \to \infty} P(\mathbf{X}(t)|_D \in D_1)$ denote the probability of active promoter at stationarity. Then, whenever the initial probability equals that of the stationary, i.e. $P(\mathbf{X}(0)|_D \in D_1) = \pi_1$, the average protein dynamics follows the differential equation

$$\frac{d}{dt}\langle x_p(t)\rangle = \pi_1\alpha - \beta\langle x_p(t)\rangle. \tag{3}$$

When DNA is modelled with one binding site, the promoter can be in only two states, and the analytic solution to Eq. (2) is tractable. In general, as activators and repressors bind to different regions (operator sites) of the same DNA molecule, the respective number of regulatory configurations at the promoter grows combinatorially with the number of operator sites. For instance, one hypothesised mechanism in λ-phage, containing only three left and three right operators, leads to 1200 different DNA configurations [35][1]! In such cases, the simplification based on the argument of fast equilibrium is often employed, meaning that the transient protein dynamics is computed according to Eq. (3), thus neglecting the transient changes in probability distribution among the DNA regulatory configurations.

Fast equilibrium assumption is a prerequisite to applying a widely popular *statistical thermodynamics* model [38]. Assuming that the DNA regulatory configurations mix rapidly, this model allows to experimentally estimate the free energies of each promoter configuration, and then, subsequently, to derive the equilibrium constants[2] for each of the reactions [28,38]. As the absolute and precise values of kinetic rates are rarely available in practice, this method is powerful, because it allows to predict the dynamics of a genetic circuit from a scarcely available experimental data. However, the statistical thermodynamics model is applicable only when the assumption of rapid equilibrium at the promoter is valid.

In the following, we showcase a simple, experimentally realisable genetic circuit which demonstrates an interesting situation where the long transient at the promoter creates phenotypes that are qualitatively distinct from the phenotypes created when the promoter configurations start at the equilibrium. In particular, these phenotypes are long lived transients: they all converge to a single value, but so slowly, that they seem stable for an extended time period, longer than typical experiment duration.

3   Problem Statement

We focus on a single gene regulation without feedback, where activators and repressors are inputs, and the average protein expression is the output. Assuming that a fixed amount of activators and repressors are added to the system with a possible time lag, our reference scenarios are (Fig. 4 in Appendix):

- $\mathcal{X}_{A\|R}$ in which activator and repressor are introduced together,
- $\mathcal{X}_{A\to R}(\Delta)$ in which the activator is introduced Δ time units before the repressor, and
- $\mathcal{X}_{R\to A}(\Delta)$ in which the repressor is introduced Δ time units before the activator.

---

[1] Models used in this paper will count 23 and 6 distinct DNA binding configurations.
[2] The ratio between the binding and unbinding rate.

Our goal is to construct a genetic circuit with the following requirements: (i) it is realistic, that is - experimentally implementable in the sense that the mechanism can be implemented by the current technology and kinetic rate values are in realistic ranges, (ii) the scenarios provide striking differences in the shape and duration of transient protein output. To quantify the latter, we introduce two quantitative measures:

- *amplitude*, the maximum distance in phenotype of the scenario with delay from the scenario without delay, that is

$$\alpha_{|s} := \max_{t \geq t_0} |\langle x_p \rangle(t|s) - \langle x_p \rangle(t|\mathcal{X}_{A||R})|,$$

  where s refers to the scenario in question ($\mathcal{X}_{A \to R}(\Delta)$ or $\mathcal{X}_{R \to A}(\Delta)$) and $x_p(t|s)$ denotes the average protein number in a population at time t in scenario s, and
- *halflife*, the time the system takes from the moment of reaching the amplitude, to the moment when the distance from the phenotype without delay disappears, that is

$$t_{1/2|s} := \arg \min_{t \geq t_{\alpha|s}} \{t \mid |\langle x_p \rangle(t|s) - x_p(t \mid \mathcal{X}_{A||R})| < \frac{1}{2}\alpha_{|s}\},$$

where $t_{\alpha|s}$ denotes the moment when the amplitude is reached in scenario s.

In summary, the *amplitude* reflects how observable is the sensitivity to the delay among inputs, and the second measure, *halflife*, reflects how slow is the convergence to the real equilibrium after the amplitude has been observed. Long lived transients are characterised by a large amplitude relative to the basal expression and a half-life exceeding several cell division cycles.

4   Searching for Long Lived Transients

We develop and analyse models for two promoter architectures (drafted in Fig. 5):

- **Model without looping.** A basic mechanism for activation and repression is assumed: repressor R competes with RNAP, and the activator A recruits the polymerase RNAP and binds independently of the repressor and the polymerase (configurations shown in Fig. 6).
- **Model with looping.** In the model with looping (Fig. 5, right), two activators and two repressors can bind the DNA; Binding of the second activator (resp. repressor) promotes looping of the DNA in the active (resp. repressed) state, thereby excluding binding of the other TF. This small mechanistic change leads to the blow-up of the state space of the CTMC to 23 states as a composition of two sub-models (Fig. 7).

Biological Context. The CTMC for the DNA switching of our prototype circuit with looping is inspired by the very well-characterized regulatory mechanisms of the *lac* operon and of the bacteriophage lambda genetic switch [17,30].

Mathematical Context. In Markov chains, the time to be ϵ-close to equilibrium, *mixing time*, varies depending on the initial distribution, the chain connectivity, and the rate parameters. In particular, long mixing times are prominent for chains with a large spectral gap of the underlying generator matrix, and can be guaranteed for chains with large connectivity diameter, suggesting that more states and sparse connectivities generally can prolong the mixing time [21]. Still, tightly estimating bounds on the mixing time for a given chain is an open problem, beyond the scope of this manuscript. Intuitively, DNA looping architecture is a good candidate for creating large mixing times, because the looped states are quick to reach when only activators or only repressors are present, but, once entered, they are then hard to exit ('dynamically trapped states').

4.1   Model Implementation

Implementation. The models are written and analysed within the rule-based modelling framework Kappa which allows us to represent the mechanistic model concisely and to run an efficient stochastic simulation algorithm [3,11]. Source-code of the rule-based models is given in Appendix 1.B. Parameter exploration and additional output analysis were performed with Python.

Simulation. We simulated multiple samples of the stochastic model, and we statistically estimated the first two moments of protein expression. In the model with looping, we used 1000 individual cells for a time of $36000\,\mathrm{s} = 10\,\mathrm{h}$, that equals around 20 average cell doubling times, where inputs are added from time point $t_0 = 5400 = 1.5\,\mathrm{h}$ (see Table 2 for all simulation parameters).

Kinetic Rates. All model parameters are in realistic ranges taken from the literature, given in Table 1 and further explained in the Appendix. The mechanism for the activator is inspired by the λ-phage. The mechanism for repressor is inspired by the lac operon. Further values that were tested to show the generality of our approach came from other well-characterized TFs such as CRP. The chosen parameter values were found in the literature, both for the scenario without looping [6,15] and for the scenario with looping [32–34,39,40].

Parameter Search. We implemented a grid search of the viable parameter space (for different levels of eleven kinetic rate parameters, and the amounts of activator, repressor and RNAP), where we compute the average protein expression, amplitude and half-life for a subset of all parameter combinations. In our implementation, the user specifies a range for each parameter, and the models are executed, figures drawn for each possible parameter combination.

5   Results

In further text, by *phenotype*, we mean the average protein expression in a population of 1000 cells. All three scenarios $\mathcal{X}_{A||R}$, $\mathcal{X}_{A \to R}(\Delta)$ and $\mathcal{X}_{R \to A}(\Delta)$ have the

Long lived transients (reference parameter set)

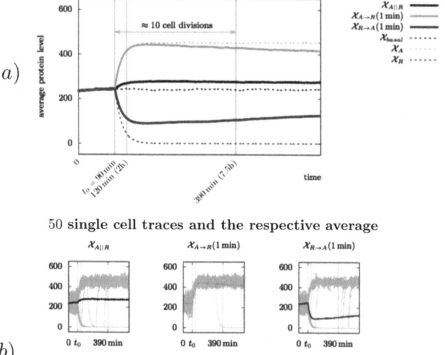

a)

50 single cell traces and the respective average

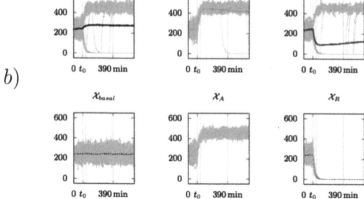

b)

Fig. 1. A small delay in arrival times of TFs can give rise to qualitatively opposite, stable transient phenotypes for a long period of time (a) Average protein level for a population of 1000 cells, in three input scenarios (full lines) and three reference scenarios (dotted lines). (b) 50 single cell traces (grey lines) and the respective average, for each of the six modes.

same phenotypes eventually. As a reference, we also analyse the scenarios where no TFs are input ($\mathcal{X}_{basal}$), only activators ($\mathcal{X}_A$) or only repressors are input ($\mathcal{X}_R$). We investigated the phenotype in the three scenarios for a large range of parameter combinations. (there are $2^{11} > 2000$ combinations when only two values for each parameter are set). We choose one parameter set as the *reference parameter set* (shown in Fig. 1a), where the phenotypes are symmetric with respect to the $\mathcal{X}_{A||R}$ scenarios in the sense that the protein expression deviates in the same amount from the phenotype of $\mathcal{X}_{A||R}$ and the rate of reaching the phenotype of $\mathcal{X}_{R||A}$ is of the same scale.

A Small Delay in Arrival Times of TFs Can Give Rise to Qualitatively Opposite, Stable Transient Phenotypes for a Long Period of Time. In Fig. 1a, we plot the observable phenotype – the mean of protein expression for a given population of cells – in the three regimes of interest (full lines) and the three reference regimes (dotted lines). Three distinct transient phenotypes are observed:

- for $\mathcal{X}_{A||R}$, the expression is close to the level of basal expression,
- the transient regime for the input $\mathcal{X}_{A \to R}(1 \, \text{min})$ shows high expression for multiple average cell doubling times, while
- the transient regime for the input $\mathcal{X}_{R \to A}(1 \, \text{min})$ shows low expression for multiple average cell doubling times.

The transient for the input $\mathcal{X}_{A||R}$ lasts roughly for one average cell doubling time (30 min), while both phenotypes for $\mathcal{X}_{A \to R}(1 \, \text{min})$ and $\mathcal{X}_{R \to A}(1 \, \text{min})$ last well over 10 average doubling times. Therefore, the delay in arrival times of TFs can result in long lived transient regimes with qualitatively opposite phenotypes (both differing significantly from the equilibrium phenotype), depending on which TF arrives first. Moreover, each of the phenotypes seems stable at the time-scale of multiple cell lifetimes. In other words, the small delays, hence two different histories of input, produce substantially different routes to the equilibrium, and the routes are so slow that they appear as steady state behavior at the timescale of most experiments.

In Fig. 1b, we see that individual cells exhibit 'all-or-none' behaviour: an individual cell either has high or low expression and the phenotype depends on whether the cell entered the active looped state or the repressed looped state. The expected time that a cell spends in one looped state is long. The protein expression for 50 randomly chosen single cells is displayed in Fig. 1b for each of the three regimes. In regime $\mathcal{X}_{A||R}$, an individual cell either has high expression at around 400 proteins or low expression, being fully inhibited. The noise around the low expression value is not observable in the plot, because the low expression is fully inhibited most of the time. If the DNA unloops and subsequently loops towards a different regulatory state, eg from looped repressed to looped active state (or vice-versa), the protein expression will change from low to high expression (or vice-versa). In the taken time window (10 h), three (out of 50) displayed traces switching from the high to low expression level and one trace switching from low to high expression level. As expected, the average expression in a given population (thick line in respective color) follows a continuous line; It is saturated at around 270 protein molecules. In regime $\mathcal{X}_{A \to R}$, all of the displayed 50 cells enter the active looped state before the repressors are input, but, due to the slow unlooping, the high expression profile is long-preserved, resulting in slow switching towards the low-expression state, and hence long transient time towards the average expression. In regime $\mathcal{X}_{R \to A}$, even though repressors are input first, some cells are activated, but most of the cells are repressed. Similarly as in the profile $\mathcal{X}_{A \to R}$, since DNA unlooping from the repressed state is slow, the transient of the average protein expression is also slow. The reference scenarios - $\mathcal{X}_{basal}$, $\mathcal{X}_A$ and $\mathcal{X}_R$ show the expected behaviour.

Long Lived Transients are Robust to Changes in Kinetic Parameters. Are long lived transients a consequence of system regulatory architecture or a careful tuning of kinetic parameters? To tackle this question, we chose six different parameter combinations, listed in Table 1, and we reproduce the plot shown in Fig. 1a for each of the parameter combinations, each for three time delays - 1 min, 5 min and 15 min. $\mathbf{p}_1$ is the reference parameter set (the plot shown in Fig. 1). Results, shown in Fig. 8, confirm that the long lived transients are preserved with the chosen parameter changes. Higher unlooping rate for either activators or repressors results in shortening the transient and moving the average expression level to lower and higher value respectively ($\mathbf{p}_2$ and $\mathbf{p}_3$). While decreasing the number of activators does not change the phenotype much ($\mathbf{p}_4$), decreasing the number of repressors results in complete dominance of activation effect when both TFs are input simultaneously ($\mathbf{p}_5$). Still, the delay of activator input shows full repression profile for a long period of time. When RNAP rates are scaled so that the binding and unbinding rates are both ten times slower, the duration of transients shortens and the three input regimes show the same output after much shorter time ($\approx$10 h, $\mathbf{p}_6$).

To quantify the effect of long-lived transients, in Fig. 12 (up, model with looping), we see that the reference parameter set ($\mathbf{p}_1$) has a halflife longer than 20 hours no matter if the delay occurs in favour of the activator or repressor. The halflife decreases significantly in cases when the unlooping rate is decreased (one at a time - $\mathbf{p}_2$ and $\mathbf{p}_3$), or when RNAP binding and unbinding rate is scaled down ($\mathbf{p}_6$), while the change in the number of activators/repressors reflect more on the amplitude than on the halflife ($\mathbf{p}_4$, $\mathbf{p}_5$).

Long Lived Transients Are Not Observed in the Model Without Looping. We next inquire how changes in regulatory architecture affect the behaviour, i.e. is DNA looping essential for observing the long lived transients? We repeated the experiments on a model without looping. Phenotypes for six parameter combinations, listed in Table 1[3], are each plotted for three time delays (Fig. 10). For all parameter combinations, the amplitude and duration of transient regimes is clearly correlated with the duration of delay - the longer delays induce longer transient regimes. The transient phase is significantly shorter than in the model with looping (notice the different time-scale than in Figs. 1 and 8), but they still can last for several cell doubling times (for delays of 15 min up to 2.5 h or 5 average doubling times). However, they are not long lived transients, as the shape of transients clearly reveals that the steady-state regime is going to be reached later on, that is, the transients in this model would not be easily confused with the steady state. The observations above are indicating that looping is essential

---

[3] Notice that these six parameter combinations are different than those used for the model with looping.

for creating the effect of long lived transients. $\mathbf{p}_1$ is the reference parameter set, which we choose so that the level of expression when both activators and repressors are input is close to basal (the TFs neutralise each-other's effect overall). As expected, decreasing the recruitment by the activator results in lower stationary expression ($\mathbf{p}_2$), increasing the number of repressors results in stronger repression ($\mathbf{p}_3$), weakening the repressor binding results in higher expression ($\mathbf{p}_4$), weak binding of repressor in combination is not affected by decreasing the recruitment by activator simultaneously ($\mathbf{p}_5$) and weak binding of repressor in combination with more repressor molecules results in low expression ($\mathbf{p}_6$).

In the summary of characteristics of long lived transients for the model without looping (shown in Fig. 12 down), we see that, for a delay of 5 min, all parameter combinations achieve the amplitude at comparable scale as that in the case of model with looping. However, the maximal halflife in all tested parameter points is 15 min, a 100-fold difference with respect to the halflife of long lived transients in the model with looping, confirming that adding the looped configurations was essential for the effect of long-lived transients.

Phenotypes in Long Lived Transients can be Modulated by the Delay Between Inputs. We now comment on the dependency on the delay. In Fig. 11, the phenotypes in scenario $\mathcal{X}_{A \to R}(\Delta)$ are observably equivalent for all chosen values of delay. In particular, they transiently reach the same protein expression value as the scenario $\mathcal{X}_A$ where only activator is present. Therefore, this scenario seems to be independent of delay timing between TFs as long as the delay occurs in favour of the activator.

On the other hand, the difference between phenotypes in scenario $\mathcal{X}_{R \to A}(\Delta)$ is different for delay $\Delta = 1$ min than for delays $\Delta \in \{5 \text{ min}, 15 \text{ min}\}$. While for all three delays, the effect of long lived transients can be observed (the slope of approaching the limit value is small), the phenotype (protein expression around which the transients seem to stabilise) is different. It appears that, unlike delays longer than 5 min, the delay of 1 min is not long enough for the population to repress protein expression to a value as low as in the scenario $\mathcal{X}_R$ (where only repressor is present). In other words, the lowest gene expression value for delay of 1 min is never as small as in the scenario $\mathcal{X}_R$. To investigate the dependency of the transient phenotype on the delay, we simulated the scenario for several delay values between 1 sec and 5 min, namely $\Delta \in \{1s, 20s, 40s, 60s, 120s, 180s\}$ and we computed the amplitude for the scenario $\mathcal{X}_{R \to A}$. The plot in Fig. 11 demonstrates that the amplitude approaches the value of $\mathcal{X}_R$ scenario exponentially fast with increasing delay time.

Plotting the phenotypes for scenario $\mathcal{X}_{A \to R}(\Delta)$ for delays between 1 s and 1 min shows that the same activated gene expression levels are observed even for delays as small as 1 s (plots not shown). The explanation for different sensitivity of transient phenotypes to the delay in scenarios $\mathcal{X}_{A \to R}(\Delta)$ and $\mathcal{X}_{R \to A}(\Delta)$ are the different mechanisms implementing the activation and repression. When activator is input first, it quickly binds both operator sites and the probability of being in the looped active state almost instantaneously increases to the maximum value (as fast as within 1 s), and then starts decreasing only very slowly towards the equilibrium as soon as the repressor is present as well. On the other hand, when the repressor is input first, it does not bind both operator sites as quickly, because it is competing with the abundant RNAP, even while the activator is not in the system. Only if there is enough time for the repressor to reach the looped repressed state with a probability nearly as high as in case of repressors only, the maximally repressed expression level will be observable. Otherwise, as soon as the activator is in the system, the probability of being in the looped repressed state starts shifting slowly towards the equilibrium point, and, consequently, the protein expression in the population starts increasing.

Long Lived Transients in Protein Expression Follow the Long Lived Transients (Mixing Times) in Promoter Activity. In Fig. 9, we plot the probability of the active regulatory configuration of the promoter for six different parameter combinations (listed in Table 1). Plots show the expected agreement with those in Fig. 8.

6 Discussion

Given that the processes of transcription and translation of a gene into a protein introduce intrinsic time delays and that there is no global synchronization among the arrival times of different molecular species at molecular targets, the simultaneous arrival of TFs in genetic circuits is a strong assumption. We subjected this assumption to a perturbation analysis, where the perturbed parameters are the relative arrival times of the TFs (different to the usual choices of perturbation parameters being the kinetic rates). We simulated a simple and realistic genetic circuit with two inputs and we showed that, in presence of small perturbations in the arrival of inputs (shorter than 1 min), the circuit can exhibit three qualitatively distinct phenotypes which are stable for as long as any typical experiment would last (longer than 20 cell doubling-times). This has wide implications.

First, while our showcase example was constructed with the goal of demonstrating that long lived transients can appear in gene regulation, there are reasons to believe that many other gene regulatory schemes also exhibit long lived transients and implement multiple phenotypes by modulating the timing of inputs. To see this, consider that the number of potential phenotypes grows factorially with the number of inputs per gene as it is determined by the number of possible input orderings, meaning that, for instance, only 5 inputs would require us to analyse $5! = 120$ different input scenarios. Moreover, our analysis indicates that long lived transients are possible in promoters with many configurations and certain states that are easy to reach but hard to exit. For instance, genomic regulation of the development of sea urchin embryo shows potential for long lived transients. The relevance of transient TF production has already been determined in this system [2,44]: multiple TFs regulate a single gene which in turn has multiple targets, and there is clear differentiation between upstream and downstream components in the network. Therefore, considering long lived transients might clear up some puzzling observations like the discrepancy of TF interactions between endogenous promoters and minimal promoters controlled by three Endo16 regulatory modules [43].

Secondly, our proof-of-concept case study suggests that any modelling approach which assumes perfectly synchronous arrival of TFs or assumes rapid equilibrium at the promoter, may fail to explain a variety of phenotypes and raise false conclusions. To illustrate this point, think of an experimentalist who observes the system which seems equilibrated, but is a long lived transient (e.g. in our case study, a delay in favour of the activator occurs). Assuming that what she sees is an equilibrium, following the approach of statistical thermodynamics, she would proceed by estimating the free energies of binding configurations, but these estimates would be wrong, as the real equilibrium is much further away. Moreover, the obtained model would explain a single phenotype, and not the variety of quasi-stable phenotypes such as the ones we see in our showcase example. In summary, one cannot ignore the order of stimulating a cell, even when the GRN under consideration is assumed to be feedback-free. Similarly, one cannot assume what is observed towards the end of the life cycle is close to equilibrium even when the system seems relatively stable, e.g. growth at steady state in bulk, even when the stimulation was completed very early on. This opens further important questions such as how can an experimentalist who observes a stable phenotype for the chosen experiment duration, distinguish between a long lived transient and a real equilibrium? One immediate insight is the critical importance of experimentally measuring the kinetic rates as accurately as possible, and taking the timing of inputs into account.

Finally, our case study opens the possibility that GRNs are exploiting event timing to perform desired behaviours - it suggests that the cell does not compute with equilibrium dynamics - as is widely assumed in the field (with the exception of 'well behaved' limit cycle behaviours or pulsatile behaviour [22]), but uses the transients to react to stimuli and to memorise events. The DNA may be encoding more behaviours and thus phenotypes than an understanding based on the conventional input to output mapping suggests. In particular, as our analysis of delay timing between TFs shows (Fig. 11), a whole range of different stable gene expression levels can be encoded in the event timing of inputs. More broadly, this aspect may provide an explanation to why an organism can display so many more phenotypes, though the number of genes is limited, as the complexity of the organism increases, e.g. number of genes in bacteria and human vary by a factor of only 4!

Our primary goal was to show that a simple gene regulation without feedback, with realistic parameters, can exhibit long lived transients. We hypothesised that the promoter architecture with looping will have the desired feature, and, in order to find the feature, we performed a search over the 11-dimensional parameter space, which allowed us to display and discuss a range of parameterisations showing interesting behaviour. One of the compelling questions for future work is formalisation and computation of robustness of a given promoter architecture wrt. property of long-lived transients, as well as its sensitivity to a specific (group of) parameters. To this end, we believe that the ideas of parameter synthesis for stochastic chemical reaction networks, extensively studied in [4, 9, 10] (where the properties of the CTMC assigned to general biochemical reaction networks are expressed in continuous signalling logic (CSL)), would be a useful starting point.

Related Work. The consequences of combined effects of time delay and intrinsic noise on gene regulation has been studied in [45]. In more recent works [23], the authors elucidate the importance of relative timing of TF activation in combinatorial gene regulation with pulsatile signals. Like Lin et al. [23], our work shows that relative timing between TFs may be used by the cell to implement responses to different environments and therefore has to be taken into consideration for modelling gene expression patterns. However, while the authors in [23] suggest that the phenotypes differ in pulsatile regulation patterns, our study reveals the existence of long lived transients. From a dynamical system point of view, the effect of long lived transients that we present here can be seen through the prism of general theoretical frameworks such as proposed in [31, 42], where the authors discuss how to detect and automatically compute the meta-stable states from only the topology and timescales of the network; It would be interesting to see how precisely these methods could be used to detect the long lived transients we showcase in this paper.

Of relevance for synthetic biology, our construction based on looping suggests a way to implement memory units, though they may be leaky, in the sense that the signal is slowly being lost. In a broader context, cellular memory refers to systems whose present phenotype is dependent on the history of input stimuli and therefore the trajectory by which it has been reached [8]. The molecular mechanisms associated with such memory effects are usually based on feedback loops (e.g. the E. coli lac operon), DNA methylation patterns (e.g. temperate phage, pilus synthesis, cell differentiation) or inversions catalysed by site-specific recombinases (e.g. the Salmonella Hin system or the E. coli Fim system) [8, 29]. The long lived transient behaviour observed in our simulations differs from the mentioned memory mechanisms as it is purely relying on dynamical trapping of the transcriptional state. Different to the usual references to cellular memory, the long lived transients require no stabilisation of the phenotype through strong (covalent) modification of the DNA or any kind of feedback of the output on the promoter state (which is generally considered necessary for cellular memory).

The nature of the observed long lived transient states confer an epigenetic nature to these states. Methylation of histones is widely used in eukaryotic gene regulation as a modulator of gene activity that confers memory and stability to gene expression states. However, unlike methylation that requires a sleuth of specialised proteins that expend energy in order to form covalent bonds of methyl groups to histones, the long lived transients arise simply as a dynamical property of the system.

Acknowledgements. We are very grateful to Moritz Lang, Tiago Paixao and Jakob Ruess, for their feedback during the manuscript preparation.

Appendix 1.A Parameter Values

Table 1 lists the parameter ranges used for our case study example. We next explain the choice of each of the parameters with respect to their biological context.

1.A.1 Stochastic Scaling Constant
The stochastic scaling of rates and concentrations is done with a standard scaling rate for *E. coli* cell $N = 10^9$ [26].

1.A.2 Protein Production and Degradation
The protein production is taken $0.5\,\text{molec.s}^{-1}$ ([41], caption of Fig. 2) and the degradation rate is taken $0.001\,\text{s}^{-1}$ (corresponding to the halflife of 12 min, consistent with [26]).

1.A.3 RNAP Rates
On rate, off rate and number of RNAP molecules are consistent with the orders of values reported in [5,14,35].

1.A.4 Activator
The activation mechanism is inspired by the activation of the PRM promoter in the lysogenic state by protein CI in the regulation of λ-phage: CI competes with Cro to bind to the promoter sites, and, when bound, it recruits RNAP (increases PRM activity). The mechanism with looping, explained at mechanistic detail level in [35], contains three left and three right operators, leading to 1200 different DNA binding states. We model a mechanism with two states for the activator without looping ('bound' or 'not bound') and with four binding states for the

Table 1. Parameter combinations tested in the model with looping.

Parameter set	RNAP			A binding		R binding		unloop A	unloop R	# A	# R
ID[a]	on[b]	off	ifA	on	off	on	off	u_A	u_R	x_A	x_R
Reference set (p_1)	$10^4 N^{-1}$	$1.6 \cdot 0.01$	$9\times$	$8.8 \cdot 10^7 N^{-1}$	0.0264	$8.8 \cdot 10^7 N^{-1}$	0.016	1000	1000	275	350
increase unloop A (p_2)	–	–	–	–	–	–	–	–	100	–	–	–
increase unloop R (p_3)	–	–	–	–	–	–	–	–	–	100	–	–
decrease # A (p_4)	–	–	–	–	–	–	–	–	–	–	10	–
decrease # R (p_5)	–	–	–	–	–	–	–	–	–	–	–	10
downscale RNAP rates (p_6)	$10\times$	$10\times$	–	–	–	–	–	–	–	–	–	–

[a] The stochastic scalling of rates and concentrations is done with $N = 10^9$. The choice of this and other parameters is detailed in the main text of the appendix.
[b] All on-rates are given in units $\text{molec.}^{-1}\text{s}^{-1}$, off-rates in units s^{-1}. The unlooping rate is specified relative to the unbinding of the respective transcription factor - eg. it means that the unlooping rate is 1000 times smaller than the unbinding of the TF A. x_A, x_R are given in molecule numbers.

activator with looping (see Fig. 5). The on-rate, off-rate well as the number of activators is taken from [35] (page 82) When activator is bound, the recruitment of RNAP is increased by factor 10 or 50 ([34] and [27] respectivelly) (Table 3).

1.A.5   Repressor

The repression mechanism is inspired by the well-studied transcriptional regulation, there is a word missing after transcription of the lac operon, the repressor LacI. We take the binding and unbinding rates for the repressor from ([39], Fig. 4).

1.A.6   Looping Rates

The stability of the looped state is incorporated in the model by scaling down the unlooping rate. We choose the scaling factors of 100 and 1000 based on the computation of the ratio of dissociation rates for the models with and without looping ([40], Table 1; parameter a in [39]). The mechnism proposed in, eg. [39] suggests that the looping increases the binding rate (due to increased local concentration of TFs), while leaving the unbinding rate unchanged. As the scaled on-rates may exceed theoretical limit for diffusion-limited reactions, in our model, we incorporate the same effect by leaving the binding rate identical, and scaling down the unlooping rate.

Table 2. Parameter combinations tested for the model without looping.

Parameter set ID	RNAP			A binding		R binding		# A	# R
	on	off	ifA	on	off	on	off	x_A	x_R
Reference set ($\mathbf{p_1}$)	$10^4 V^{-1}$	$1.6 \cdot 0.01$	$49\times$	$8.8 \cdot 10^7 V^{-1}$	0.0264	$8.8 \cdot 10^7 V^{-1}$	0.016	275	10
low recruitment by A ($\mathbf{p_2}$)	$-$	$-$	$9\times$	$-$	$-$	$-$	$-$	$-$	$-$
increase # R ($\mathbf{p_3}$)	$-$	$-$	$-$	$-$	$-$	$-$	$-$	$-$	350
weak R binding ($\mathbf{p_4}$)	$-$	$-$	$-$	$-$	$-$	$-$	0.19	$-$	$-$
low recruitment by A, weak R binding ($\mathbf{p_5}$)	$-$	$-$	$9\times$	$-$	$-$	$-$	0.19	$-$	$-$
weak R binding, increase # R ($\mathbf{p_6}$)	$-$	$-$	$-$	$-$	$-$	$-$	0.19	$-$	350

Table 3. Simulation parameters: all models were run for three different delays and in six different regimes.

Input time	Total time	Time delays	Input schemes	Simulation points	# samples
5400 s	36000 s	60 s, 300 s, 900 s	$\{X_{basal}, X_{both}, X_A, X_R, X_{AR}, X_{RA}\}$	2000	1000

Appendix 1.B   Kappa Models

```
####### MODEL 1 (no looping) ############


#### Signatures
%agent: D(a,d) # Declaration of agent representing DNA with two binding
    sites: 'a' for
binding the activator and 'd' for binding the Polymerase or the repressor
%agent: RNAP(d) # Declaration of Polymerase with binding site named 'd'
%agent: A(a) # Declaration of actovatpr A with binding site named 'a'
%agent: R(d) # Declaration of repressor R with binding site named 'd'
%agent: P() # Declaration of protein P


#### Rules
# numbers after the '!' sign denote bond identifiers
# for bimolecular reactions, the rate is scaled with the average number
    of molecules in
the cell 'N' in order to convert from units 'per Mol per sec' to 'per
    molecule per sec')


# POLYMERASE
RNAP(d), D(d) -> RNAP(d!1), D(d!1) @ 'on_rnap' # RNAP binds, bimolecular
    reaction
RNAP(d!1), D(d!1) -> RNAP(d), D(d) @ 'off_rnap' # RNAP unbinds


# PROTEIN
RNAP(d!1), D(d!1) -> RNAP(d!1), D(d!1), P() @ 'p_on' # P is expressed
    when RNAP is bound
P() -> @ 'p_off'


# ACTIVATION
A(a), D(a) <-> A(a!1), D(a!1) @ 'on_a', 'off_a' # A binds to D
A(a!1), D(a!1,d), RNAP(d) -> A(a!1), D(a!1,d!2), RNAP(d!2) @
    'on_rnap_if_a' # A recruits
RNAP, that is, if A is bound, RNAP binds with larger affinity


# INHIBITION
R(d), D(d) <-> R(d!1), D(d!1) @ 'on_b', 'off_b' # repressor binds the
    'd' site of D;
Since 'd' is also the site for binding RNAP, when the repressor binds to
    site 'd', it
prevents the RNAP from binding and hence inhibits the protein expression.
```

```
#### Variables
## rates

%var: 'N' 10^9 # the average number of molecules in the cell
%var: 'on_rnap' 10^4/'N' # division because the reaction is bimolecular
%var: 'off_rnap' 1.6*0.01
%var: 'on_rnap_if_a' 49*'on_rnap'
%var: 'on_b' 8.8*10^7/'N'
%var: 'off_b' 0.016
%var: 'on_a' 8.8*10^7/'N'
%var: 'off_a' 0.0264
%var: 'p_on' 0.5
%var: 'p_off' 0.001
%var: 'rnap0' 1500
%var: 'a_add' 275
%var: 'b_add' 10

%var: 'p0' 240 # initial number of proteins
%var: 'b0' 0 # initial number of B molecules
%var: 'a0' 0

#### Observables

%obs: 'protein' P()
%obs: 'd_active' D(d!1),RNAP(d!1)

##### Perturbation

#%mod: [T]= 5400 do $ADD 'a_add' A(a)
%mod: [T]= 5400 do $ADD 'b_add' R(d)

#### Initial conditions
%init: 1 D(d,a)
%init: 'rnap0' RNAP(d)
%init: 'b0' R(d)
%init: 'a0' A(a)
%init: 'p0' P()
```

```
####### MODEL 2 (with looping) ############


#### Signatures
%agent: D(a1,a2,d,b2,loop~0~1) # Declaration of agent representing DNA
    with four binding
sites: 'a1' and 'a2' for binding the activators, 'd' and 'b2' for
    binding the repressor
(both) or Polymerase (site 'd'), and site 'loop' which indicates whether
    the DNA is looped
```

or not.
%agent: RNAP(d) # Declaration of Polymerase with binding site named 'd'
%agent: A(a) # Declaration of actovatpr A with binding site named 'a'
%agent: R(d) # Declaration of repressor B with binding site named 'd'
%agent: P() # Declaration of protein P

Rules
numbers after the '!' sign denote bond identifiers
for bimolecular reactions, the rate is scaled with the average number
    of molecules in
the cell 'N' in order to convert from units 'per Mol per sec' to 'per
    molecule per sec')

POLYMERASE
RNAP(d), D(d) -> RNAP(d!1), D(d!1) @ 'on_rnap' # RNAP binds, bimolecular
    reaction
RNAP(d!1), D(d!1) -> RNAP(d), D(d) @ 'off_rnap' # RNAP unbinds

PROTEIN
RNAP(d!1), D(d!1) -> RNAP(d!1), D(d!1), P() @ 'p_on' # P is expressed
    when RNAP is bound
P() -> @ 'p_off'

ACITIVATION

#A binds to the site 'a1' or site 'a2' of DNA whenever it is not looped
A(a), D(a1,a2,loop~0) <-> A(a!1), D(a1!1, a2, loop~0) @ 'on_a', 'off_a'
A(a), D(a1,a2,loop~0) <-> A(a!1), D(a1, a2!1, loop~0) @ 'on_a', 'off_a'

# lopping is immediate when the second activator binds
A(a!1), D(a1!1,a2,loop~0), A(a) <-> A(a!1), D(a1!1,a2!2,loop~1), A(a!2)
    @ 'loop_a', 'unloop_a'
A(a!1), D(a1,a2!1,loop~0), A(a) <-> A(a!1), D(a1!2,a2!1,loop~1), A(a!2)
    @ 'loop_a', 'unloop_a'

if A is bound to site 'a1', it recruits RNAP
A(a!1), D(a1!1,d), RNAP(d) -> A(a!1), D(a1!1,d!2), RNAP(d!2) @
    'on_rnap_if_a'

INHIBITION
R binds to site 'd' or site 'b2' of DNA whenever it is not looped
By binding to site 'd', repressor inhibits the binding of RNAP to the
    same site, and
hence inhibits the expression of the protein indirectly
R(d), D(d,b2,loop~0) <-> R(d!1), D(d!1,b2,loop~0) @ 'on_b', 'off_b'

```
R(d), D(d,b2,loop~0) <-> R(d!1), D(d,b2!1,loop~0) @ 'on_b', 'off_b'

# looping is immediate when the second repressor binds
R(d!1), D(d!1,b2,loop~0), R(d) <-> R(d!1), D(d!1,b2!2,loop~1), R(d!2) @
    'loop_b', 'unloop_b'
R(d!1), D(d,b2!1,loop~0), R(d) <-> R(d!1), D(d!2,b2!1,loop~1), R(d!2) @
    'loop_b', 'unloop_b'

#### Variables
## rates
%var: 'N' 10^9 # the average number of molecules in the cell
%var: 'on_rnap' 10^4/'N'
%var: 'off_rnap' 1.6*0.01
%var: 'on_rnap_if_a' 9*'on_rnap'
%var: 'on_b' 8.8*10^7/'N'
%var: 'off_b' 0.19
%var: 'on_a' 8.8*10^7/'N'
%var: 'off_a' 0.0264
%var: 'p_on' 0.5
%var: 'p_off' 0.001
%var: 'rnap0' 1500 # initial number of RNAP molecules
%var: 'a_add' 275
%var: 'b_add' 350
%var: 'unloop_a' 'off_a'/1000
%var: 'unloop_b' 'off_b'/1000

%var: 'p0' 240 # initial number of proteins
%var: 'b0' 0 # initial number of B molecules
%var: 'a0' 0

%var: 'loop_a' 'on_a'
%var: 'loop_b' 'on_b'

#### Observables

%obs: 'protein' P()
%obs: 'd_active' D(d!1),RNAP(d!1)
##### Perturbation

#%mod: [T]= 5400 do $ADD 'a_add' A(a)
%mod: [T]= 5400 do $ADD 'b_add' R(d)

#### Initial conditions
%init: 1 D(d,a1,a2,b2,loop~0)
%init: 'rnap0' RNAP(d)
%init: 'b0' R(d)
%init: 'a0' A(a)
%init: 'p0' P()
```

Appendix 1.C Supplementary Theory and Proofs

1.C.1   Deterministic Limit

In the continuous, deterministic model of a chemical reaction network, the state $\mathbf{z}(t) = (z_1, \ldots, z_n)(t) \in \mathbb{R}^n$ is represented by listing the concentrations of each species. The dynamics is given by a set of differential equations in form

$$\frac{d}{dt} z_i = \nu_{ij} \sum_{j=1}^{r} k_j \prod_{i=1}^{n} z_i(t)^{a_{ij}}, \tag{4}$$

where k_j is a deterministic rate constant, computed from the stochastic one and the volume N according to $k_j := c_j N^{|\mathbf{a}_j|-1}$ ($|\mathbf{x}|$ denotes the 1-norm of the vector $\mathbf{x}$). The deterministic model is a limit of the stochastic model when all species in a reaction network are highly abundant [19]: by scaling the species multiplicities with the volume: $Z_i(t) = X_i(t)/N$, adjusting the propensities accordingly, in the limit of infinite volume $N \to \infty$, the scaled process $\mathbf{Z}(t)$ follows an ordinary differential Eq. (4).

1.C.2   Expected Output in the Transient

The CME implies that the expectation of the marginal distribution of $\{X_t\}$ satisfies the equations

$$\frac{d}{dt} \mathsf{E}(\mathbf{X}_t) = \sum_{j=1}^{r} \boldsymbol{\nu}_j \mathsf{E}(\lambda_j(\mathbf{X}_t)). \tag{5}$$

To check (5), observe a transition from $\mathbf{x}$ to $\mathbf{x} + \boldsymbol{\nu}_j$. The term $\lambda_j(\mathbf{x}) \mathsf{P}(\mathbf{X}_t = \mathbf{x})$ appears exactly once when summing up for the state $\hat{\mathbf{x}} = \mathbf{x}$ as the outflow probability, and exactly once when summing up for the state $\hat{\mathbf{x}} = \mathbf{x} + \boldsymbol{\nu}_j$, as the inflow probability. This gives the term $(\mathbf{x} + \boldsymbol{\nu}_j) - \mathbf{x} = \boldsymbol{\nu}_j \cdot \lambda_j(\mathbf{x}) p^{(t)}(\mathbf{x})$. It is worth noting that, upon scaling the rate constants, the equations for $\mathsf{E}(\mathbf{X}_t)$ are equivalent to (4) only if all rate functions are linear, that is, when all reactions are unimolecular.

1.C.3   Proof for Lemma 1

We first notice that the process $\mathbf{X}(t)|_{\mathsf{D}}$ is indeed Markovian, because all states of $\mathbf{X}(t)$ projected to the same state in $\mathbf{X}(t)|_{\mathsf{D}}$ are behaviourally indistinguishable (bisimulation equivalent), due to rates between lumped states not depending on protein count. From (5), it follows that

$$\frac{d}{dt} \langle x_{\mathsf{P}}(t) \rangle = -\beta \langle x_{\mathsf{P}}(t) \rangle + \sum_{j=1}^{r} 1 \cdot \mathsf{E}(\alpha \cdot x_{\mathsf{D}1j}(t)) = -\beta \langle x_{\mathsf{P}}(t) \rangle + \alpha \sum_{\text{all } j} \langle x_{\mathsf{D}1j}(t) \rangle,$$

where $\langle x_{\mathsf{D}1j}(t) \rangle$ denotes the expected value of being in one of the active promoter configurations. The latter equals (2), since in every reachable state $\mathbf{x} \in (D_0 \uplus D_1) \times \{0, 1, \ldots\}$, exactly one DNA configuration takes value 1.

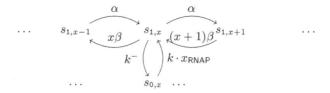

Fig. 2. Transitions of the CTMC underlying basal gene expression. The state space $S \cong \{0,1\} \times \{0,1,2,\ldots\}$, such that $s_{1,x}$ denotes an active configuration (where the RNAP is bound to the DNA) and $x \in \mathbf{N}$ protein molecules.

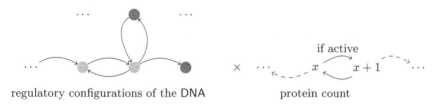

regulatory configurations of the DNA   protein count

Fig. 3. Each binding configuration of the DNA can be active (green, polymerase bound) or inactive (gray, polymerase not bound). Protein count can increase only when the DNA configuration is active. (Color figure online)

Appendix 1.D Supporting Figures

See Fig. 13.

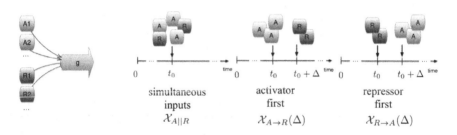

Fig. 4. Searching for long lived transients in gene regulation without feedback: three modelled scenarios. We demonstrate that small delays Δ can raise qualitatively different phenotypes, which are stable for cell lifetime.

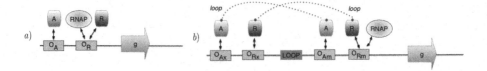

Fig. 5. Two prototype GRNs and their promoter logic: (a) Model without looping: regulatory architecture (promoter logic), (b) Model with looping: regulatory architecture (promoter logic). Mechanistic models are listed in (Appendix 1.B).

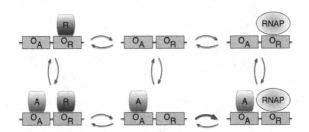

Fig. 6. Model without looping: the CTMC regulating six different DNA configurations. Thicker blue line denotes that the recruitment of RNAP is faster when the activator is bound. (Color figure online)

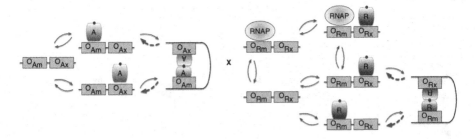

Fig. 7. Model with looping: CTMC regulating the DNA configurations has 23 different states. It is naturally represented as a composition of two sub-models: (left) the switching among configurations with respect to activator binding to its main and auxiliary binding sites (O_{Am} and O_{Ax} respectively, and (right) the switching among configurations with respect to repressor binding to its main and auxiliary binding sites (O_{Am} and O_{Ax} respectively. The unlooping rates (thicker blue lines) are typically much weaker than the TF unbinding. Any combination of the states in the two sub-models can be observed (reachable), except the state where both repressor and activator are looped. (Color figure online)

The effect of kinetic parameters on the shape and duration of transients for a model with looping

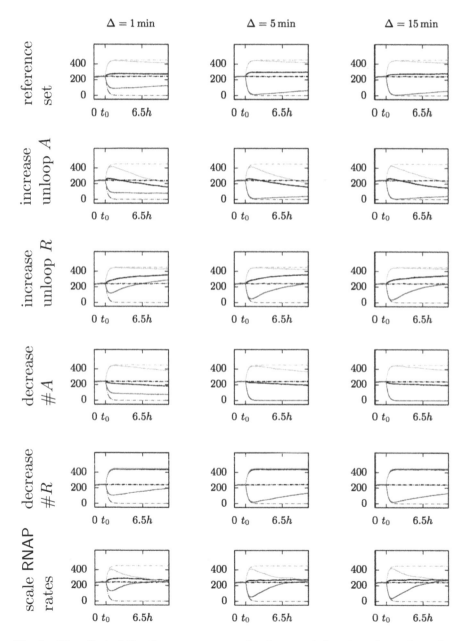

Fig. 8. The effect of kinetic parameters on the shape and duration of transients for a model with looping (for six parameter values listed in Table 1 and time delays of 1 min, 5 min and 15 min respectively).

Long lived transients in protein expression are due to long lived transients in promoter activity

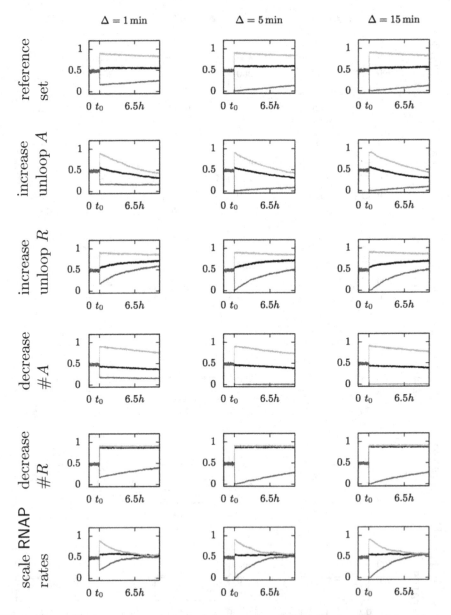

Fig. 9. Long lived transients in protein expression follow the long lived transients (mixing times) in promoter activity. We plot the statistically inferred probability of promoter logic being in the active state for six different parameter combinations, listed in Table 1 and time delays of 1 min, 5 min and 15 min respectively.

Long lived transients in protein expression are not observed in the model without looping

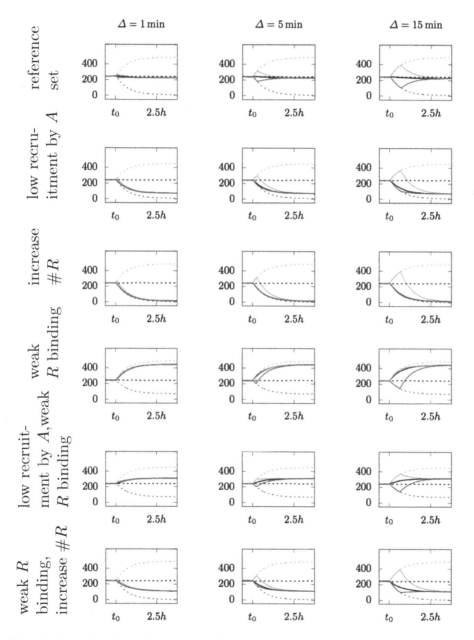

Fig. 10. Long lived transients are not observed in the model without looping (for six parameter values listed in Table 1 and time delays of 1 min, 5 min and 15 min respectively).

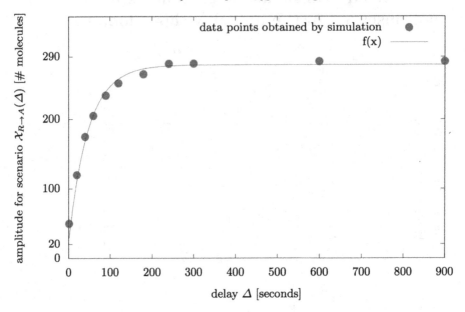

Fig. 11. In the reference parameter set, when there is no delay ($\Delta = 0$), the phenotype in scenario $\mathcal{X}_{R \to A}(\Delta)$ is equal to the one in scenario $\mathcal{X}_{A\|R}$, visibly different than the phenotype $\mathcal{X}_R$ (290 protein molecules). The difference of from the scenario $\mathcal{X}_{A\|R}$ (the characteristic we formally termed *amplitude* – see Sect. 5) exponentially grows as the delay increases, that is, it quickly approaches the phenotype of scenario $\mathcal{X}_R$. The difference of $\mathcal{X}_{R \to A}(\Delta)$ from $\mathcal{X}_R$ becomes observably negligible already for delays larger than $\Delta = 5\,\mathrm{min} = 300\,\mathrm{s}$ (difference of 10 molecules, 0.035% of the initial difference). We obtained the dependency by fitting the data obtained by simulating the system for $\Delta \in \{1, 20, 40, 60, 120, 180, 240, 300, 600, 900\}$.

Characteristics of long lived transients

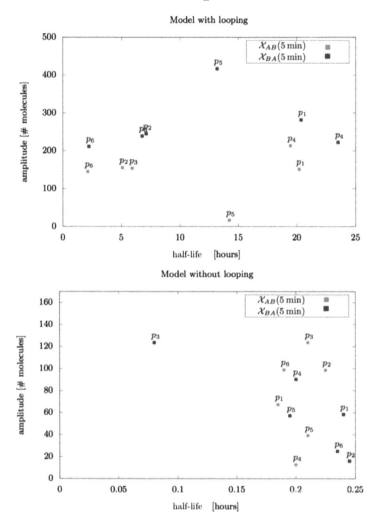

Fig. 12. For chosen parameter sets (Tables 1 and 2) and for a delay $\Delta = 5$ min, we plot the amplitude and the half-life (defined in Sect. 2).

Characteristics of long lived transients

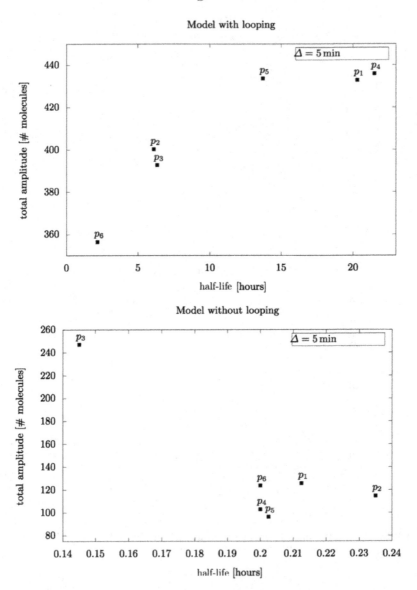

Fig. 13. We define a more global measure of the effect of delay in arrival times of TFs: first, the range of values that can be observed with a delay $\Delta = 5\,\text{min}$ by the measuring the maximum distance between phenotypes the scenarios $\mathcal{X}_{A \to R}$ and $\mathcal{X}_{R \to A}$ $\beta := \max_{t \geq t_0} |x_p(t|\mathcal{X}_{A \to R}) - x_p(t|\mathcal{X}_{R \to A})|$, and secondly, the halflife of this range $t_{1/2} := \arg\min_{t \geq t_\beta} \{t \mid |x_p(t|\mathcal{X}_{A \to R}) - x_p(t \mid \mathcal{X}_{R \to A})| < \frac{1}{2}\beta\}$.

References

1. Bintu, L., et al.: Transcriptional regulation by the numbers: applications. Curr. Opin. Genet. Devel. **15**(2), 125–135 (2005)
2. Bolouri, H., Davidson, E.H.: Modeling transcriptional regulatory networks. Bioessays: News Rev. Mol. Cell. Devel. Biol. **24**(12), 1118–1129 (2002)
3. Boutillier, P., et al.: The Kappa platform for rule-based modeling. Bioinformatics **34**(13), i583–i592 (2018)
4. Brim, L., Češka, M., Dražan, S., Šafránek, D.: Exploring parameter space of stochastic biochemical systems using quantitative model checking. In: Sharygina, N., Veith, H. (eds.) CAV 2013. LNCS, vol. 8044, pp. 107–123. Springer, Heidelberg (2013). https://doi.org/10.1007/978-3-642-39799-8_7
5. Brunner, M., Bujard, H.: Promoter recognition and promoter strength in the Escherichia coli system. EMBO J. **6**(10), 3139 (1987)
6. Buchler, N.E., Gerland, U., Hwa, T.: On schemes of combinatorial transcription logic. Proc. Nat. Acad. Sci. **100**(9), 5136–5141 (2003)
7. Byers, R., Hansell, R., Madras, N.: Stability-like properties of population models. Theor. Popul. Biol. **42**(1), 10–34 (1992)
8. Casadesús, J., D'Ari, R.: Memory in bacteria and phage. BioEssays **24**(6), 512–518 (2002)
9. Češka, M., Dannenberg, F., Paoletti, N., Kwiatkowska, M., Brim, L.: Precise parameter synthesis for stochastic biochemical systems. Acta Informatica **54**(6), 589–623 (2017)
10. Ceska, M., Šafránek, D., Dražan, S., Brim, L.: Robustness analysis of stochastic biochemical systems. PLoS ONE **9**(4), e94553 (2014)
11. Danos, V., Feret, J., Fontana, W., Harmer, R., Krivine, J.: Rule-based modelling of cellular signalling. In: Caires, L., Vasconcelos, V.T. (eds.) CONCUR 2007. LNCS, vol. 4703, pp. 17–41. Springer, Heidelberg (2007). https://doi.org/10.1007/978-3-540-74407-8_3
12. Gillespie, D.: Exact stochastic simulation of coupled chemical reactions. J. Phys. Chem. **81**, 2340–2361 (1977)
13. Guet, C.C., Elowitz, M.B., Hsing, W., Leibler, S.: Combinatorial synthesis of genetic networks. Science **296**(5572), 1466–1470 (2002)
14. Harada, Y., Funatsu, T., Murakami, K., Nonoyama, Y., Ishihama, A., Yanagida, T.: Single-molecule imaging of RNA polymerase-dna interactions in real time. Biophys. J. **76**(2), 709–715 (1999)
15. Hermsen, R., Tans, S., Ten Wolde, P.R.: Transcriptional regulation by competing transcription factor modules. PLoS Comput. Biol. **2**(12), e164 (2006)
16. Jacob, F.: Evolution and tinkering. Science **196**(4295), 1161–1166 (1977)
17. Jacob, F., Monod, J.: Genetic regulatory mechanisms in the synthesis of proteins. J. Mol. Biol. **3**(3), 318–356 (1961)
18. Kurtz, T.G.: Solutions of ordinary differential equations as limits of pure jump Markov processes. J. Appl. Probab. **7**(1), 49–58 (1970)
19. Kurtz, T.G.: Limit theorems for sequences of jump Markov processes approximating ordinary differential processes. J. Appl. Probab. **8**(2), 344–356 (1971)
20. Kwok, R.: Five hard truths for synthetic biology. Nature **463**(7279), 288–290 (2010)
21. Levin, D.A., Peres, Y., Wilmer, E.L.: Markov Chains and Mixing Times. American Mathematical Society (2009)

22. Levine, J.H., Lin, Y., Elowitz, M.B.: Functional roles of pulsing in genetic circuits. Science **342**(6163), 1193–1200 (2013)
23. Lin, Y., Sohn, C.H., Dalal, C.K., Cai, L., Elowitz, M.B.: Combinatorial gene regulation by modulation of relative pulse timing. Nature **527**(7576), 54–58 (2015)
24. Marchisio, M.A., Stelling, J.: Automatic design of digital synthetic gene circuits. PLoS Comput. Biol. **7**(2), e1001083 (2011)
25. McAdams, H.H., Arkin, A.: It's a noisy business! genetic regulation at the nanomolar scale. Trends Genet. **15**(2), 65–69 (1999)
26. Milo, R., Jorgensen, P., Moran, U., Weber, G., Springer, M.: Bionumbers–the database of key numbers in molecular and cell biology. Nucleic Acids Res. **38**(suppl 1), D750–D753 (2010)
27. Müller-hill, B.: Lac Operon. Wiley Online Library (1996)
28. Myers, C.J.: Engineering Genetic Circuits. CRC Press (2009)
29. Nashun, B., Hill, P.W., Hajkova, P.: Reprogramming of cell fate: epigenetic memory and the erasure of memories past. EMBO J. **34**(10), 1296–1308 (2015)
30. Ptashne, M.: A Genetic Switch: Phage Lambda Revisited, vol. 3. Cold Spring Harbor Laboratory Press Cold Spring Harbor, New York (2004)
31. Radulescu, O., Swarup Samal, S., Naldi, A., Grigoriev, D., Weber, A.: Symbolic dynamics of biochemical pathways as finite states machines. In: Roux, O., Bourdon, J. (eds.) CMSB 2015. LNCS, vol. 9308, pp. 104–120. Springer, Cham (2015). https://doi.org/10.1007/978-3-319-23401-4_10
32. Saiz, L., Rubi, J.M., Vilar, J.M.G.: Inferring the in vivo looping properties of DNA. Proc. Nat. Acad. Sci. U.S.A. **102**(49), 17642–17645 (2005)
33. Saiz, L., Vilar, J.M.: DNA looping: the consequences and its control. Curr. Opin. Struct. Biol. **16**(3), 344–350 (2006). Nucleic acids/Sequences and topology Anna Marie Pyle and Jonathan Widom/Nick V Grishin and Sarah A Teichmann
34. Saiz, L., Vilar, J.M.: Stochastic dynamics of macromolecular-assembly networks. Mol. Syst. Biol. **2**(1) (2006)
35. Santillán, M., Mackey, M.C.: Why the lysogenic state of phage λ is so stable: a mathematical modeling approach. Biophys. J. **86**(1), 75–84 (2004)
36. Schnoerr, D., Sanguinetti, G., Grima, R.: Approximation and inference methods for stochastic biochemical kinetics–a tutorial review. J. Phys. A: Math. Theor. **50**(9), 093001 (2017)
37. Segal, E., Widom, J.: From dna sequence to transcriptional behaviour: a quantitative approach. Nat. Rev. Genet. **10**(7), 443–456 (2009)
38. Shea, M.A., Ackers, G.K.: The OR control system of bacteriophage lambda: a physical-chemical model for gene regulation. J. Mol. Biol. **181**(2), 211–230 (1985)
39. Vilar, J.M., Leibler, S.: DNA looping and physical constraints on transcription regulation. J. Mol. Biol. **331**(5), 981–989 (2003)
40. Vilar, J.M., Saiz, L.: Dna looping in gene regulation: from the assembly of macromolecular complexes to the control of transcriptional noise. Curr. Opin. Genet. Devel. **15**(2), 136–144 (2005)
41. Vilar, J.M., Saiz, L.: Suppression and enhancement of transcriptional noise by DNA looping. Phys. Rev. E **89**(6), 062703 (2014)
42. Vivek-Ananth, R., Samal, A.: Advances in the integration of transcriptional regulatory information into genome-scale metabolic models. Biosystems **147**, 1–10 (2016)

43. Yuh, C.H., Bolouri, H., Davidson, E.H.: Cis-regulatory logic in the endo16 gene: switching from a specification to a differentiation mode of control. Devel. (Cambridge, England) **128**(5), 617–629 (2001)
44. Zeller, R.W., Griffith, J.D., Moore, J.G., Kirchhamer, C.V., Britten, R.J., Davidson, E.H.: A multimerizing transcription factor of sea urchin embryos capable of looping DNA. Proc. Nat. Acad. Sci. **92**(7), 2989–2993 (1995)
45. Zhu, R., Salahub, D.: Delay stochastic simulation of single-gene expression reveals a detailed relationship between protein noise and mean abundance. FEBS Lett. **582**(19), 2905–2910 (2008)

A Logic-Based Learning Approach to Explore Diabetes Patient Behaviors

Josephine Lamp[1]($\boxtimes$), Simone Silvetti[3], Marc Breton[2], Laura Nenzi[4], and Lu Feng[1]

[1] Department of Computer Science, University of Virginia, Charlottesville, VA, USA
{jl4rj,lu.feng}@virginia.edu
[2] Center for Diabetes Technology, University of Virginia, Charlottesville, VA, USA
mb6nt@virginia.edu
[3] Esteco S.p.A., Trieste, Italy
simone.silvetti@gmail.com
[4] University of Trieste, Trieste, Italy
lnenzi@units.it

Abstract. Type I Diabetes (T1D) is a chronic disease in which the body's ability to synthesize insulin is destroyed. It can be difficult for patients to manage their T1D, as they must control a variety of behavioral factors that affect glycemic control outcomes. In this paper, we explore T1D patient behaviors using a Signal Temporal Logic (STL) based learning approach. STL formulas learned from real patient data characterize behavior patterns that may result in varying glycemic control. Such logical characterizations can provide feedback to clinicians and their patients about behavioral changes that patients may implement to improve T1D control. We present both individual- and population-level behavior patterns learned from a clinical dataset of 21 T1D patients.

Keywords: Signal Temporal Logic · Learning · Type I Diabetes

1 Introduction

Type I Diabetes (T1D) is a chronic disease in which the body's ability to synthesize insulin is destroyed, as the patient's immune system attacks the insulin-producing cells of the pancreas [10]. Insulin is an important hormone used by cells to absorb glucose for energy production. 425 million people worldwide have Diabetes (Type I and Type II), including 1,106,500 children and adolescents living with T1D [15]. Intensive insulin therapy effectively reduces the risk of long-term complications of T1D (such as nerve or kidney damage) in which patients are required to inject or infuse insulin throughout the day to replace the normal pancreas function. Unfortunately, this means the burden of managing T1D falls to patients as they are required to manage a variety of behavioral factors (e.g., insulin injection, exercise, eating) that affect T1D. Studies have found that such factors affect a patient's overall glycemic control: e.g., exercise

© Springer Nature Switzerland AG 2019
L. Bortolussi and G. Sanguinetti (Eds.): CMSB 2019, LNBI 11773, pp. 188–206, 2019.
https://doi.org/10.1007/978-3-030-31304-3_10

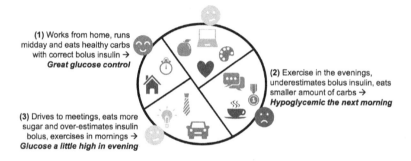

Fig. 1. Hypothetical patient behaviors resulting in different glycemic control outcomes.

may lower blood glucose values while carbs from meals increase blood glucose levels [1,23]. Figure 1 shows a set of hypothetical patient behaviors that may result in varying glycemic control. For example, on days when a patient exercises in the evening and underestimates the insulin absorption amount, they may have poor glycemic control (hypoglycemia) the next morning. Characterization of these behaviors can be used by clinicians to counsel their patients on strategies to optimize glycemic control using predictive recommendations (e.g., if you exercise late at night, make sure you eat a snack before you go to bed to avoid morning hypoglycemia). However, it is challenging to accurately identify T1D patient behavior patterns due to inherently messy patient data and the individual variability of patient behavior and physiology.

In this paper, we present a logic-based learning approach to address these challenges and explore T1D patient behaviors. Our approach takes advantage of the expressiveness and explainability of Signal Temporal Logic (STL) [19] and uses STL learning [21] to learn a set of STL formulas that characterize both individual- and population-level T1D patient behaviors. We argue that STL is a suitable representation of patient behavior patterns, because it can capture the temporal relations of Diabetes patient actions and glycemic outcomes. In addition, STL formulas are easily explainable to clinicians and patients. We apply our approach to learn STL formulas representing T1D patient behaviors from a clinical dataset including a variety of patient physiological and behavioral data, such as Continuous Glucose Monitors (CGM) sensor readings, heart rate, step count and activity intensity recorded by Fitbit, insulin pump injection records, self-reported meals and blood glucose finger pricks (SMBG). We envision that the learned STL formulas can provide clinically-relevant insights for clinicians and patients to develop behavioral change strategies to improve glycemic control.

The rest of the paper is organized as follows: Sect. 2 introduces the background of STL and learning techniques. Section 3 describes our approach to learn STL formulas for characterizing T1D patient behaviors. Sections 4 and 5 present our key findings about individual- and population-level patient behaviors, respectively. Section 6 summarizes related work, and Sect. 7 draws conclusions and discusses future research directions.

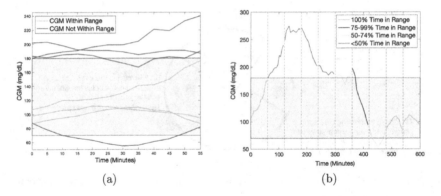

Fig. 2. (a) Example CGM trajectories that satisfy (green trajectories) or violate (red trajectories) the STL formula $\square(cgm \geq 70 \wedge cgm \leq 180)$. (b) An example illustrating the labeling mechanism of patient data. The CGM trajectory is chopped into several one-hour chunks divided by the vertical dashed blue lines. Each chunk is assigned with one of the four labels based on the percentage of time that the CGM value is within the target grey region. (Color figure online)

2   Preliminaries

In this section, we briefly introduce background on Signal Temporal Logic (STL) and STL learning techniques. Formally, the syntax of an STL formula φ is defined as follows:

$$\varphi ::= \mu \mid \neg\varphi \mid \varphi \wedge \varphi \mid \square_{(u,v)}\varphi \mid \Diamond_{(u,v)}\varphi \mid \varphi \mathbf{U}_{(u,v)}\varphi,$$

where μ is a signal predicate in the form of $g(\tau) > 0$ with a signal variable $\tau \in \mathcal{X}$ and function $g : \mathcal{X} \to \mathbb{R}$. The temporal operators $\square$, $\Diamond$, and $\mathbf{U}$ denote "always", "eventually," and "until", respectively. The bounded interval (u, v) denotes the time interval of temporal operators and can be omitted if the interval is $[0, +\infty)$. For example, we can specify a diabetes management rule "continuous glucose monitoring signal should always be between 70 and 180" [24] using a STL formula $\square(cgm \geq 70 \wedge cgm \leq 180)$.

The satisfaction of a formula is verified over a signal trajectory. For example, the formula $\square(cgm \geq 70 \wedge cgm \leq 180)$ can be verified over the time series of CGM signals shown in Fig. 2(a). STL considers two different semantics (Boolean and quantitative) to describe the satisfaction of a formula. The Boolean semantics checks if a trajectory satisfies a STL formula. For example, some CGM signals shown in Fig. 2(a) violate the STL formula $\square(cgm \geq 70 \wedge cgm \leq 180)$ because their CGM values go under 70 or above 180. The quantitative semantics returns a real-valued *robustness metric* that can be interpreted as a measure of the satisfaction [9]. Signal trajectories exhibiting weakening robustness with respect to a given property can be said to be moving toward a state of violation. We refer to [2,4,9,11,13] for a more detailed description of STL and its semantics.

STL learning provides techniques to infer STL formulae and parameters from signal trajectories. STL learning goes beyond property specification and allows for the automated identification of interesting behaviors that may not initially be apparent to the human eye. Nenzi et al. [21] present a STL learning method that learns the best set of STL formulas to discriminate between a two-label dataset of trajectories (e.g., regular and anomalous). This method uses a bi-level optimization process: it learns the STL formula structure using a discrete optimization of a genetic algorithm, and then synthesizes the parameters for the formulas using the Gaussian Process Upper Confidence Bound algorithm. We expand upon the STL learning tool developed in [21] to learn STL formulas representing T1D patient behaviors. However, since our clinical dataset has four labels (shown in Fig. 2(b)) rather than two labels, we need to adapt the tool for our problem. In addition, our goal is not to learn STL formulas that best discriminate between data with different labels. Instead, we are interested in learning STL formulas that can characterize patient behaviors that fall under the same label. We present our approach of learning STL formulas from T1D patient behaviors in the next section.

3   Methodology

We first describe the clinical dataset, then present our approach of learning individual- and population-level patient behaviors as illustrated in Fig. 3.

3.1   Clinical Dataset Description

Our dataset was collected during the observation period leading up to inpatient clinical trials in 2016–2017 at the Center for Diabetes Technology at the University of Virginia. The dataset contains 21 patients, ages ranging from 17 to 55, with an average age of 36 ± 10.4. Each patient has about 2 months of consecutive data. The data includes blood glucose readings from a Continuous Glucose Monitor (recorded in a variable named CGM), different types of insulin injections called boluses (total bolus, meal bolus, basal bolus, and correction bolus), meal carbs, patient-recorded blood glucose values from a finger prick (SMBG) and recordings of hypoglycemia (SMBG-Hypo). The data also contains exercise data recorded from a Fitbit including Heart Rate (HR), step count, calories, distance (in miles), and a Fitbit calculated activity level (in range of 1 to 4, with 1 being equivalent to little activity, and 4 being equivalent to intense activity.)

3.2   STL Learning for Individual Patient Behaviors

The approach of learning for individual behaviors is shown in Fig. 3 in the top yellow flowchart. We first pre-processed the data, and then added a multi-class labeling mechanism for our unlabeled patient data using CGM time in range, based on medical domain knowledge. Next, we fed our data and labels into our STL learning tool, to output STL formulas that classify specific patient behaviors. Finally, our results were validated for clinical insights. Each of these steps is explained in greater detail below.

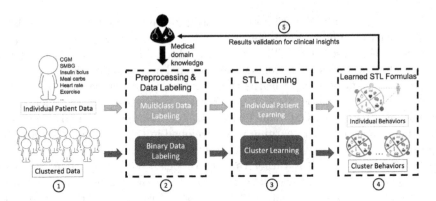

Fig. 3. Approach overview for learning STL formulas representing individual- (top yellow flowchart) and population-level (bottom blue flowchart) patient behavior patterns. (Color figure online)

Data Pre-processing. As our clinical data was messy and sampled at different rates, the first step in our methodology was to pre-process the data. We combined all data variables into a single file, and aligned them on a five minute sampling rate, to match the set sampling rate of the CGM. The variables HR and steps were sampled at more frequent rates (data was recorded a couple times per minute,) and we used a sliding average to compute the HR value, and summed the total steps in the time frame to align with each five minute interval. In addition, we added a detector to indicate when patients were exercising. For the purposes of our approach, we determined that a patient was exercising when the Fitbit Activity Level was ≥ 3, and/or when the patient had ≥ 3000 steps in 30 min, following approaches used to detect exercise in [5,20]. Finally, the data was layered into one hour time chunks to be fed into our STL Learning algorithm. We choose a one hour time chunk such that we would have enough data points (12 points) per layer for interesting learning to happen, but also small enough to provide detailed granularity within each patient's data.

Labeling. Next, we added labels for each one hour time chunk by hand. We used CGM Time in Range [12,18]—the percentage of time a patient spends in a well controlled blood glucose range (between 70 and 180 mg/dL)—as our labeling mechanism. This metric is commonly used by clinicians to determine how well controlled patients are, and as such served as an appropriate labeling mechanism for our data. We developed 4 sets of labels based on the total percentage of time the patient was in a well-controlled range: 100%, 75–99%, 50–74%, and <50% time in range. These thresholds were chosen based on clinical advice, and to evenly stratify the labels across our data points. Since our STL Learning algorithm cannot handle mutli-class labels, we had to create 4 label sets for each of these classes with binary indicators. In essence, for each label set, the hour chunk of data was given a positive label if it met the correct time in range (i.e. 100%) and a negative label if not. An example labeling scheme for a patient is shown in Fig. 2(a). For instance, for the first label class, for each one hour layer, if 100% of the CGM data points are in well controlled ranges then the label is a +1, and if it is anything else, then it is a −1 label. This is repeated for every hour chunk of the patient's data. For the second labeling class (75–99% label,) a +1 label is given if 75–99% of the CGM data points are in well controlled ranges for the hour time chunk, and −1 label if not, and so on for the rest of the data and label classes.

STL Learning and Validation. Once we had developed our four labeling classes, we fed our dataset and each of the four labeling sets into the STL learning tool [21] described in Sect. 2. For example, we fed the dataset with our 100% time in range labels, then with our 75–99% labels, etc. The tool works by generating formulas and picking the set that best separates our two classes (the positive and negative labelled classes). The tool then outputs these sets of formulas with the accuracy and misclassification rate (MCR). We define accuracy as $\frac{\text{True Positives} + \text{True Negatives}}{\text{Total}}$ and MCR as 1 - accuracy. In our case, we end up with 4 different final formula sets for each of our labeling classes. These formulas represent specific *rules* that classify particular patient behaviors with positive and negative labels. A formula is considered a good candidate for characterizing data with a given label if it separates the +1 and −1 classes with a high accuracy and low MCR. For instance, if we are classifying data using the 100% labels, a returned formula is good if it has a high percentage of data instances correctly classified in the positive label (+1, meaning they belong to the 100% class.)

3.3 STL Learning for Clustered Population Behaviors

The approach for learning population behaviors is shown in Fig. 3 in the bottom blue flowchart. First, we cluster the patient data into four population groups based on the overall percentage of time patients are in a well controlled CGM range. Next, the data is pre-processed and labeled, and the STL learning tool is used to learn formulas representative of our patient clusters. Four sets of formulas (for each of our clusters) are outputted and our results are validated for clinical insights at a population level.

Clustering. The first thing we did was divide our patient data into clusters based on how well controlled they were for the entire time period of data (approx. 2 months per patient), based on the average CGM time in range. We had 4 clusters, grouped by best controlled patients to worst: Cluster 1 had patients that were well controlled >79% of the time, Cluster 2 had patients that were well controlled 70–79% of the time, Cluster 3 had patients that were well controlled 60–69% of the time, and Cluster 4 had patients that were well controlled <60% of the time. We clustered patients in this way to ensure a relatively even distribution of patients per cluster (∼5 patients per cluster). Figure 4(a) shows a plot of the different patient clusters with the percentage of time their blood glucose is high (>180 mg/dL) vs the percentage of time their blood glucose is low (<70 mg/dL). We then pre-processed and chopped the data into 1 h time chunks, following the same methodology used for individual patients.

Labeling. Next, we labeled each of our four clusters using a binary methodology: For each hour time chunk, if patients were 75–100% controlled, a positive label

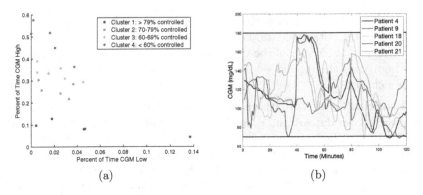

Fig. 4. (a) Clusters of Patient Data plotted for percentage of time patients are in a high blood glucose range (>180 mg/dL) vs in a low blood glucose range (<70 mg/dL), and (b) Sample patient trajectories of Cluster 1 (well controlled >79% of the time).

was added (+1), and if they were <75% a negative label was added (−1). It is important to note here that although patients were clustered based on their overall average time in range (i.e. Cluster 1 is for patients who had >79% average time in range,) the patients within each cluster are not always in those set ranges, and there may be periods where they are more or less controlled than their average. As a result, it is necessary to label each time chunk individually based on the actual percentage of time they are in range for that *specific time chunk*. For each cluster we generated one labeling set.

STL Learning and Validation. We then fed each cluster and its binary label set into the STL Learning tool individually, to output four formula sets, representative of each of the clusters patients' behaviors with accuracy and MCR metrics. Similar to the STL learning for individual patients, our outputted formulas are representative of *rules* that characterize the population level behaviors of the cluster. For example, Fig. 4(b) shows some sample CGM trajectories of Cluster 1 patients, and the learned STL formula that characterizes these trajectories is $\Box(cgm \geq 70 \wedge cgm \leq 180)$.

4   Learning Results for Individual Behaviors

In the following, we present our key findings about *personalized* STL formulas (rules) learned from individual patients' data using the methodology described in Sect. 3.2.

4.1   Personalized Bounds from Repeated Rules

One of the first interesting things we found were repeating formulas for different patients that had the same STL formula structure, but different personalized parameters, representative of patient bounds for specific variables. We identified repeated rules for CGM, HR, basal bolus and total bolus. The structure of such rules is shown as follows,

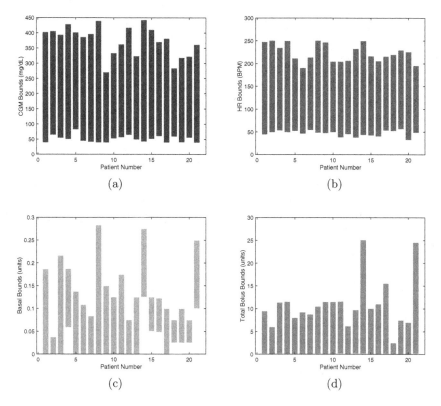

Fig. 5. Individual patient bounds for CGM (a), Heart Rate (b), Basal Bolus (c) &
Total Bolus (d) found from Repeated Rules (see Rule 1) for each patient's 2-months of
data.

$$\varphi = \Box_{[0,1]}(x \geq \alpha \wedge x \leq \beta) \tag{1}$$

where the time interval bound is within 1 h, x is the signal variable (e.g., cgm),
and α and β are parametric lower and upper bounds of the signal variable.
Figure 5 shows personalized parameter values learned for different patients. Since
these rules encompass the range of specific bounds patients have for different
data variables (i.e. CGM and HR), they are good classifiers for our data, and
therefore show up repeatedly for each patient. This is supported by the fact that
these rules generally (with the exception of HR bounds), have high accuracy
rates (see Table 5 in the Appendix). Since HR is highly variable even just for an
individual patient, it is not surprising that their accuracy is not extremely high.
However, we include the bounds in our results for all patients, since these rules
did show up repeatedly, and were accurate for some patients. We will explain the
significance and use of the personalized bounds for each specific variable next.

Identifying CGM bounds as in Fig. 5(a) allows for an understanding of the
range of blood glucose values patients may have within a specific time period (in
this case within a 2 month time period for our data). This may be relevant to note

to help clinicians tailor treatment options, especially if a patient consistently has very large CGM ranges over periods of many months: the clinician may find it useful to find out the source of such wide variability, as well as determine other options that might help the patient reduce such large hypo- or hyper-glycemic occurrences. Visualizing HR bounds as shown in Fig. 5(b) provides an overview of the minimum and maximum heart rate values a patient experiences. Although not clinically significant, it can be used as a quick, ballpark idea of patients' maximum heart rates, as well as their normal resting heart rates. One of the most interesting and clinically relevant bounds we are able to identify is personalized basal insulin bounds for patients, as shown in Fig. 5(c). These are very useful for determining the appropriate basal rates for individual patients. Currently, such bounds are estimated based on clinical expertise and then changed over time (using a guess-and-check method) after conferring with patients. Being able to determine the proper ranges for patients in an automated way over time is a great advantage of this approach, and can help clinicians and patients save time. The fourth type of bound we are able to identify is total bolus bounds, as shown in Fig. 5(d). These bolus amounts are also variable by patient, and include total meal, correction and basal bolus amounts. Although not quite as helpful as the basal bolus bounds, they still provide clinicians with an overview of the range of bolus amounts patients may have for certain time periods.

In addition, we also identified repeated rules for meal carbs and time bounds when eating occurs, as well as exercise intensity and timing on a patient-by-patient basis. These rules were identified across all 4 label classes by comparing learned rules from similar time periods. For instance, we compared all of the rules for a single patient in the morning time (i.e. between 7:00 and 11:00) and noticed repeating rules that identified meal and exercise times for that individual patient. These rules are defined as follows,

$$\varphi = \Diamond_{[\alpha,\beta]}(x \leq \kappa \wedge y \geq \lambda) \tag{2}$$

where α and β are the time bound parameters, x and y are the variables the bounds are generated for (meal, HR, steps, or activity level), and κ and λ are parameters. As mentioned before, we do not focus on the use of rules to discriminate between different classes, but rather on the *types* of behaviors we can classify within and across different label classes. In this case, these rules allow for an understanding of when patients are eating and exercising. As an example, we identified this rule for Patient 1 with an accuracy rate of 70.09%, indicating the patient consistently eats a meal of 10 to 65 carbs between 18:01 and 19:37.

$$\varphi = \Diamond_{[18:01,19:37]}(meal \leq 65 \wedge meal \geq 10) \tag{3}$$

In another example, we identified this rule for Patient 15 (accuracy 88%) indicating the patient consistently exercises at a moderate intensity level (Fitbit Activity Level indicator of 3 or above) between 17:59 and 19:00:

$$\varphi = \Diamond_{[17:59,19:00]}(HR \leq 212 \wedge activityLevel \geq 3) \tag{4}$$

4.2 Unique Formula Relationships

We also generated a variety of unique formulas that enable learning about the specific relationships between different variables (e.g. exercise and CGM) for individual patients. These relationships were identified across all our label classes, and as such may indicate behaviors resulting in better/worse glycemic control. However, in this section we only elucidate the *types* of rules we identify, and we will discuss the implications for good and bad control in Sect. 4.4.

Meals and CGM. We are able to identify relationships between eating and the resulting change in blood glucose values within all four of our labeling classes. An example formula for Patient 3 showing that the CGM changes within 35 min after the patient eats from the 75–99% class is shown below (MCR = 9.09%):

$$\varphi = ((meal \geq 1) \ \mathbf{U}_{[13:15,13:50]} \ (cgm \geq 120)) \tag{5}$$

These rules are useful to quickly understand how meals affect a patient's blood glucose and the specific time range these effects occur.

Meals and Meal Bolus. We can identify relationships for the amount of meal bolus given for various meals. For example, we find the following rule for Patient 11 (MCR 13.63%) in the 50–74% class, which states that they will have a meal bolus of $\leq$ to 0.8 units given a meal $\geq$ to 23 carbs:

$$\varphi = \Box_{[8:09,9:09]}(mealBolus \leq 0.8 \wedge meal \geq 23) \tag{6}$$

These rules provide some insight into the bolus levels for patients based on their carb amount. This is useful to understand to help patients tune and identify the correct amounts of bolus they should infuse based on the carbs they eat.

Exercise and CGM. Similar to meals and CGM, we can identify specific relationships about the effect exercise has on patient blood glucose levels. For example, we identify the following formula for Patient 1 (MCR = 17.143%) in the 100% class, which states that the patient's CGM value is greater than 120 mg/dL whenever the patient has an activity level of 3 or greater.

$$\varphi = \Box_{[20:31,21:14]}(cgm \geq 120 \wedge activityLevel \geq 3) \tag{7}$$

These rules are useful to quickly understand how exercise may affect a patient's blood glucose and the specific time range these effects occur.

Eating and Exercise. We can also identify different instances of eating and exercise. These include eating before exercise as shown in Rule 8 for Patient 20 (MCR = 0%) from the <50% class and eating during actual periods of exercise, as shown in Rule 9 for Patient 21 (MCR = 5%) from the <50% class.

$$\varphi = ((meal \geq 1) \ \mathbf{U}_{[18:17,18:32]} \ (activityLevel \geq 2)) \tag{8}$$

$$\varphi = \Diamond_{[12:52,13:07]}(activityLevel \geq 2 \wedge meal \geq 10) \tag{9}$$

These rules are interesting to identify as they provide insights for clinicians into the strategies specific patients use to help keep their blood glucose in the proper ranges before (or during) exercise. For instance, some may eat a small snack before they begin their workout to help prevent hypoglycemia, and others may begin their workout, then realize they are becoming hypoglycemic and eat a snack during a break in the workout to prevent this.

Exercise and Basal Bolus. Finally, we are able to identify basal adjustments before or during the start of exercise, such as the one shown in the rule below for Patient 11 (MCR = 7%) from the 75–99% class:

$$\varphi = ((basalBolus \leq 0.0345)\ \mathbf{U}_{[10:48,11:22]}\ (activityLevel \geq 3)) \tag{10}$$

Similar to the rules identified for eating and exercise, these rules provide insight into decisions patients make to manage their blood glucose before exercise.

4.3   Behavioral Interventions

We were able to generate rules that identify specific behavioral interventions patients engage in, across our four label classes. As a reminder, the focus of our approach is to characterize behaviors within labeled classes, and these rules provide insights into when patients are intervening in their T1D management, by double checking their blood glucose values and/or making corrections to their bolus levels. These rules are interesting as they indicate how proactive individuals are in monitoring and adjusting aspects of their glycemic control. Patients double check their blood glucose through a finger prick for SMBG values, and these formulas provide information about the circumstances under which patients may check their blood glucose. For instance, they may occur at regular time intervals, or around other events such as hyper- or hypo-glycemia as in Rule 11 (Patient 4, MCR = 0%), or before exercise as in Rule 12 (Patient 8, MCR = 14.5%).

$$\varphi = \Diamond_{[14:09,14:29]}(cgm \geq 195\ \&\ smbg \geq 200) \tag{11}$$

$$\varphi = ((smbg \geq 82)\ \mathbf{U}_{[10:36,11:59]}\ (activityLevel \geq 3)) \tag{12}$$

In addition, our rules identify correction bolus times and amounts, such as those in Rule 13 (Patient 13, MCR = 12.12%) and 14 (Patient 4, MCR = 5%).

$$\varphi = ((basalBolus \leq 0.04)\ \mathbf{U}_{[8:15,11:48]}\ (corrBolus \geq 0.459)) \tag{13}$$

$$\varphi = \Diamond_{[16:58,17:55]}(totalBolus \leq 2.105 \wedge corrBolus \geq 4.07) \tag{14}$$

4.4   Occurrences of Good and Bad Control

Using our unique relationships, we were able to identify specific instances that patient behaviors may have resulted in good or bad control, based on which class label the rule was identified in. We identified many different types of rules classifying these behaviors, but due to space constraints we provide 6 total rules

Table 1. Example rules capturing good and bad instances of control

Class label	Patient	Formula	MCR
Good: 100%	1	$\varphi = \Diamond_{[13:24,15:22]}(smbgHypo \geq 1 \wedge meal \geq 49)$	0%
Good: 75–99%	2	$\varphi = \Diamond_{[12:00,12:55]}(smbgHypo \geq 1 \wedge totalBolus \leq 7.18)$	0.2%
Good: 100%	21	$\varphi = ((activityLevel \leq 4)\ \mathbf{U}_{[10:36,11:59]}\ (corrBolus \geq 5.9))$	5%
Bad: 50–74%	5	$\varphi = \Diamond_{[15:00,17:41]}(cgm \leq 68 \wedge basalBolus \leq 0.011)$	13.18%
Bad: <50%	7	$\varphi = \Diamond_{[11:55,13:02]}(activityLevel \geq 4 \wedge cgm \leq 65)$	1.8%
Bad: <50%	15	$\varphi = ((meal \leq 44)\ \mathbf{U}_{[21:09,23:37]}\ (cgm \geq 210))$	6.36%

with their MCR in Table 1. For instance, in the case of good control we identified periods where patients were hypoglycemic and ate a meal (to raise their blood sugar,) were hyperglycemic and added a meal bolus (to lower blood sugar), and where the correct amounts of correction boluses were taken. For incidents of bad control, we identified periods where patients exercised but their blood glucose was too low (and no corrective actions were taken,) instances where incorrect bolus amounts for meals were taken and instances of incorrect basal or bolus adjustments. These rules are very helpful on a personalized level to help patients identify and correct behaviors that result in bad glycemic control.

4.5   Example Use Case

We next present a sample use case of our learned rules. Using the formulas generated for occurrences of good and bad control, we can identify the specific basal bolus amounts appropriate for different exercise intensity levels for a specific patient (i.e. Patient 21). Table 2 shows the minimum and maximum basal bounds for each activity level, and the Misclassification Rate for the good and bad formulas (MCR Good and MCR Bad), and the good and bad classification formulas used to derive each of the basal range bounds are shown below (the name of each φ indicates the label class and the activity level.) We define the 75–99% and 100% labels as the "good class" and the 50–74% and <50% labels as the "bad class". For instance, in the first row of Table 2 we can see that the basal range is between 0.066 and 0.072 for an activity level of 4. We reference Rule 15, that states that the basalBolus is below 0.072 units at the start of intense exercise (activity level 4,) and Rule 16, that states that bad control occurs when exercise activity level is 4 and the basal bolus is less than 0.065 (meaning we need a higher basal rate than this for good control.) From these we can derive the basal rate bounds: an upper rate bound of 0.072 from our good classification formula, and a lower bound of 0.066 from our bad classification formula.

Table 2. Proper basal ranges for exercise intensity for patient 21

Act. level	Basal range	Formulas used	$\mathbf{MCR}_{\text{Good class}}$	$\mathbf{MCR}_{\text{Bad class}}$
4	0.066–0.072	15, 16	14.84%	0%
3	0.073–0.077	17, 18	16.23%	10.12%
2	0.078–0.089	19	0%	N/A
1	0.09–0.1	20, 21	26.35%	1.8%

Formulas Used to Derive Table 2:

$$\varphi_{good4} = \Box_{[9:00,11:01]}(basalBolus \leq \mathbf{0.072})\mathbf{U}_{[9:10,11:01]}(activityLevel \geq 4) \quad (15)$$

$$\varphi_{bad4} = \Box_{[9:00,11:05]}(activityLevel \geq 4 \wedge basalBolus \leq \mathbf{0.065}) \quad (16)$$

$$\varphi_{good3} = \Box_{[9:00,11:00]}(activityLevel \leq 3 \wedge basalBolus \leq \mathbf{0.072}) \quad (17)$$

$$\varphi_{bad3} = \Box_{[9:02,10:59]}(activityLevel \geq 3 \wedge basalBolus \geq \mathbf{0.078}) \quad (18)$$

$$\varphi_{good2} = \Box_{[8:58,11:00]}(activityLevel \leq 2 \wedge basalBolus \leq \mathbf{0.089}) \quad (19)$$

$$\varphi_{good1} = \Box_{[8:55,10:57]}(activityLevel \leq 1 \wedge basalBolus \geq \mathbf{0.091}) \quad (20)$$

$$\varphi_{bad1} = \Box_{[8:55,11:05]}(basalBolus \leq \mathbf{0.122})\ \mathbf{U}_{[9:10,11:01]}(activityLevel \geq 1) \quad (21)$$

5   Learning Results for Population Behaviors

We now present results of population-level patient behaviors learned using the methodology in Sect. 3.3. There are several interesting key findings. First, the most controlled patients had the most number of SMBG occurrences (double checks of their blood glucose) as shown in Table 3. These occurrences were drawn from our STL formulas generated related to SMBG, an example of which is displayed in Rule 22. As mentioned before, Cluster 1 contains the best controlled patients, and Cluster 4 contains the worst controlled patients. This finding indicates that the best controlled patients double check their blood glucose much more frequently, which may result in better overall control of their T1D. This makes sense, because patients who are more actively engaged in verifying the status of their blood glucose (and other factors of their glycemic control,) are more proactive in making the necessary changes (i.e. adding a correction bolus) in order to ensure their blood glucose stays within the proper ranges. Alternatively, patients who have worse control tend to check their blood glucose values less often, meaning they may not be as aware of specific blood glucose changes that require some adjustment to the management of their T1D. The following is an example SMBG rule used to derive Table 3 for a 24 h time period for Cluster 1 (accuracy = 100%):

$$\varphi = \Diamond_{[12:00,12:00]}(smbg \geq 55 \wedge cgm \leq 400) \quad (22)$$

Table 3. Average SMBG count by cluster

Cluster number	Average count of SMBG checks
1 (best controlled)	85.00
2	68.80
3	59.67
4 (worst controlled)	50.60

Table 4. Average number and amount of correction boluses by cluster

Cluster number	Number of correction boluses	Correction bolus amount
1 (best controlled)	8.80	17.14
2	11.80	18.16
3	12.17	23.98
4 (worst controlled)	14.80	32.30

Second, from our rules we identified that as we go from the best controlled cluster (Cluster 1) to the worst (Cluster 4), we have an increased count of correction boluses per patient. This is shown in the second column of Table 4, and some sample rules that we derived these values from is shown in Rule 23. Moreover, not only do patients with worse control have an increased count of the correction boluses, they also have an increased average *amount* of actual correction bolus units taken per correction bolus occurrence. This is shown in the third column of Table 4. These findings indicate that patients who have worse control tend to need to correct their bolus levels more often, and change (i.e. increase) their actual correction bolus amounts more drastically than better controlled patients. These findings also make sense, because less controlled patients may take more of a reactive approach, (e.g. they only intervene in their control when a specific incident such as hyper- or hypo-glycemia occurs), resulting in an increased need to correct their bolus levels, and by larger unit amounts at each intervention. The following is an example Correction Bolus rule used to derive Table 4 for a 24 h time period for Cluster 4 (accuracy $= 100\%$):

$$\varphi = \Diamond_{[23:59,23:59]}(corrBolus \geq 10) \tag{23}$$

We were not able to identify any other specific formulas that made sense and that provided a good characterization between the clusters. Although different rules relating CGM or exercise to other components (i.e. basal bolus) were generated, these rules cannot be used for the entire cluster population. These types of formulas and their parameters should be very specific to individuals, and therefore cannot be generalized, even across a small cluster of patients.

6   Related Work

Learning Diabetes Patient Behaviors. A couple of works have looked at learning patient behaviors for T1D patients at a population level. Chen et al. [8] developed an "eat, trust, check" framework to model and evaluate patient insulin pump behaviors using a machine learning approach. Hoyos et al. [14] used an incremental learning approach to infer the behavior of autonomous glucose measurements and parameters for population groups of T1D patients. In addition, Cameron et al. [6] developed a model predictive controller for regulating blood glucose based on cgm readings and meal behaviors, and Paoletti et al. [22] presented a model predictive controller to administer insulin based on patient behavior (i.e. meal and exercise events). These approaches do not include behavior types beyond meals/exercise as our approach does (such as our SMBG checks) and do not employ STL Learning, so they are not able to express the range of different behaviors and personalized level of formulas that our methodology can. Moreover, Chatterjee et al. [7] designed a sensor-based at home system for T1D patients that records patient activity throughout the day to promote patient behavior change. This approach provides alerts about more high-level activities (eating and sedentary behavior), and does not provide as specific of information (such as about behavioral interventions related to SMBG) as in our approach.

STL Learning for Behavior Detection. In terms of STL Learning, a variety of papers have developed new methodologies to learn STL formula structures and their parameters for anomaly detection and behavior identification in applications such as naval surveillance and medical contexts. Kong et al. [17] developed an offline supervised learning approach that uses machine learning to detect anomalous and normal behaviors. Formula structures and parameters are synthesized using a gradient descent optimization guided by robustness and hinge loss functions in their machine learning algorithm. This work suffers from a long computational complexity, due to the time needed to optimize for the graph structure and a lack of explainability due to the ML algorithm. Our approach is explainable and facilitates greater clinician trust in the outcome of our results. In addition, Klimek [16] and Bombara et al. [3] used tree structures to generate their STL formulas and parameters. Klimek employed an online learning approach in which graph models were used to reason about objects and events, and logical truth trees were outputted to represent the formulas and their behavioral meanings. Bombara et al. use a decision tree framework and a misclassification rate optimization method to build binary decision trees representative of STL formulas and their parameters to categorize anomalous vs normal behaviors. The strict structure of the tree algorithms imposes some restrictions on the flexibility and diverse types of STL formula structures that can be outputted. As a result, these structures are not optimal choices for T1D patient data, as they lack the expressivity needed to classify diverse patient behaviors. Moreover, the formulas generated from the decision tree are long and not very human readable.

7   Discussion and Conclusion

Conclusion. In this paper, we presented an approach to learn STL formulas that characterize individual- and population-level T1D patient behaviors with varying glycemic control and applied it to a clinical dataset with 21 T1D patients' data. Our learning results provide some clinically-relevant insights for clinicians and patients to develop behavioral change strategies to improve glycemic control.

Tool Limitations. Our results are constrained by the limitations of the STL learning tool [21] in several ways. First, our patient data contain many null values. For example, patients only eat at discrete time periods, and times the patient was not eating were null. However, the tool cannot handle null values, so we had to fill all of these instances with zeros: this changes the semantic meaning of the data points, and may cause a bias in how the formula parameters are being generated (for instance, when data points are averaged to get specific parameter bounds). Second, since the tool cannot handle multi-class classification, we had to use four different sets of labels with binary indicators to cover our different classes. This may have caused some overlap in our resulting formulas. In addition, since the tool relies on a supervised classification approach, we had to supply labels to guide the learning. However, this may have resulted in missing some behavior sets that still have an effect on T1D glycemic control (but may not have a direct relationship with CGM time in range). Moreover, the tool relied on having an evenly split distribution of data labels, which proved challenging for our unevenly distributed patient data. Finally, the tool can only learn from raw data streams for short time periods, (and can not, for instance, calculate CGM rate of change or other advanced relationships), and as such we were only able to learn fairly simple rules for short time chunks. As a result, we were unable to study longer term T1D effects (e.g. multiple hour meal-bolus relationships).

Future Work. We would like to address the limitations and improve upon the capabilities of the tool, as well as integrate our patient behavior identification approach into a closed loop feedback system (e.g., implemented in a smartphone application or other wearable), which will provide real-time feedback about behaviors that have negative impact on glycemic control.

Acknowledgements. The authors would like to graciously thank the UVA Center for Diabetes Technology for providing the clinical datasets and Basak Ozaslan, Jack Corbett, Jonathan Hughes and Dr. José García-Tirado for their clinical insights and valuable discussions. Research partially supported by the Austrian National Research Networks RiSE/ShiNE (S11405) and ADynNet (P28182) of the Austrian Science Fund (FWF).

Appendix

Table 5. Accuracy rates for repeated rules

Patient	CGM %	HR %	Basal %	Bolus %
1	88.61	51.25	86.94	90
2	93.88	97.05	93.24	100
3	90.58	77.02	100	93.72
4	96.10	56.25	98.48	98.48
5	87.10	94.17	100	95.28
6	88.19	51.25	100	97.92
7	93.61	83.55	96.26	94.40
8	87.08	78.89	99.72	100
9	88.09	88.78	100	93.77
10	95.45	97.47	100	95.36
11	86.76	86.46	100	87.05
12	93.75	55.17	100	95.63
13	95.59	61.25	100	96.12
14	94.40	79.33	96.38	100
15	86.86	93.50	88.24	91.29
16	89.38	75.47	90.90	89.13
17	87.38	100	100	93.07
18	89.54	71.09	90.66	89.65
19	90.29	63.38	90.15	89.88
20	89.54	62.43	91.11	89.99
21	86.86	66.99	89.88	88.35

References

1. American Diabetes Association: 13. children and adolescents: standards of medical care in diabetes–2019. Diab. Care **42**(Suppl. 1), S148–S164 (2019)
2. Bartocci, E., Bortolussi, L., Sanguinetti, G.: Data-driven statistical learning of temporal logic properties. In: Legay, A., Bozga, M. (eds.) FORMATS 2014. LNCS, vol. 8711, pp. 23–37. Springer, Cham (2014). https://doi.org/10.1007/978-3-319-10512-3_3
3. Bombara, G., Vasile, C.I., Penedo, F.: A decision tree approach to data classification using signal temporal logic, pp. 1–10 (2016)
4. Bufo, S., Bartocci, E., Sanguinetti, G., Borelli, M., Lucangelo, U., Bortolussi, L.: Temporal logic based monitoring of assisted ventilation in intensive care patients. In: Margaria, T., Steffen, B. (eds.) ISoLA 2014. LNCS, vol. 8803, pp. 391–403. Springer, Heidelberg (2014). https://doi.org/10.1007/978-3-662-45231-8_30

5. Bumgardner, W.: The average steps per minute for different exercises. https://www.verywellfit.com/pedometer-step-equivalents-for-exercises-and-activities-3435742

6. Cameron, F., Niemeyer, G., Bequette, B.W.: Extended multiple model prediction with application to blood glucose regulation. J. Process Control **22**(8), 1422–1432 (2012)

7. Chatterjee, S., Byun, J., Dutta, K., Pedersen, R.U., Pottathil, A., Xie, H.: Designing an Internet-of-Things (IoT) and sensor-based in-home monitoring system for assisting diabetes patients: iterative learning from two case studies. Eur. J. Inf. Syst. **27**(6), 670–685 (2018)

8. Chen, S., Feng, L., Rickels, M.R., Peleckis, A., Sokolsky, O., Lee, I.: A Data-Driven Behavior Modeling and Analysis Framework for Diabetic Patients on Insulin Pumps Recommended Citation, Technical report (2015). http://repository.upenn.edu/cis_papersrepository.upenn.edu/cis_papers/791

9. Deshmukh, J., Donzé, A., Ghosh, S., Jin, X., Juniwal, G., Seshia, S.: Robust online monitoring of signal temporal logic, pp. 1–26, July 2017

10. Prevention: Type 1 diabetes for Disease Control, C.C., August 2018. https://www.cdc.gov/diabetes/basics/type1.html

11. Donzé, A., Maler, O.: Robust satisfaction of temporal logic over real-valued signals. In: Chatterjee, K., Henzinger, T.A. (eds.) FORMATS 2010. LNCS, vol. 6246, pp. 92–106. Springer, Heidelberg (2010). https://doi.org/10.1007/978-3-642-15297-9_9

12. Fabris, C., Patek, S.D., Breton, M.D.: Are risk indices derived from CGM interchangeable with SMBG-based indices? J. Diab. Sci. Technol. **10**(1), 50–59 (2016)

13. Fainekos, G.E., Pappas, G.J.: Robustness of temporal logic specifications for continuous-time signals. Theor. Comput. Sci. **410**(42), 4262 – 4291 (2009). https://doi.org/10.1016/j.tcs.2009.06.021. http://www.sciencedirect.com/science/article/pii/S0304397509004149

14. Hoyos, J.D., Bolanos, F., Vallejo, M., Rivadeneira, P.S.: Population-based incremental learning algorithm for identification of blood glucose dynamics model for type-1 diabetic patients. In: Proceedings on the International Conference on Artificial Intelligence (ICAI), pp. 29–35. The Steering Committee of The World Congress in Computer Science, Computer (2018)

15. IDF: IDF diabetes atlas 8th edition 2017 (2017). https://diabetesatlas.org/

16. Klimek, R.: Behavior recognition and analysis in smart environments for context-aware applications, October 2015 (2016). https://doi.org/10.1109/SMC.2015.340

17. Kong, Z., Jones, A., Belta, C.: Temporal logics for learning and detection of anomalous behavior. IEEE Trans. Autom. Control **62**(3), 1210–1222 (2017). https://doi.org/10.1109/TAC.2016.2585083

18. Kovatchev, B.P.: Metrics for glycaemic control-from HbA 1c to continuous glucose monitoring. Nat. Rev. Endocrinol. **13**(7), 425 (2017)

19. Maler, O., Nickovic, D.: Monitoring temporal properties of continuous signals. In: Lakhnech, Y., Yovine, S. (eds.) FORMATS/FTRTFT -2004. LNCS, vol. 3253, pp. 152–166. Springer, Heidelberg (2004). https://doi.org/10.1007/978-3-540-30206-3_12

20. Marshall, S.J., et al.: Translating physical activity recommendations into a pedometer-based step goal: 3000 steps in 30 minutes. Am. J. Prev. Med. **36**(5), 410–415 (2009)

21. Nenzi, L., Silvetti, S., Bartocci, E., Bortolussi, L.: A robust genetic algorithm for learning temporal specifications from data. In: McIver, A., Horvath, A. (eds.) QEST 2018. LNCS, vol. 11024, pp. 323–338. Springer, Cham (2018). https://doi.org/10.1007/978-3-319-99154-2_20

22. Paoletti, N., Liu, K.S., Smolka, S.A., Lin, S.: Data-driven robust control for type 1 diabetes under meal and exercise uncertainties. In: Feret, J., Koeppl, H. (eds.) CMSB 2017. LNCS, vol. 10545, pp. 214–232. Springer, Cham (2017). https://doi.org/10.1007/978-3-319-67471-1_13
23. Riddell, M.C., et al.: Exercise management in type 1 diabetes: a consensus statement. Lancet Diab. Endocrinol. **5**(5), 377–390 (2017). https://doi.org/10.1016/S2213-8587(17)30014-1
24. Young, W., Corbett, J., Gerber, M.S., Patek, S., Feng, L.: DAMON: a data authenticity monitoring system for diabetes management. In: 2018 IEEE/ACM Third International Conference on Internet-of-Things Design and Implementation (IoTDI), pp. 25–36. IEEE (2018)

Reachability Design Through
Approximate Bayesian Computation

Mahmoud Bentriou, Paolo Ballarini$^{(\boxtimes)}$, and Paul-Henry Cournède

MICS, CentraleSupélec, Université Paris-Saclay, 91190 Gif-sur-Yvette, France
{mahmoud.bentriou,paolo.ballarini,paul-henry.cournede}@centralesupelec.fr

Abstract. Time-bounded reachability problems are concerned with assessing whether a model's trajectories traverse a given region of the state-space within given time-bounds. In the case of stochastic models reachability is associated with a measure of probability which depends on the model's parameters. In this paper we propose a methodology that, given a reachability specification (for a parametric stochastic model), allows for computing a *reachability related* probability distribution on the parameter space, i.e. a distribution that allows for identifying regions of the parameter space for which there is a non-null probability to match the considered reachability specification. The methodology relies on the characterisation of *distance* between a model's trajectory and a reachability specification which we show being assessable by using a hybrid automaton as a monitor of a model's trajectory. An automata-based adaptation of the Approximated Bayesian Computation method is then introduced to estimate the reachability distribution on the parameter space.

Keywords: ABC methods · Hybrid automata · Parameter estimation

1 Introduction

Approximate Bayesian computation (ABC) algorithms have gained in popularity over the last decade and are applied for parameter inference in many modeling fields, including systems biology [16,21,26,29] and cancer research [24]. They proved powerful in many cases when classical Bayesian parameter inference methods are difficult to implement. A first formulation can be found in the population genetics field [25] which was motivated by the dimension of the studied models. ABC permits to approximate the posterior distribution of a model without evaluating the likelihood function in complex models, when the computation cost is too high or even impossible. ABC methods are likelihood-free and only rely on model simulations: simply speaking, only parameters for which simulated summary statistics are close to observed ones are preserved while the others are dismissed. This mechanism, generally used iteratively, allows for progressively converging towards regions of the parameter space with higher probability density.

© Springer Nature Switzerland AG 2019
L. Bortolussi and G. Sanguinetti (Eds.): CMSB 2019, LNBI 11773, pp. 207–223, 2019.
https://doi.org/10.1007/978-3-030-31304-3_11

The initial idea of our work relies on the similarity of this concept with that of logic driven parameter inference, i.e. an emerging area notably in systems biology. Specifically, parameter synthesis driven by temporal logic aims at identifying the regions of a model's parameter space that better fulfill a given temporal logic specification [13, 14]. We propose a likelihood-free ABC algorithm that relies on general hybrid automata to constrain the particle selection in the original ABC algorithm in order to ensure that specific logical properties of the system are satisfied by the simulations. We show that sequential versions of ABC can speed up the exploration of the subset of parameters where a temporal logic specification is satisfied thanks to a formal definition of the distance of a trajectory from the property. The paper is organised as follows: Sect. 2 introduces background material about stochastic models, reachability problems, the hybrid automata specification language and ABC statistical methods. In Sect. 3 the notion of distance of a trajectory from a spatio-temporal region is introduced and a novel ABC framework based on such distance measure is developed aimed at finding the parameter subspace such that the probability of reaching this region is positive. The novel ABC framework is demonstrated through a number of experiments in Sect. 4, while some conclusive remarks and future perspectives are discussed in Sect. 5.

2   Background

We briefly introduce the background material the remainder of the paper relies upon, namely: the basics about the class of continuous-time Markov chain models, the basics about temporal logic and reachability problems, the basics about the Hybrid Automata Specific Language and the ABC method (whose automata-extension we introduce in Sect. 3).

2.1   Continuous-Time Markov Chains

We consider continuous-time Markov chains (CTMCs) [17] as a framework for modelling networks of biochemical reactions. A CTMC $\mathcal{M}$ is a kind of stochastic process which accounts for dense time elapsing and that enjoys the *memoryless property*, i.e. the probability of observing a transition from a source state to a target state (within a given delay) depends entirely on the source state and not on the *history* that led to it. A CTMC model for a biochemical network consisting of n species interacting through m reaction channels is characterised by:

- A state space $S \subseteq \mathbb{N}^n$ whose elements are vectors $\mathbf{X} = [X_1, \ldots, X_n] \in S$ where X_i is the population, in terms of number of molecules, of the i-th species. State space S is associated with an initial state probability distribution $\pi_0 : S \to [0, 1]$ which, whenever π_0 concentrates the probability mass in a single state $s_0 \in S$, is simply denoted s_0.
- A set $R_1, \ldots, R_m$ of reaction channels where each R_j is characterised by a pair $R_j : (\nu_j, \eta_j)$ with $\nu_j = [\nu_{1j}, \ldots, \nu_{nj}]$ the *stoichiometric vector*, representing

the amount of change on each species determined by the occurrence of R_j, and $\eta_j = \eta_j(\mathbf{X}, \theta)$ the kinetic rate expressing the rate of an exponential distribution governing the occurrence of R_j as a function of the state $\mathbf{X}$ and of the parameters θ of the model.

– a d-dimensional vector of parameters $\theta = [\theta_1, \ldots, \theta_d]$ which affect the kinetic rate of the reaction channels.

Observe that the kinetic rate of a reaction channel i.e. $\eta_j = \eta_j(\mathbf{X}, \theta)$ depends both on the state $\mathbf{X}$ of the CTMC as well as on the parameters θ. To highligh the fact that the dynamics underlying a given CTMC model $\mathcal{M}$ may vary considerably depending on the considered value of θ we adopt the notation $\mathcal{M}_\theta$.

Paths of a CTMC. Given a CTMC model $\mathcal{M}_\theta = (S, R, s_0)$ we denote $Path_{\mathcal{M}_\theta}(s_1)$ the set of (possibly infinite) paths originating in state $s_1 \in S$ where a path from s_1 is a (possibly infinite) sequence $\sigma = s_1 \xrightarrow{t_1} s_2 \xrightarrow{t_2} \ldots \xrightarrow{t_{n-1}} s_n \ldots$ with $t_i \in \mathbb{R}_{>0}$ being the sojourn-time in state $s_i \in S$. For $\sigma \in Path_{\mathcal{M}_\theta}(s_1)$ a path, $i \in \mathbb{N}$ and $t \in \mathbb{R}_{>0}$, we denote $\sigma[i] = s_i$ the i-th state of σ, $\delta(\sigma, i) = t_i$ the sojourn-time of σ in the i-th state and $\sigma@t$ the state of σ at time t. It can be easily shown that a CTMC model $\mathcal{M}_\theta$ induces a probability mesure on the space of the trajectories $Path_{\mathcal{M}_\theta}(s_0)$. A measurable subset of trajectories of $Path_{\mathcal{M}_\theta}(s_0)$ may be referred to as an event of $\mathcal{M}_\theta$. Notice that trajectories of a CTMC are càdlàg (i.e. step) functions of time.

2.2   Temporal Logic and Reachability Problems

In temporal logic reasoning [3,9] the term *reachability problem* identifies the class of problems interested in establishing whether a given model *reaches* (i.e. enters), at some point during its execution, a certain region of its state-space usually associated with some state condition φ. If the considered model inherently quantifies time elapsing (like with CTMCs) then one may also consider *time-bounded reachability* whereby the focus is on establishing whether the desired region of the state-space is entered within a time-interval $[t_1, t_2] \subset \mathbb{R}_{\geq 0}$. Temporal logic formalisms are equipped with operators for expressing reachability problems. Here we briefly recall the basics of the Metric Interval Temporal Logic (MITL [22]), one of many temporal logic languages (e.g. [2,18]) that allow for stating time-bounded reachability problems for CTMC models.

MITL Temporal Logic. MITL formulae are terms of the following grammar:

$$\varphi ::= \top \mid \mu \mid \neg\varphi \mid \varphi_1 \wedge \varphi_2 \mid \varphi_1 \, \mathbf{U}^{[t_1, t_2]} \, \varphi_2$$

where $\top$ stands for the **true** formula, μ denotes an atomic proposition (i.e. an inequality built on top of model's state-variables), $\neg$ and $\wedge$ are the basic negation and conjunction connectives of propositional logic and $\mathbf{U}^{[t_1, t_2]}$ is the time-bounded *until* temporal operator with $[t_1, t_2] \subseteq \mathbb{R}_{\geq 0}$ being the bounding interval. The truth of a MITL formula is defined w.r.t. to a function of time, such as, e.g. a path σ of a CTMC model. For t an instant of time we say that $\sigma@t$ satisfies φ, denoted $\sigma@t \models \varphi$. For example a time-bounded until formula is satisfied from time t of path σ, denoted $\sigma@t \models \varphi_1 \mathbf{U}^{[t_1,t_2]} \varphi_2$, if and only if there exists $t' \in [t_1, t_2]$ such that $\sigma@(t + t') \models \varphi_2$ and $\forall t'' < t', \sigma@(t + t'') \models \varphi_1$. As usual we consider two derivations of the time-bounded until operator: the time-bounded *eventuality* $\mathbf{F}^{[t_1,t_2]}\varphi \equiv \top \, \mathbf{U}^{[t_1,t_2]}\varphi$, which stands for *"at some point within $[t_1, t_2]$ φ is satisfied"* and the time-bounded *globally* $\mathbf{G}^{[t_1,t_2]}\varphi \equiv \neg\mathbf{F}^{[t_1,t_2]}\neg\varphi$ which stands for *"φ is always satisfied within $[t_1, t_2]$"*. In the remainder, unless otherwise stated, we restrict our focus to the non-nested fragment of MITL, i.e. we consider only formulae such that the operands of a temporal modality is always an atomic proposition μ. Although a clear limitation in terms of expressiveness this constraint still allows us to treat most common reachability problems.

Model Checking CTMCs. Model checking of a (MITL) formula φ against a CTMC model $\mathcal{M}_\theta$ involves assessing the probability that $\mathcal{M}_\theta$ satisfies φ, denoted $Pr(\varphi|\mathcal{M}_\theta)$, which, roughly speaking, boils down to adding up the probability of each path that starting in the initial state s_0 at time $t = 0$ satisfies φ, i.e. $Pr(\varphi|\mathcal{M}_\theta) = Pr(\{\sigma|\sigma@0 \models \varphi, \sigma \in Path_{\mathcal{M}_\theta}(s_0)\}|\mathcal{M}_\theta)$. $Pr(\varphi|\mathcal{M}_\theta)$ may be assessed either exactly through *numerical* model checkers [10,19] (although these are affected by the state-space explosion problem, hence they are limited to models of reasonable size) or being estimated through *statistical* model checkers [5,20,27,31] (through which the estimates of $Pr(\varphi|\mathcal{M}_\theta)$ are obtained by statistical inference based on trajectory samples of arbitrary size).

2.3 Hybrid Automata Specification Language

The Hybrid Automata Specification Language [4] (HASL) is a formalism that allows for expressing sophisticated performance measures of stochastic models (CTMCs included) and assess them through a statistical model checking approach. The expressive power of HASL relies on the use of a linear hybrid automaton (LHA) as a machinery to filter trajectories sampled from a given model $\mathcal{M}$. LHA generalise timed automata (TA), in that they may be equipped with generic real-valued variables which include but are not limited (as with TA) to clocks, hence allowing for computing useful statistics during the analysis of a model's trajectory which opens up to plentiful applications. In this paper we apply the HASL formalism to reachability problems by developing HASL specifications for measuring the distance of trajectories from a given reachability region (see Sect. 3.1) associated to a reachability problem expressed in MITL terms. We briefly recall the nature of LHA referring the reader to [4] for more details. An HASL specification consist of a linear hybrid automaton (LHA) which is defined as an n-tuple:

$$\mathcal{A} = \langle E, L, \Lambda, \textit{Init}, \textit{Final}, X, \textit{flow}, \rightarrow \rangle$$

where: E is a finite alphabet of events; L is a finite set of locations; $\Lambda : L \rightarrow Prop$, a location labelling function (*Prop* being the set of atomic proposition built on top of variables X); *Init* is a subset of L called the initial locations; *Final* is a subset of L called the final locations; $X = (x_1, ... x_n)$ a n-tuple of data variables; $flow : L \mapsto Ind^n$ is a function which associates each location with an n-tuple of indicators with the i-th indicator, denoted $flow_i$, representing the rate at which variable x_i evolves; $\rightarrow \subseteq L \times ((\mathsf{Const} \times 2^E) \uplus (\mathsf{IConst} \times \{\sharp\})) \times \mathsf{Up} \times L$, a set of edges, where the notation $l \xrightarrow{\gamma, E', U} l'$ means that $(l, \gamma, E', U, l') \in \rightarrow$, with Const the set of constraints, whose elements are boolean combinations of inequalities of the form $\sum_{1 \le i \le n} \alpha_i x_i + c \prec 0$ where α_i and c are constants), $\prec \in \{=, <, >, \le, \ge\}$, whereas IConst is the set of left-closed constraints. Selection of a model's trajectories through an automaton $\mathcal{A}$ is achieved through *synchronization* of $\mathcal{M}_\theta$ with $\mathcal{A}$, i.e. by letting $\mathcal{A}$ synchronise its transitions with the transitions of the trajectory σ being sampled. To this aim an LHA admits two kinds of transitions: *synchronising* transitions (associated with a subset $E \subseteq \Sigma$ of event names, with ALL denoting Σ), which may be traversed when an event (in E) is observed on σ, and *autonomous* transitions (denoted by $\sharp$) which are traversed autonomously (and have priority over synchronized transitions), on given conditions, typically to update relevant statistics or to terminate (accept) the analysis of σ. For example, autonomous transition $l_0 \xrightarrow{\sharp, \top, \{n := x_O, d := \infty\}} l_1$ of automaton $\mathcal{A}_F$ in Fig. 2 is fired unconditionally (constraint $\top$) and updates variables n and d to the value of species O in current state and ∞ respectively. On the other hand transition $l_3 \xrightarrow{ALL, \top, \{n := x_O\}} l_1$ is fired only when any event (i.e. ALL) is observed on the synchronising trajectory. Since automata-based formalisms are at least as expressive as temporal logic based on classical temporal modalities (see [12]) in the remainder we denote $\mathcal{A}_\varphi$ the HASL automaton equivalent to a MITL formula φ (i.e. $\mathcal{A}_\varphi$ accepts a trajectory σ of a CTMC model $\mathcal{M}_\theta$ if and only if $\sigma \models \varphi$).

2.4 The ABC Method

Approximate Bayesian Computation (ABC) methods are concerned with estimating the posterior distribution of a model's parameters θ based on some observed (experimental) data y_{exp}. Considering a prior distribution on the parameters $\pi(.)$, observations $y_{exp} \in \mathcal{Y}$ and likelihood function $p(.|\theta)$ of a model, the objective of Bayesian estimation is to determine the posterior distribution:

$$\pi(\theta|y_{exp}) = \frac{p(y_{exp}|\theta)\pi(\theta)}{\int_{\theta'} p(y_{exp}|\theta')\pi(\theta') d\theta'} \tag{1}$$

In complex models, the likelihood function $p(y_{exp}|\theta)$ may be too expensive to compute or even intractable, which hinders the determination of the posterior distribution by classical methods. ABC algorithms were designed to handle these situations. We refer to [23, 28] for interesting and pretty complete surveys of ABC or rejection-sampling methods. To illustrate the general idea, we give below a simple likelihood-free algorithm:

Algorithm 1. Simple ABC

Require: $y_{exp}, \epsilon, \rho, \eta$
Ensure: $(\theta_i)_{1 \leq i \leq N}$ drawn from $\pi_{ABC,\epsilon}$
    for $i = 1 : N$ **do**
        repeat
            $\theta' \sim \pi(.)$
            $y' \sim p(.|\theta')$
        until $\rho(\eta(y'), \eta(y_{exp})) \leq \epsilon$
        $\theta_i \leftarrow \theta'$
    end for

---

where $\eta : \mathcal{Y} \to \mathcal{S} \subset \mathbb{R}^{k_1}$ is a function that computes summary statistics on the observations and $\rho : \mathcal{S} \times \mathcal{S} \to \mathbb{R}^+$ is a distance in the space of summary statistics. The choice of summary statistics is a crucial point in ABC (see for example [1]). The resulting samples $(\theta_i, y_i)_{1 \leq i \leq N}$ are drawn from the joint distribution:

$$\pi_{ABC,\epsilon}(\theta, y | y_{exp}) \propto \mathbb{1}_{A_{\epsilon,y_{exp}}}(y) p(y|\theta) \pi(\theta)$$

where $A_{\epsilon,y_{exp}} = \{y'/\mathbb{1}_{\rho(\eta(y'),\eta(y_{exp}))\leq\epsilon}\}$. The marginal distribution of $\pi_{ABC,\epsilon}(.,.)$ is:

$$\pi_{ABC,\epsilon}(\theta | y_{exp}) \propto \int_y \pi_{ABC,\epsilon}(\theta, y | y_{exp}) dy$$

When η is the identity function, as ϵ tends to zero, we get:

$$\lim_{\epsilon \to 0} \pi_{ABC,\epsilon}(\theta | y_{exp}) \propto \lim_{\epsilon \to 0} \int_y \pi_\epsilon(\theta, y | y_{exp}) dy \propto \lim_{\epsilon \to 0} \int_y \mathbb{1}_{A_{\epsilon,y_{exp}}}(y) p(y|\theta) \pi(\theta) dy$$

$$\propto \int_y \delta_{y_{exp}} p(y|\theta) \pi(\theta) dy \propto p(y_{exp}|\theta) \pi(\theta)$$

Therefore $\pi_{ABC,\epsilon}$ approximates the posterior distribution, all the better when ϵ is smaller.

Algorithm 2. ABC Population Monte-Carlo

Require: N: number of particles, $y_{exp}, (\epsilon_i)_{1 \leq i \leq M}, \rho, \eta$
Ensure: $(\theta_j)_{1 \leq j \leq N}$ drawn from π_{ABC,ϵ_M}
    Iteration $i = 1$: find $(\theta_j^{(1)})_{1 \leq j \leq N}$ with algorithm ABC 1
  $\omega_j \leftarrow \frac{1}{N}$
  for $i = 2 : M$ **do**
    for $j = 1 : N$ **do**
        repeat
            Take θ'_j from $(\theta_j^{(i-1)})_{1 \leq j \leq N}$ with probabilities $(\omega_j)_{1 \leq j \leq N}$
            $\theta_j^{(i)} \sim K(.|\theta'_j)$
            $y' \sim p(.|\theta_j^{(i)})$
        until $\rho(\eta(y'), \eta(y_{exp})) \leq \epsilon_i$
        $\omega_j \leftarrow \dfrac{\pi(\theta_j^{(i)})}{\sum_{j'=1}^{N} \omega_{j'}^{(i-1)} K(\theta_j^{(i)} | \theta_{j'}^{(i-1)})}$
    end for
    Normalize $(\omega_j)_j$
  end for

2.5   ABC Population Monte-Carlo

The chosen value of ϵ is crucial for the performance of Algorithm 1: a small ϵ is needed to achieve a good approximation, however this may result in high rejection rate leading to cumbersome computations. To overcome this issue the more elaborate Algorithm 2, known as ABC population Monte-Carlo (ABC-PMC), has been proposed [6]. It is an SMC based approach [11] through which a population of N particles is iteratively sampled with increasing accuracy until the targeted level of accuracy ϵ_M, is obtained. At the first iteration particles are initialised through the simple ABC Algorithm 1 using a large enough ϵ_1 to limit the computation cost. Then, at each step i, $i = 2, \ldots, M$, the particles are moved by a transition kernel $K(.|.)$ (for example a Gaussian one [11]) until they match the next level, tighter, approximation constraint ϵ_i. At iteration M, we finally get N particles that fulfill the desired approximation ϵ_M. Some ad-hoc strategies are proposed to find a proper sequence $(\epsilon_i)_{1 \le i \le M}$ ensuring an efficient convergence towards the posterior distribution.

3   ABC for Reachability Design

In this section we present the core contribution of the paper namely: we formalise the notion of distance of a model's trajectories from a *reachability region*, we introduce the HASL specifications for measuring the trajectories distance w.r.t. a given reachability problem and finally we introduce an adaptation of the ABC method to the reachability problem, i.e. we adapt ABC so that the convergence is guided by the distance from a reachability region.

3.1   Reachability Distances

In order to adapt procedures for searching the parameters space driven by a reachability problem we introduce different notions of distance of a model's trajectory from a (set of) reachability region(s) associated with a reachability problem.

Definition 1 (Time-bounded reachability region). *Given an n-dimensional CTMC population model $\mathcal{M}_\theta$ with state-space $S \subseteq \mathbb{N}^n$ and a simple (time-bounded) reachability formula, either $\varphi \equiv F^{[t_1,t_2]}\mu$ or $\varphi \equiv G^{[t_1,t_2]}\mu$ (with $[t_1, t_2] \subset \mathbb{R}_{\ge 0}$) we define $Rreg(\mathcal{M}_\theta, \varphi) \subset S \times [t_1, t_2]$ the time-bounded reachability region of $\mathcal{M}_\theta$ w.r.t. φ as the hyperrectangle:*

$$Rreg(\mathcal{M}_\theta, F^{[t_1,t_2]}\mu) = Rreg(\mathcal{M}_\theta, G^{[t_1,t_2]}\mu) = \{((s,t) \mid s \models \mu \wedge t \in [t_1,t_2]\}$$

Definition 2 (Distance from an eventual region $F^{[t_1,t_2]}\mu$). *Given a trajectory $\sigma \in Path^{\mathcal{M}_\theta}(s_0)$ of a CTMC $\mathcal{M}_\theta$ and an eventual reachability property $\varphi \equiv F^{[t_1,t_2]}\mu$ we define the distance $d(\sigma, F^{[t_1,t_2]}\mu))$ from region $Rreg(\mathcal{M}_\theta, F^{[t_1,t_2]}\mu)$ as the minimal euclidean distance of any point of σ that occurs within $[t_1, t_2]$ from S_μ, the subset of states of $\mathcal{M}_\theta$ that fulfils μ.*

$$d(\sigma, F^{[t_1,t_2]}\mu) = argmin_{t \in [t_1,t_2]} d_e(\sigma@t, S_\mu)$$

where $d_e(s, S_1) = argmin_{s' \in S_1} \sqrt{\sum_{i=1}^{n}(s_i - s'_i)^2}r$ denotes the euclidean distance of point $s \in S$ from subset $S_1 \subseteq S$ with $S \subseteq \mathbb{N}^n$ an n-dimensional space.

Definition 3 (Distance from a global region $G^{[t_1,t_2]}\mu$). *Given a trajectory $\sigma \in Path_{\mathcal{M}_\theta}(s_0)$ of a CTMC $\mathcal{M}_\theta$ and a global reachability property $\varphi \equiv G^{[t_1,t_2]}\mu$ we define the distance $d(\sigma, G^{[t_1,t_2]}\mu))$ from region $Rreg(\mathcal{M}_\theta, G^{[t_1,t_2]}\mu)$ as the sum of the euclidean distance from S_μ of any point of σ that occurs within $[t_1,t_2]$*

$$d(\sigma, G^{[t_1,t_2]}\mu) = \int_{t_1}^{t_2} d_e(\sigma@t, S_\mu)dt$$

Observe that, in agreement with the semantics of the eventual (F) and the global (G) modalities, the distance of a CTMC trajectory σ from an eventual formula $d(\sigma, F^{[t_1,t_2]}\mu) = 0$ if and only if σ has at least one point traversing region $S_\mu \times [t_1,t_2]$ while, on the other hand, the distance from a global formula $d(\sigma, G^{[t_1,t_2]}\mu) = 0$ if and only if all points of σ fall in $S_\mu \times [t_1,t_2]$ (see Fig. 1).

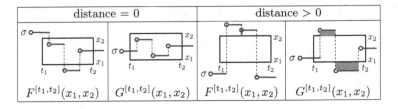

Fig. 1. Examples of trajectories with zero-distance (left) and positive distance (right) from an eventual, respectively a global, region (positive distances are depicted in gray). (Color figure online)

The following proposition states an intuitively trivial, yet relevant, aspect relating a trajectory's distance of from a region with the satisfaction of the corresponding MITL formula.

Proposition 1. *For $\sigma \in Path_{\mathcal{M}_\theta}(s_0)$ a path of a CTMC $\mathcal{M}_\theta$ and φ a MITL reachability formula of kind $\varphi = F^{[t_1,t_2]}\mu$ or $\varphi = G^{[t_1,t_2]}\mu$ then*

$$\sigma \models \varphi \Longleftrightarrow d(\sigma, \varphi) = 0$$

Proof. Trivial.

3.2   HASL Specifications for Measuring Reachability Distance

Given a (global or eventual) reachability formula φ referred to a model $\mathcal{M}_\theta$ we show how to set up an LHA for measuring the distance of a trajectory of $\mathcal{M}_\theta$ from region $Rreg(\mathcal{M}_\theta, \varphi)$. For the sake of simplicity in Fig. 2 we show automata $\mathcal{A}_F$, respectively $\mathcal{A}_G$, referred to a mono-dimensional region corresponding to atomic formula $\mu \equiv x_1 \leq x_O \leq x_2$ (where x_O denotes the population of an observable quantity O of $\mathcal{M}_\theta$ and $x_1 < x_2 \in \mathbb{N}$). Distance automata for n-dimensional regions are just adaptation of those in Fig. 2.

Distance Automaton $\mathcal{A}_F$. Automaton $\mathcal{A}_F$ in Fig. 2 is designed for measuring the (average) distance (Definition 2) of trajectories of a CTMC model $\mathcal{M}_\theta$ from the region associated with $F^{[t_1,t_2]}(x_1 \le x_O \le x_2)$, i.e. the region corresponding to quantity O being $x_O \in [x_1, x_2]$ within time $t \in [t_1, t_2]$.

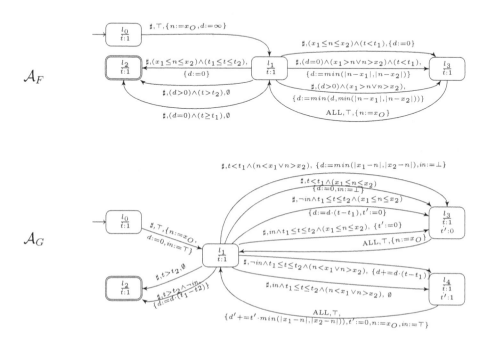

Fig. 2. Distance automata for regions $F^{[t_1,t_2]}(x_1 \le x_O \le x_2)$ and $G^{[t_1,t_2]}(x_1 \le x_O \le x_2)$

It uses 3 variables: d (computed distance), t (current time along the trajectory) and n (population of the observed quantity x_O). Initially ($l_0 \to l_1$) the distance is initialised to $d := \infty$ and the initial value of the observed species stored in $n :=$ x_O. Once in l_1 if the trajectory is inside the region (and this include even initially with $t = 0$ in case $t_1 = 0$ too) then transition $l_1 \xrightarrow{(x1 \le n \le x_2) \wedge (t_1 \le t \le t_2), \{d := 0\}} l_2$ fires and computation stops setting $d := 0$. On the other hand if, while in l_1, the trajectory has not entered the region the distance d must be updated which is dealt by 3 autonomous transitions $l_1 \to l_3$. In case O has entered $[x_1, x_2]$ before the considered time-window ($l_1 \xrightarrow{(x1 \le n \le x_2) \wedge (t < t_1), \{d := 0\}} l_3$) the distance is set to $d := 0$. This is because, since CTMC trajectories are *càdlàg* functions of time[1], if the next reaction occurs at time $t \ge t_1$ then it is certain that the current trajectory has at least one point within the considered region hence the trajectory

---

[1] i.e. right continuous with left limits, see Fig. 1.

is accepted ($l_1 \xrightarrow{(d=0)\wedge(t\geq t_1)} l_2$) with distance $d = 0$. On the other hand if O has not entered $[x_1, x_2]$ then d is set to the distance of the current point from $[x_1, x_2]$, if the previous point was in $[x_1, x_2]$ ($l_1 \xrightarrow{(d=0)\wedge(x_1>n\vee n>x_2)} l_3$), or to the minimum between the previous value of d and the distance of the current point from $[x_1, x_2]$ if the previous point was not in $[x_1, x_2]$ (i.e. $l_1 \xrightarrow{(d>0)\wedge(x_1>n\vee n>x_2)} l_3$). The transition $l_3 \xrightarrow{ALL,\top,\{n:=x_O\}} l_1$ is traversed whenever a novel point is added to the trajectory and brings the automaton back to l_1 so that the distance can be updated accordingly. Finally transition $l_1 \xrightarrow{\natural,(d>0)\wedge(t>t_2),\emptyset} l_2$ halts the computation as soon as the trajectory exit the temporal region $[t_1, t_2]$ not having traversed $[x_1, x_2]$: at that point the distance of the trajectory is already stored in d and needs no update (even in case of a trajectory consisting of a single point).

Distance Automaton $\mathcal{A}_G$. Automaton $\mathcal{A}_G$ (Fig. 2) is similar to $\mathcal{A}_F$ only that the computed distance d corresponds with the integral of the distance of points which falls outside the region within time window $[t_1, t_2]$. It uses the same variables as $\mathcal{A}_F$ plus an extra timer t', for measuring the duration of a segment falling outside the region within $[t_1, t_2]$, and a boolean flag in, which is set to $\perp$ (i.e. `false`) if the trajectory have no segment originating in the region within $[t_1, t_2]$. After variables are initialised ($l_0 \to l_1$) analysis begins in l_1: for events occurring at $t < t_1$ the distance is set to either $d = 0$ ($l_1 \to l_3$ top arc), if $\sigma@t \in [x_1, x_2]$, or to the distance of $\sigma@t$ from $[x_1, x_2]$ otherwise ($l_1 \to l_3$ midway arc). This is because if the next point of σ happens at $t > t_2$ then the final distance is given by $d \cdot (t_2 - t_1)$ ($l_1 \to l_2$ bottom arc). Conversely for events occurring at $t \in [t_1, t_2]$ the distance is either incremented with the surface underlying the segment (of duration t') laying outside $[x_1, x_2]$, if $\sigma@t \notin [x_1, x_2]$ (sequence $l_1 \to l_4 \to l_1$), or is left unchanged if $\sigma@t \in [x_1, x_2]$ (sequence $l_1 \to l_3 \to l_1$).

Example 1 (Enzymatic reaction system). We consider a simple model of an enzymatic reaction (ER) system whereby a *substrate* species S is converted into a *product* P through mediation of an *enzyme* E. The dynamics is given by chemical equations (2) which depend on the parameters $\theta = \{k_1, k_2, k_3\}$, i.e. the kinetic rate constants of reactions R_1, R_2, R_3:

$$R_1 : E + S \xrightarrow{k_1} ES \qquad R_2 : ES \xrightarrow{k_2} E + S \qquad R_3 : ES \xrightarrow{k_3} E + P \qquad (2)$$

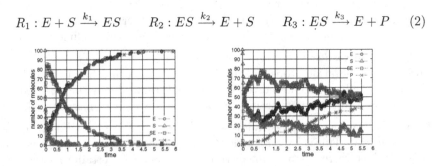

Fig. 3. Trajectories of the ER system with $\theta = (1, 1, 1)$ (left) and $\theta = (0.1, 1, 0.1)$ (right).

We assume *mass-action* as the law for kinetic rates of the 3 reactions (i.e. the actual rate of each reaction is given by the product of the abundances of the reactants times the kinetic rate constant). Figure 3 shows two (4-dimensional) trajectories sampled from the CTMC model $\mathcal{M}_\theta$ of the enzymatic reaction system with initial state $(E_0, S_0, ES_0, P_0) = (100, 100, 0, 0)$ and parameters $\theta = (1,1,1)$ (left) and $\theta = (0.1, 1, 0.1)$ (right). The dynamics of the ER system (Fig. 3) is such that the totality of the substrate (initially $S_0 = 100$) is converted into the product at a speed dependent on parameters θ. With $\theta = (1,1,1)$ the totality of S is converted within $T \sim 5$ whereas if we slow down by a ten-fold both formation of the ES complex and synthesis of P (i.e. $\theta = (0.1, 1, 0.1)$)) we have that only about 30% of S has been converted within $T \sim 5$.

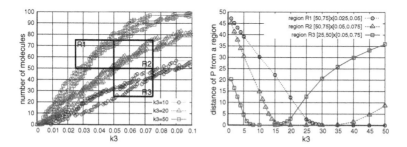

Fig. 4. P-projected trajectories of the CTMC model of ER with different spatio-temporal regions (left), and distances computed thtoguh $\mathcal{A}_F$ (right). (Color figure online)

Testing Distance Automata. To test automaton $\mathcal{A}_F$ we run a few experiments on the CTMC model of the ER system using the statistical model checker Cosmos [5]. Figure 4 (left) shows batches of (the projection over the P dimension of) trajectories of the ER model corresponding to parameter sets $\theta_1 : (1,1,50)$, $\theta_2 = (1,1,20)$ and $\theta_3 = (1,1,10)$ together with regions associated with formulae $\varphi_1 : F^{[0.025,0.05]}(50 \leq x_P \leq 75)$ (r1), $\varphi_2 : F^{[0.05,0.075]}(50 \leq x_P \leq 75)$ (r2) and $\varphi_3 : F^{[0.05,0.075]}(25 \leq x_P \leq 50)$ (r3). It is evident that trajectories for $\mathcal{M}_{\theta_1}$ (purple) are more likely to traverse r_1, those for $\mathcal{M}_{\theta_2}$ (green) to traverse r2 and those for $\mathcal{M}_{\theta_3}$ (blue) to traverse r3. Such intuition is confirmed by plots in Fig. 4 (right) which depict (the average value of the) distance of trajectories from regions r1, r2 and r3 measured with Cosmos, in function of k_3, using specific instances of $\mathcal{A}_F$ i.e. $\mathcal{A}_{\varphi_1}$, $\mathcal{A}_{\varphi_2}$ and $\mathcal{A}_{\varphi_3}$. We observe that e.g. the measured distance from region r1 monotonically decreases as k_3 increases and gets equal to zero for $k_3 \geq 30$ while the distance from region r3 is null when $10 \leq k_3 \leq 15$ whereas it grows as we increases k_3.

3.3   ABC with Reachability Distances

We consider the problem of defining an ABC framework for exploring the parameter space of a CTMC model $\mathcal{M}_\theta$ so that the probability of $\mathcal{M}_\theta$ to satisfy a reachability formula φ (either $\varphi \equiv F^{[t_1,t_2]}\mu$ or $\varphi \equiv G^{[t_1,t_2]}\mu$) is positive. The intuition behind this idea is that the exploration of the parameter space can be driven efficiently by taking into account the notion of distance of a trajectory $\sigma \in Path_{\mathcal{M}_\theta}(s_0)$ from the reachability region corresponding to φ (see Sect. 3.1). We point out that with the distance driven ABC the estimation of the posterior distribution ($\pi_{\varphi-ABC}$) is no longer computed as a limit approximation (i.e. $\lim_{\epsilon \to 0} \hat{\pi}_{ABC,\epsilon}(\cdot|y_{exp})$), as with classical ABC (Algorithm 1), but rather as an estimation of the exact posterior distribution, since trajectories are accepted exclusively if their distance is zero.

Simple ABC with Reachability Distance. We define a modified version of Algorithm 1 adapted to reachability distance:

Algorithm 3. ABC driven by $\mathcal{A}_\varphi$ automaton

Require: $\pi(.)$
Ensure: $(\theta_i)_{0 \leq i \leq N}$ drawn from $\pi_{\varphi-ABC}$
    for $i = 1 : N$ **do**
        repeat
            $\theta' \sim \pi(.)$
            $\sigma' \sim \mathcal{M}_{\theta'}$
        until $d(\sigma', \varphi) = 0$
        $\theta_i, \sigma_i \leftarrow \theta', \sigma'$
    end for

Here we draw a parameter θ' from the prior $\pi(.)$, we simulate a path σ' according to the CTMC $\mathcal{M}_{\theta'}$, and accept θ' if its distance from φ is $d(\sigma', \varphi) = 0$. (i.e. if $\sigma' \models \varphi$). Hence, $(\theta_i, \sigma_i)_i$ are drawn according to a density $\pi_{\varphi-ABC}$:

$$\pi_{\varphi-ABC}(\theta_i, \sigma_i) \propto \mathbb{1}_{d(.,\varphi)=0}(\sigma_i)p(\sigma_i|\theta_i)\pi(\theta_i) \propto \mathbb{1}_{C_{\theta_i,\varphi}}(\sigma_i)p_{\mathcal{M}_{\theta_i}}(\sigma_i)\pi(\theta_i)$$

where $C_{\theta,\varphi}$ the set of paths of $\mathcal{M}_\theta$ that satisfies φ, $Pr_{\theta,s_0}(C_{\theta,\varphi}) = Pr(\varphi|\mathcal{M}_\theta)$ and $p_{\mathcal{M}_{\theta_i}}$ is the density related to Pr_{θ_i,s_0}. The marginal of θ is given by:

$$\pi_{\varphi-ABC}(\theta_i) \propto \int_{\sigma \in Path_{\mathcal{M}_\theta}(s_0)} \mathbb{1}_{C_{\theta_i,\varphi}}(\sigma)p_{\mathcal{M}_{\theta_i}}(\sigma)\pi(\theta_i)d\sigma \propto \pi(\theta_i)\int_{\sigma \in C_{\theta_i,s_0}} dPr_{\theta_i,s_0}(\sigma)$$

$$\propto Pr_{\theta_i,s_0}(C_{\theta_i,s_0})\pi(\theta_i) = Pr(\varphi|\mathcal{M}_\theta)\pi(\theta_i)$$

With an uniform prior $(\theta_i)_i \sim \pi_{\varphi-ABC}(\theta_i) \propto Pr(\varphi|\mathcal{M}_\theta)$. The $\varphi - ABC$ density of θ is then proportional to the probability that φ is satisfied by $\mathcal{M}_\theta$.

ABC-PMC with Reachability Distance. Following the same approach we introduce the adapted ABC-PMC algorithm with distance automaton which allows for a smaller *runtime* than Algorithm 3. Observe that if with Algorithm 3 we do not really exploit the notion of continuous distance (i.e. we only accept/reject trajectories depending on whether their distance is null, i.e. if they satisfy φ, which is a simple Monte-Carlo approach), with Algorithm 4 we actually use the distance to rank paths and accept those parameters whose corresponding paths are closer (better ranked) than others, even if they don't necessarily satisfy φ. In this Sequential Monte-Carlo based version, the decreasing sequence of ϵ is set automatically. For the first iteration, we randomly sample parameters from the prior, simulate the paths and compute the empirical $\alpha - quantile$ of the distances of the simulated paths from φ for ϵ_1. Then, at each iteration we find parameters so that the simulated paths satisfy the acceptance condition with the current ϵ and then take the empirical $\alpha - quantile$ for the new ϵ until ϵ is equal to zero.

Algorithm 4. ABC Population Monte-Carlo driven by $\mathcal{A}_\varphi$ automaton

---

Require: N: number of particles, $\pi(.)$ prior, $d(.,.)$ the distance from $\mathcal{A}_\varphi$, $\alpha \in (0,1)$
Ensure: $(\theta_j)_{1 \leq j \leq N}$ drawn from $\pi_{\varphi-ABC}$

    $i \leftarrow 1$
    $(\theta_j^{(1)})_{1 \leq j \leq N} \sim \pi(.)$
    $\forall j \in 1, \ldots, N, \sigma_j \sim \mathcal{M}_{\theta_j}$
    $\epsilon \leftarrow quantile(\alpha, d(\sigma_j, \varphi))_{1 \leq j \leq N}$
    $(\omega_j)_{1 \leq j \leq N}^{(1)} \leftarrow \frac{1}{N}$
    while $\epsilon > 0$ **do**
        for $j = 1 : N$ **do**
            repeat
                Take θ_i' from $(\theta_j^{(i-1)})_{1 \leq j \leq N}$ with probabilities $(\omega_j)_{1 \leq j \leq N}$
                $\theta_j^{(i)} \sim K(.|\theta_i')$
                $\sigma' \sim p(.|\theta_j^{(i)})$
                $d_j \leftarrow d(\sigma', \varphi)$
            until $d_j \leq \epsilon$
            $\omega_j \leftarrow \dfrac{\pi\left(\theta_j^{(i)}\right)}{\sum\limits_{j'=1}^{N} \omega_{j'}^{(i-1)} K(\theta_j^{(i)}|\theta_{j'}^{(i-1)})}$
        end for
        Normalize $(\omega_j^{(i)})_j$
        $\epsilon \leftarrow quantile(\alpha, (d_j)_{1 \leq j \leq N})$
        $i \leftarrow i + 1$
    end while

---

4   Experiments

Table 1 reports about the evaluation[2] of the posterior distribution $\pi_{\varphi-ABC}$ obtained by application of the automaton-driven adaptation of the ABC method

---

[2] Experiments were performed on HPC resources from the "Mésocentre" (http://mesocentre.centralesupelec.fr/) through a prototype tool (available at https://gitlab.centralesupelec.fr/2017bentrioum/abc-automaton) written in Julia [7] and based on the ADJUSTIN' modelling platform [30].

Table 1. Posterior distributions computed through the ABC-distance method for different reachability regions of the enzymatic reaction model.

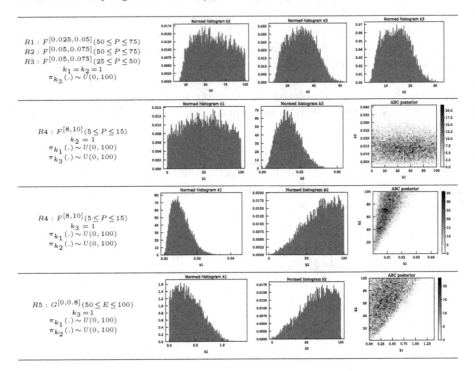

(Algorithms 3 and 4) to different reachability formulae referred to the ER model. Specifically we considered a few examples of F (eventual) reachability formulae as well as one example of G (global) reachability property. The first line of Table 1 depicts the marginal distribution of k_3 computed w.r.t. to regions R_1, (left) R_2 (center) and R_3 (right) where $R1, R2, R3$ correspond with the reachability formulae $\varphi_1 : F^{[0.025, 0.05]}(50 \leq P \leq 75)$, $\varphi_2 : F^{[0.05, 0.75]}(50 \leq P \leq 75)$ respectively $\varphi_3 : F^{[0.05, 0.75]}(25 \leq P \leq 50)$ (see Fig. 4). The marginal for $R1$ shows a rather uniform profile with about the 95% credibility interval that φ is satisfied for $k_1 \in [20, 100]$, which is in agreement with the average distance measure (Fig. 4 right). The marginals for $R2$ and $R3$, instead, gather about the 95% credibility interval on smaller intervals $k_3 \in [15, 50]$ ($R2$) resp. $k_3 \in [5, 25]$ ($R3$), again in line with measured distance (Fig. 4 right). The 2nd and 3rd row of Table 1 refer to the evaluation of posterior for the "eventual" region $R4$ for product P (i.e. $\varphi_4 : F^{[8,10]}(5 \leq P \leq 15)$) while the 4th row refers to the "global" region $R5$ for the enzyme E (i.e. $\varphi_4 : G^{[0,08]}(50 \leq E \leq 100)$), obtained by searching the parameter space w.r.t two (out of three) parameters. The rectangular profile of the joint posterior in the 2nd row (computed with $k_2 = 1$ and $\pi_{k_3}(.), \pi_{k_1}(.) \sim U(0, 100)$) indicates that k_3 is the "dominant" parameter w.r.t region $R4$ as the uniform-like profile of the marginal of k_1 indicates that k_1 has little effect on reaching

$R4$. The triangular profile of the joint posterior in 3rd and 4th row (computed with $k_3 = 1$ and $\pi_{k_1}(.), \pi_{k_2}(.) \sim U(0, 100)$) indicates that only very low values of k_1 ($k_1 \leq 0.015$ for $R4$, $k_1 \leq 1$ for $R5$) combined with rather high-values of k_2 (i.e. $k_1 \in [40, 100]$ for $R4$, $k_1 \in [50, 100]$ for $R5$) results in trajectories entering $R4$, resp. never leaving $R5$, which means the algorithm caught the correlation between the parameters. This is intuitively correct in both cases, in fact $R4$ corresponds to a very low synthesis of P which is not compatible with fast creation of the ES complex (i.e. only very small k_1 are not ruled out) and even the compensation effect obtained by fast decomplexion (i.e. large k_2) won't suffices for trajectories to stay in $R4$. Similarly $R5$ bounds the speed of the initial decrease of E (which initially is $E_0 = 100$), to 50 within $t \leq 0.8$ which again is compatible only with slow ES complexation and cannot be compensated by fast decomplexation.

Remarks. Results have been obtained by running Algorithm 4 with sample size $N = 10000$, $\alpha = 0.5$. There are no notable differences of performance between Algorithms 3 and 4 for regions R1, R2, R3 because of the large distributions. However Algorithm 3 is not worth considering for R4 and R5: one can see with the considered priors the probability to get a couple of parameters in the second distribution is about $\frac{90 \times 0.03}{\frac{2}{100 * 100}} \approx 10^{-4}$ which gives infinitesimal probability for drawing from the prior $N = 10000$ particles in the distribution and this simple computation doesn't even take into account a parameter in the obtained distribution could produce paths that don't satisfy φ. By adding several transitional step with Algorithm 4 the problem becomes feasible. Results for R4 required about 4×10^5 simulations of the model.

Related work. Several important results have been obtained in recent times w.r.t. studying the satisfaction of a temporal specification φ w.r.t. to a parametric CTMC model $\mathcal{M}_\theta$. In [8] authors outline conditions under which the satisfaction of φ is a differentiable function of parameters θ and set up a Bayesian framework for efficiently estimating the probability $Pr(\varphi | \mathcal{M}_\theta)$. In [15] authors introduce a framework for identifying the regions of the parameter space that either comply with a *threshold* CSL problem (i.e. such that $Pr(\varphi | \mathcal{M}_\theta) \sim r$, with $\sim \in \{<, \leq, \geq, >\}$ and $r \in [0, 1]$) or maximise the probability of a CSL specification.

5   Conclusion

We presented a new methodology for efficient exploration of a CTMC parameter space so that it has a positive probability to satisfy a time-bounded reachability specification φ. The methodology is an adaptation of the ABC method which, based on the notion of distance from a reachability region, outputs a density which is proportional to the probability of satisfying φ with uniform priors. Through experiments on a enzymatic reaction model we showed both the efficiency (i.e. the adapted ABC-PMC algorithm allows for treating computationally costly reachability problems which simple ABC cannot deal with) as well as the consistency of the proposed approach. Future works include the extension to a larger fragment of reachability problems (e.g. Until formulae and nested operators) as well as to other categories of problems (beyond reachability), i.e. those for which a notion of trajectory distance can be conceived.

References

1. Nunes, M.A., Balding, D.J. On optimal selection of summary statistics for approximate Bayesian computation. Stat. Appl. Genet. Mol. Biol. **9** (2010). Article no. 34
2. Baier, C.: On algorithmic verification methods for probabilistic systems. Habilitation thesis, Fakultät für Mathematik & Informatik, Universität Mannheim (1998)
3. Baier, C., Katoen, J.-P.: Principles of Model Checking. MIT Press, Cambridge (2008)
4. Ballarini, P., Barbot, B., Duflot, M., Haddad, S., Pekergin, N.: HASL: a new approach for performance evaluation and model checking from concepts to experimentation. Perform. Eval. **90**, 53–77 (2015)
5. Ballarini, P., Djafri, H., Duflot, M., Haddad, S., Pekergin, N.: COSMOS: a statistical model checker for the hybrid automata stochastic logic. In: Proceedings of the 8th International Conference on Quantitative Evaluation of Systems (QEST 2011), pp. 143–144. IEEE Computer Society Press, September 2011
6. Beaumont, M.A., Cornuet, J.-M., Marin, J.-M., Robert, C.P.: Adaptive approximate Bayesian computation. Biometrika **96**(4), 983–990 (2009)
7. Bezanson, J., Edelman, A., Karpinski, S., Shah, V.: Julia: a fresh approach to numerical computing. SIAM Rev. **59**(1), 65–98 (2017)
8. Bortolussi, L., Milios, D., Sanguinetti, G.: Smoothed model checking for uncertain continuous-time Markov chains. Inf. Comput. **247**, 235–253 (2016)
9. Clarke, E.M., Grumberg, O., Peled, D.A.: Model Checking. MIT Press, Cambridge (1999)
10. Dehnert, C., Junges, S., Katoen, J.-P., Volk, M.: A storm is coming: a modern probabilistic model checker. In: Proceedings of the 29th International Conference on Computer Aided Verification (CAV 2017) (2017)
11. Del Moral, P., Doucet, A., Jasra, A.: Sequential Monte Carlo samplers. J. R. Stat. Soc. B **68**(3), 411–436 (2006)
12. Donatelli, S., Haddad, S., Sproston, J.: Model checking timed and stochastic properties with CSL^{TA}. IEEE Trans. Softw. Eng. **35**, 224–240 (2009)
13. Donzé, A.: Breach, a toolbox for verification and parameter synthesis of hybrid systems. In: International Conference on Computer Aided Verification, pp. 167–170 (2010)
14. Haghighi, I., Jones, A., Kong, Z., Bartocci, E., Gros, R., Belta, C.: SpaTeL: a novel spatial-temporal logic and its applications to networked systems. In: Proceedings of the 18th International Conference on Hybrid Systems: Computation and Control, pp. 189–198 (2015)
15. Ceska Jr., M., Dannenberg, F., Paoletti, N., Kwiatkowska, M., Brim, L.: Precise parameter synthesis for stochastic biochemical systems. Acta Inf. **54**(6), 589–623 (2017)
16. Koutroumpas, K., Ballarini, P., Votsi, I., Cournède, P.H.: An infinite mixture models approach. In: Bioinformatics, Bayesian parameter estimation for the Wnt pathway (2016)

17. Kulkarni, V.G.: Modeling and Analysis of Stochastic Systems, 3rd edn. Chapman & Hall/CRC Texts in Statistical Science. CRC Press, Boca Raton (2016)
18. Kwiatkowska, M., Norman, G., Parker, D.: Stochastic model checking. In: Bernardo, M., Hillston, J. (eds.) SFM 2007. LNCS, vol. 4486, pp. 220–270. Springer, Heidelberg (2007). https://doi.org/10.1007/978-3-540-72522-0_6
19. Kwiatkowska, M., Norman, G., Parker, D.: PRISM: probabilistic symbolic model checker. In: Field, T., Harrison, P.G., Bradley, J., Harder, U. (eds.) TOOLS 2002. LNCS, vol. 2324, pp. 200–204. Springer, Heidelberg (2002). https://doi.org/10.1007/3-540-46029-2_13
20. Legay, A., Sedwards, S., Traonouez, L.-M.: Plasma lab: a modular statistical model checking platform. In: Margaria, T., Steffen, B. (eds.) ISoLA 2016. LNCS, vol. 9952, pp. 77–93. Springer, Cham (2016). https://doi.org/10.1007/978-3-319-47166-2_6
21. Lenive, O., Kirk, P.D.W., Stumpf, M.P.H.: Inferring extrinsic noise from single-cell gene expression data using approximate Bayesian computation. BMC Syst. Biol. **10**, 81 (2016)
22. Maler, O., Nickovic, D.: Monitoring temporal properties of continuous signals. In: Lakhnech, Y., Yovine, S. (eds.) FORMATS/FTRTFT-2004. LNCS, vol. 3253, pp. 152–166. Springer, Heidelberg (2004). https://doi.org/10.1007/978-3-540-30206-3_12
23. Marin, J.-M., Pudlo, P., Robert, C.P., Ryder, R.J.: Approximate Bayesian computational methods. Stat. Comput. **22**(6), 1167–1180 (2012)
24. Plagnol, V., Tavaré, S.: Approximate Bayesian computation and MCMC. In: Niederreiter, H. (ed.) Monte Carlo and Quasi-Monte Carlo Methods 2002, pp. 99–113. Springer, Heidelberg (2004). https://doi.org/10.1007/978-3-642-18743-8_5
25. Pritchard, J.K., Seielstad, M.T., Perez-Lezaun, A., Feldman, M.W.: Population growth of human Y chromosomes: a study of Y chromosome microsatellites. Mol. Biol. Evol. **16**(12), 1791–1798 (1999)
26. Ratmann, O., Andrieu, C., Wiuf, C., Richardson, S.: Model criticism based on likelihood-free inference, with an application to protein network evolution. Proc. Natl. Acad. Sci. **106**, 10576–10581 (2009)
27. Sen, K., Viswanathan, M., Agha, G.: VESTA: a statistical model-checker and analyzer for probabilistic systems. In: Second International Conference on the Quantitative Evaluation of Systems (QEST 2005), pp. 251–252, September 2005
28. Sisson, S.A., Fan, Y., Beaumont, M.A.: Overview of approximate Bayesian computation (1), 1–66 (2018)
29. Toni, T., Welch, D., Strelkowa, N., Ipsen, A., Stumpf, M.P.H.: Approximate Bayesian computation scheme for parameter inference and model selection in dynamical systems. J. R. Soc. Interface **6**(31), 187–202 (2009)
30. Viaud, G.: Statistical methods for the genotypic differentiation of plants using growth models. Université Paris-Saclay, Theses (2018)
31. Younes, H.L.S.: Ymer: a statistical model checker. In: Etessami, K., Rajamani, S.K. (eds.) CAV 2005. LNCS, vol. 3576, pp. 429–433. Springer, Heidelberg (2005). https://doi.org/10.1007/11513988_43

Fast Enumeration of Non-isomorphic Chemical Reaction Networks

Carlo Spaccasassi, Boyan Yordanov, Andrew Phillips, and Neil Dalchau[⊠]

Biological Computation group, Microsoft Research, Cambridge CB1 2FB, UK
{t-caspac,yordanov,aphillip,ndalchau}@microsoft.com

Abstract. Chemical reaction networks (CRNs) have been applied successfully to model a wide range of phenomena and are commonly used for designing molecular computation circuits. Often, CRNs with specific properties (oscillations, Turing patterns, multistability) are sought, which entails searching an exponentially large space of CRNs for those that satisfy a property. As the size of the CRNs being considered grows, so does the frequency of isomorphisms, by up to a factor $N!$, where N is the number of species. Accordingly, being able to *generate* sets of non-isomorphic CRNs within a class can lead to large computational savings when carrying out global searches. Here, we present a bijective encoding of bimolecular CRNs into novel vertex-coloured digraphs called Complex-Species graphs. The problem of enumerating non-isomorphic CRNs can then be tackled by leveraging well-established computational methods from graph theory [20]. In particular, we extend Nauty, the graph isomorphism tool suite by McKay [22]. Our method is highly parallelisable and more efficient than competing approaches, and a software package (**genCRN**) is freely available for reuse. Non-isomorphs are generated directly by **genCRN**, alleviating the need to store intermediate results. We provide the first complete count of all 2-species bimolecular CRNs and extend previous known counts for classes of CRNs of special interest, such as mass-conserving and reversible CRNs.

1 Introduction

Chemical reaction networks (CRNs) are widely recognised as a convenient formalism for modelling and analysing a broad range of biochemical systems [1,17]. In recent years, they have also been used for designing synthetic systems with specified behaviours, such as distributed consensus networks [9], oscillators [34] and feedback control circuits [27]. CRNs provide a convenient abstraction for modelling synthetic biological systems, while also supporting a mapping to biological implementations in both molecular [33] and genetic [28] circuits.

CRNs also support a broad range of analysis methods, which can be used to check the desired properties of a system prior to its implementation. In particular, a promising approach is to encode a CRN as a graph and analyse its properties using graph-theoretic methods. A CRN is essentially a map from a

© Springer Nature Switzerland AG 2019
L. Bortolussi and G. Sanguinetti (Eds.): CMSB 2019, LNBI 11773, pp. 224–247, 2019.
https://doi.org/10.1007/978-3-030-31304-3_12

multiset of *reactant* species to a multiset of *product* species, which can be can be written as

$$\alpha_{ji} A_i \xrightarrow{k_j} \beta_{ji} A_i, \quad j \in \{1, \dots, M\}, i \in \{1, \dots, N\} \tag{1}$$

where N denotes the number of species, M denotes the number of reactions, α_{ji} and β_{ji} denote the multiplicity of species A_i in the reactants and products of reaction j, respectively, and $k_j \in \mathbb{R}_+$ denotes the rate constant of reaction j. In a pair of landmark papers [15,16], Feinberg encoded CRNs as *complex graphs* – where each vertex represents a complex and each directed edge represents a reaction – and related the *deficiency* of a CRN to the existence of positive steady states. Alternative graph encodings have also been developed, including *species-reaction* (SR) graphs, which are directed bipartite graphs whose vertices are either species or reactions. These SR graphs are used to check for the existence of multiple equilibria that can be determined from network structure alone [11,12]. Graph-theoretic properties have also been developed for detecting oscillations [24] and Turing instabilities [23], and for assessing concentration robustness [32].

As the number of species N increases, the number of possible reactions grows such that the number of bimolecular CRNs grows as $\mathcal{O}(2^{N^4})$ (see Lemma 4 in Appendix B). As a result, the design space of CRNs with more than a few species is intractable to explore systematically [26], and the design of CRNs with specified behaviours remains largely an artisanal process. One approach to exploring the design space of CRNs more efficiently is to filter out CRNs that are *isomorphic* and therefore exhibit identical behaviour. In principle, the enumeration of these non-isomorphic CRNs can make it possible to exhaustively explore an otherwise intractable space, since as the number of species N increases, the number of isomorphic CRNs also increases substantially. More generally, enumerating non-isomorphic CRNs can be used to determine which CRNs satisfy a property in a *complete* sense, such as determining the complete set of 2-species CRN oscillators [4], or the smallest CRN admitting bistability [35].

The problem of enumerating non-isomorphic CRNs is related to the problem of enumerating non-isomorphic graphs, which is NP-hard and, worse still, considered to be a pathology of computer science research [30]. Several methods for working with graph isomorphisms already exist, the most notable of which is NAUTY [22], which can efficiently compute a canonical form of a graph, find its automorphism group and its generators. NAUTY also provides enumeration tools for graphs, digraphs and vertex colouring, among others. Also related is Polya's enumeration theorem, which *counts* non-isomorphic graphs *without* constructing them [29]. The most promising method for working with isomorphic CRNs was introduced in [3], which uses NAUTY to encode a CRN as a *species-reaction Petri net* [2]. This is similar to an SR-graph, except that edges are directed and weighted. It was inspired by attempts to enumerate CRNs in [13], which also leverages NAUTY. However, species-reaction Petri nets need to be encoded as multidigraphs, which are not supported natively in NAUTY. They can be encoded in terms of digraphs, but such an encoding is not enumerable in NAUTY without also generating invalid multidigraphs. This requires storing and then filtering out non-isomorphs after enumeration, which penalises run-time. Time measurements or a software tool are not available in [3], so it is hard to quantify the number of non-isomorphs and their negative impact on performance. The maximum counts reported in these works are for the bimolecular CRN classes of size (N, M) from $(2,7)$, $(3,6)$, $(4,5)$, $(5,4)$ to $(9,4)$ and $(10,3)$, with a maximum running time of 20 days [13]; $(5,5)$ is reported in [3].

The other major challenge of checking large sets of CRNs is that storing the set of CRNs in the memory of a computer becomes impossible beyond some problem size (N, M), even when using a memory-efficient representation of the CRN. For example, the $(5,5)$ class stored in the encoding of [3] takes 64.4 gigabytes of disk space; the $(5,6)$ class of reversible CRNs takes 198 gigabytes. As such, the only way to proceed practically is to directly generate non-isomorphic CRNs using the *canonical construction path* method [21], check whether the CRN satisfies the predicate, and write those to file (or store in memory if the satisfying subset happens to be small enough).

In this paper, we present an efficient method for *generating* non-isomorphic bimolecular CRNs. Our method can determine the complete subset of CRNs of a given size that satisfy a specified property, without the need to enumerate and store in memory all non-isomorphic CRNs of that size. By creating such a generator, our method can be used to ask complete questions for larger CRN sizes than was previously possible, since memory is no longer limiting. Instead, it is limited only by the computation time of testing each non-isomorphic CRN. Our approach is based on a new graph encoding of CRNs that we name the *Complex-Species graph* (*CS-graph*), and we prove that isomorphisms of bimolecular CRNs are equivalent to isomorphisms of CS-graphs. Our method also facilitates a tighter relationship with NAUTY than previous methods, leading to efficiency benefits and high parallelisation. We are the first to report that there are precisely 535,852,102 bimolecular 2-species CRNs, and extend the counts of non-isomorphic CRNs with more than 2 species beyond what has been reported previously, including counts for $(10,5)$, $(5,6)$ and $(4,7)$ in less than a day. We provide execution times of all enumerations, and a new computational tool (genCRN) for enumerating non-isomorphic CRNs with several filters. Using genCRN, it is now possible to explore the design space of larger CRNs satisfying a given set of properties.

2   Methods

We first present the *Complex-Species graph* (CS-graph), an encoding of bimolecular CRNs into directed coloured graphs, and prove that CS-graphs faithfully encode bimolecular CRNs up to isomorphism, in the sense that two bimolecular CRNs are isomorphic if and only if their CS-graphs are isomorphic. We then explain how CS-graphs facilitate the fast enumeration of the set of all CRNs that are non-isomorphic to one another.

2.1   Complex-Species Graph Encoding

We begin with a formal definition of CRNs and CRN isomorphism. We define S to be a set of species and $\mathbb{C}$ to be the space of complexes, which is any combination of species that may appear as the reactant or product set in a reaction. A set $C \subset \mathbb{C}$ is a set of multisets, where $c \in C$ is a pair (A, m) with $A \in S$ and $m \in \mathbb{N}$.

Definition 1 (CRN). *A chemical reaction network $\mathcal{N} = (S, C, \mathcal{R})$ consists of a set of species S, a set of complexes $C \subset \mathbb{C}$, and a set of reactions $\mathcal{R} \subset C \times C$ with $(y, y) \notin \mathcal{R}$ for any $y \in C$.*

Two CRNs are isomorphic when they are identical under species renaming:

Definition 2 (CRN isomorphism). *Let $\mathcal{N}_1$ and $\mathcal{N}_2$ be chemical reaction networks. $\mathcal{N}_1$ is isomorphic with $\mathcal{N}_2$, or $\mathcal{N}_1 \cong \mathcal{N}_2$, if there exists a permutation π over S such that $\mathcal{N}_1\pi = \mathcal{N}_2$.*

Here, we have written the function application $\mathcal{N}_1\pi$ in postfix notation. Note that the reaction rates are not relevant in CRN isomorphism, so are not included in this definition and are omitted from the remainder of this paper.

Before introducing CS-graphs, we introduce a technical device to more conveniently index the elements of a set:

Definition 3 (Indexed set). *Let I and S be sets, and f be a bijection $I \to S$. The set $\{S_i\}_{i \in I} \triangleq \{S_i \mid S_i = f(i), i \in I\}$ is an indexed set, and I is the indexing set. We write S_i for $S_i = f(i)$ with $i \in I$ when f is clear from the context.*

If $\mathcal{N} = (S, C, \mathcal{R})$ is a CRN, we indicate with $\{c_i\}_{i \in I}$ the indexed set of complexes occurring in $\mathcal{N}$, and with $\{A_j\}_{j \in J}$ the indexed set of species occurring in $\mathcal{N}$. For the remainder of this section, the indexing sets I and J always index respectively the complexes and the species of a CRN; moreover, we assume that $I \cap J = \emptyset$ and $I, J \subset \mathbb{N}$. We are now ready to define the CS-graph of a bimolecular CRN:

Definition 4 (Complex-Species graph). *Let $\mathcal{N}$ be a bimolecular CRN with indexed sets $\{c_i\}_{i \in I}$ and $\{A_j\}_{j \in J}$. The Complex-Species graph $[\![\mathcal{N}]\!]_J^I$ is the quadruple $\langle V, E, \sigma, \rho \rangle$, where:*

$$
\begin{aligned}
V &= I \cup J & \text{(Vertices)} \\
E &= \{(j, i) \mid A_j \text{ occurs in } c_i\} & \text{(Edges)} \\
&\cup \{(i_1, i_2) \mid c_{i_1} \to c_{i_2} \text{ occurs in } \mathcal{N}\}
\end{aligned}
$$

$$
\sigma(i) = \begin{cases} \emptyset & \text{if } c_i = \emptyset \\ \square & \text{if } c_i = A_j \\ 2\square & \text{if } c_i = 2A_j \\ \square\square & \text{if } c_i = A_{j_1} + A_{j_2} \end{cases} \qquad \text{(Stoichiometry function)}
$$

$$
\rho(j) = A_j \qquad \text{(Labelling function)}
$$

for $i, i_1, i_2 \in I$ and $j, j_1, j_2 \in J$.

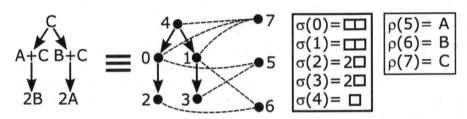

Fig. 1. Complex-Species graph of a bimolecular CRN. The set $I = \{0, 1, 2, 3, 4\}$ is the indexing set for the complexes of the CRN, while $J = \{5, 6, 7\}$ is the indexing set for its species. The concrete names of the indexes are unimportant; any disjoint set I and J can be used. Set I indexes the complexes of a CRN, the stoichiometry function σ assigns a multiplicity to each index (e.g. σ assigns heterodimer to node 0, homodimer to node 2 and monomer to node 4). An edge between two nodes in I represents a CRN reaction. Set J indexes the CRN species, with labelling function ρ assigning them concrete species names. A dashed edge from node $j \in J$ to node $i \in I$ means that species $\rho(j)$ occurs in complex i.

Figure 1 provides a visual representation of a Complex-Species graph. Notice that it is not possible to distinguish monomers from homodimers using the encoding's vertices and edges alone; this is accomplished by σ. The indexing set I in the figure is the same indexing set returned by NAUTY for that digraph. Appendix C shows an extension to CS-graphs to encode CRNs with higher molecularity.

Two CS-graphs are isomorphic when their underlying graphs are isomorphic and have the same stoichiometry:

Definition 5 (CS-graph isomorphism). *Let* $[\![\mathcal{N}_1]\!]_{J_1}^{I_1} = \langle V_1, E_1, \sigma_1, \rho_1 \rangle$ *and* $[\![\mathcal{N}_2]\!]_{J_2}^{I_2} = \langle V_2, E_2, \sigma_2, \rho_2 \rangle$. *Complex-Species graph* $[\![\mathcal{N}_1]\!]_{J_1}^{I_1}$ *and* $[\![\mathcal{N}_2]\!]_{J_2}^{I_2}$ *are isomorphic, or* $[\![\mathcal{N}_2]\!]_{J_2}^{I_2} \cong [\![\mathcal{N}_2]\!]_{J_2}^{I_2}$, *if there exist bijections* $\alpha : I_1 \to I_2$ *and* $\beta : J_1 \to J_2$ *such that:*

1. $V_1 \alpha \beta = V_2$
2. $E_1 \alpha \beta = E_2$
3. $\sigma_1 \alpha = \sigma_2$

where $\alpha\beta$ *stands for the function composition of* α *and* β.

As already pointed out, the actual indexing sets used in a CS-graph are unimportant. As a matter of fact, we can show that CS-graphs of the same CRN are all isomorphic with each other:

Lemma 1. *Let* $\mathcal{N}$ *be a bimolecular CRN. Then* $[\![\mathcal{N}]\!]_{J_1}^{I_1} \cong [\![\mathcal{N}]\!]_{J_2}^{I_2}$ *holds for any indexing sets* I_1, I_2, J_1, J_2.

Proof. The lemma is proved by explicitly constructing bijections $\alpha = \{(i_1, i_2)|c_{i_1} = c_{i_2}$ for $i_1 \in I_1, i_2 \in I_2\}$ and $\beta = \{(j_1, j_2) \mid A_{j_1} = A_{j_2}$ for $j_1 \in J_1, j_2 \in J_2\}$ that satisfy Definition 5. See Appendix B for more details.   □

Having proved this result, and since CRN isomorphism provides a permutation of species π such that two CRNs become equal, it is easy to show that CRN isomorphism implies CS-graph isomorphism:

Lemma 2. *Let $\mathcal{N}_1$ and $\mathcal{N}_2$ be bimolecular CRNs. If $\mathcal{N}_1 \cong \mathcal{N}_2$, then $[\![\mathcal{N}_1]\!]_{J_1}^{I_1} \cong [\![\mathcal{N}_2]\!]_{J_2}^{I_2}$ for any indexing sets I_1, I_2, J_1 and J_2.*

Proof. By Definition 2, there exists a permutation π over the species of $\mathcal{N}_1$ such that $\mathcal{N}_1\pi = \mathcal{N}_2$. Notice that by Lemma 1 we can deduce $[\![\mathcal{N}_1\pi]\!]_{J_1}^{I_1} \cong [\![\mathcal{N}_2]\!]_{J_2}^{I_2}$ for any indexing sets I_1, I_2, J_1 and J_2. The lemma is proved by taking $\alpha = \{(i_1, i_2) \mid c_{i_1}\pi = c_{i_2} \text{ for } i_1 \in I_1, i_2 \in I_2\}$ and $\beta = \{(j_1, j_2) \mid A_{j_1}\pi = A_{j_2} \text{ for } j_1 \in J_1, j_2 \in J_2\}$. □

When two CS-graphs are isomorphic, the indexed sets of complexes and species provide enough information to reconstruct an isomorphism π for their original CRNs:

Lemma 3. *Let $\mathcal{N}_1$ and $\mathcal{N}_2$ be bimolecular CRNs with indexing sets respectively I_1, J_1 and I_2, J_2. If $[\![\mathcal{N}_1]\!]_{J_1}^{I_1} \cong [\![\mathcal{N}_2]\!]_{J_2}^{I_2}$, then $\mathcal{N}_1 \cong \mathcal{N}_2$.*

Proof. Let $[\![\mathcal{N}_1]\!]_{J_1}^{I_1} = \langle V_1, E_1, \sigma_1, \rho_1 \rangle$ and $[\![\mathcal{N}_2]\!]_{J_2}^{I_2} = \langle V_2, E_2, \sigma_2, \rho_2 \rangle$, such that $[\![\mathcal{N}_1]\!]_{J_1}^{I_1} \cong [\![\mathcal{N}_2]\!]_{J_2}^{I_2}$. By hypothesis, $[\![\mathcal{N}_1]\!]_{J_1}^{I_1} \cong [\![\mathcal{N}_2]\!]_{J_2}^{I_2}$ implies the existence of bijections α and β that satisfy conditions 1 to 3 in Definition 5. The lemma is proved by taking $\pi = \{(A_{j_1}, A_{j_2}) \mid j_1\beta = j_2\} \circ \pi_I$ where π_I is the identity function over $\mathcal{S}$. See Appendix B for more details. □

We can now show that CS-graphs are a faithful encoding of bimolecular CRNs up to isomorphism:

Theorem 1 (Faithful encoding). *Let $\mathcal{N}_1$ and $\mathcal{N}_2$ be bimolecular CRNs with indexing sets respectively I_1, J_1 and I_2, J_2. Then $\mathcal{N}_1 \cong \mathcal{N}_2$ if and only if $[\![\mathcal{N}_1]\!]_{J_1}^{I_1} \cong [\![\mathcal{N}_2]\!]_{J_2}^{I_2}$.*

Proof. By Lemmas 2 and 3. □

2.2 Isomorph-Free Complex-Species Graphs Enumeration

Our non-isomorphic CS-graph enumeration method entails the generation of all non-isomorphic bimolecular CRNs, by virtue of Theorem 1. Our method inputs are the numbers of complexes L, reactions M and species N of the output CRNs. CS-graphs are generated through four successive enumeration stages, where each stage turns a structure generated in the previous stage into a list of more refined non-isomorphic structures (Fig. 2).

The first stage enumerates all undirected graphs with L nodes and M edges. Each undirected graph represents the topology of a CRN. The second stage orients the edges of an undirected graphs in all possible directions, including both directions at the same time (by replacing an undirected edge with two opposite directed edges). A directed edge between two nodes represents a reaction between

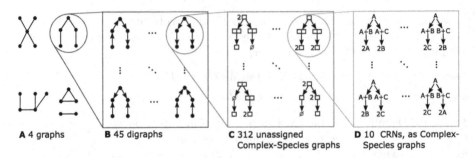

Fig. 2. Isomorph-free generation of 3-species bimolecular CRNs with 5 complexes and 4 reactions. CRN enumeration comprises four stages. It starts with generating the list of all non-isomorphic undirected graphs with 5 vertices and 4 edges (A); for each graph the list of all non-isomorphic directed graphs (B); for each directed graph the list of all non-isomorphic unassigned CS-graphs (C), and finally for each unassigned CS-graph the list of all non-isomorphic CRNs (D). The total count of elements enumerated from a single element in each stage is reported. There are 428,502 non-isomorphic CRNs against 635,040 isomorphs in total.

two complexes; two opposite edges mean a reversible reaction. For each directed graph, the third stage assigns all possible stoichiometries σ to the nodes, generating a list of *unassigned CS-graphs*, that is CS-graphs without species nodes. The fourth stage finds all possible assignments for N species nodes to an unassigned CS-graph, therefore listing non-isomorphic CRNs by Theorem 1.

The state-structured enumeration we present (Fig. 2) is based on McKay's *canonical construction path* method [21], whereby a generation of larger structures is first constructed from a previous generation of smaller non-isomorphic structures, and then filtered out by some *canonical form* function f. This function maps all structures in an isomorphism class to the same structure in that class, which is called *canonical*. Only the canonical form is retained from the generated structures.

For example, let G be a graph of size n. A new graph G' of size $n+1$ can be obtained by adding a new node to G and a new set of edges between the new node and any subset of nodes in G. The new graph G' is discarded unless G' is in canonical form, i.e. $G' = f(G')$ for a canonical function f. A simple but inefficient example of $f(G')$ is to apply all possible node permutations to G', sort the resulting graphs by lexicographic order on their edges, and return the least graph in the sorting. Starting from the empty graph, it is then possible to enumerate all graphs by iteratively constructing and filtering larger non-isomorphic structures.

2.3   Enumeration Invariants and Implementation Details

Although the generation of classes of larger structures might grow combinatorially, in practice the judicious use of graph invariants reduces this number greatly [6,21]. A notable example is the graph isomorphism tool suite by McKay, based on NAUTY [22]. NAUTY is a fast coinductive algorithm to find a graph's

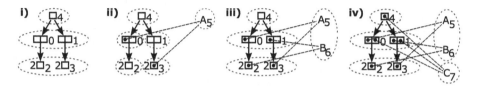

Fig. 3. Assignment of three species to an unassigned CS-graph in four steps. To simplify the visualisation, complex multiplicities have been overlaid over the complex nodes, where complex and species indices are subscript next to each node in grey. The red dashed circles indicate the existence of an *automorphism*, a permutation of nodes that maps the CS-graph to itself. For example, in step (i) the permutation $[0 \mapsto 1, 1 \mapsto 0, 2 \mapsto 3, 3 \mapsto 2, 4 \mapsto 4]$, or (01)(23)(4) in permutation cycle notation [31], returns the same CS-graph. At step (ii) the only automorphism is the identity permutation. The resulting CS-graph represents the CRN from Fig. 1; the automorphism reveals that species A and B are symmetric in the CRN. (Color figure online)

canonical labelling and its automorphism group [31]. A graph automorphism is a permutation of vertices that maps the graph onto itself. The actions of the automorphism group can generate a graph's isomorphism class very efficiently. Moreover, the generators can be used early to avoid generating non-canonical candidate structures immediately.

Graphs and directed graphs in the first and second stage of Fig. 2 are generated respectively by `geng` and `directg`, two enumeration programs available in the NAUTY tool suite. The enumeration of unassigned CS-graphs can be encoded as graph-vertex colouring problems. We use four colours, one per stoichiometry type (naught, monomer, homodimer, heterodimer). The enumeration of a coloured graph is performed by another NAUTY tool, `vcolg`. However, not all graph colourings result in a valid CRN stoichiometry: for example, a 3-species CRN cannot have 4 monomers, since there are only 3 species available to make a monomer from. We have thus modified `vcolg` to enumerate valid stoichiometries only, by providing an upper bound for the number of nodes with each specific colour. A valid CRN can have at most one naught complex, n monomers, n homodimers and $\binom{n}{2}$ heterodimers. The total number of complexes for a given maximum number of reactants p is given by the sum of the multiset coefficients:

$$L_p(n) := \sum_{i=0}^{p} \left(\!\binom{n}{i}\!\right) = \sum_{i=0}^{p} \binom{n+i-1}{i} = \binom{n+p}{p} \tag{2}$$

which for bimolecular CRNs with N species is $\binom{N+2}{2}$.

The last stage is the enumeration of CRNs from an unassigned CS-graph, for which we have developed a custom algorithm following the canonical construction path method [21]. The structures we augment are partially assigned CS-graphs, starting from an unassigned one. Larger structures are obtained by adding a new species node, together with a set of edges that assign the new species to a subset of the complex nodes in the graph, until all N have been added and all complexes are valid. We call *species assignment* the set of complex nodes targeted by the new species; for example, in step (ii) of Fig. 3, $\{0, 3\}$ is the species assignment for A.

As pointed out in [20], it is crucial to exploit graph invariants in order to curb the number of larger structures to test for canonicity. We adopt some of geng's invariants in our enumeration method when adding a new species node. In order to avoid constructing the same graph more than once by adding the same species assignments in a different order, we impose a lexicographic order on the species assignments. For example, the choice of species assignment $\{1, 2\}$ at step (iii) of Fig. 3 is allowed, because it is greater than the previous assignment $\{0, 3\}$ at step (ii); if $\{1, 2\}$ is chosen first, $\{0, 3\}$ breaks the lexicographic order and is illegal. Similarly, the cardinality of the species assignment must be equal or greater than the previous one. As for vertex colouring, not all species assignments are valid; for example, the same species cannot be assigned to two different monomer complexes, or a species to naught. Such assignments are discarded immediately.

After augmenting a CS-graph with a new species, it is tested for canonicity. The test applies all possible automorphisms α and β to the current CS-graph G: if G is the least graph of all $G\alpha\beta$ graphs by lexicographic order, then G is in canonical form, and used to assign more species to it. When the automorphism group is trivial (the only automorphism is the identity) any species assignment added in lexicographic order is already canonical.

The fourth stage does not add new complex nodes or edges, therefore the automorphisms β over complexes are either the same or they decrease after adding new species assignment, which might introduce asymmetries in the graph. For example, the species assignments of A in step (ii) of Fig. 3 introduces an asymmetry that renders the automorphism graph trivial (the only automorphism is the identity); however adding a species assignment B of the same cardinality in step (iii) restores the group. For this reason we only recompute a CS-graph's automorphism group after increasing the cardinality of its species assignments (which, as previously pointed out, are only added in increasing order).

An implementation of our method is available online at https://github.com/CSpaccasassi/genCRN for Windows and Unix systems. Our tool, called genCRN, implements the third and fourth stage of Fig. 2, and relies on inputs from geng and directg. genCRN is based on version 2.6 of NAUTY, where geng only generates graphs with a maximum size of 32 nodes. Our tool has the same limitation, it can only produce CS-graphs of size $|I \cup J| \leq 32$. Later versions of NAUTY raise this limit to 64 nodes; we leave the extension of our implementation to 64 nodes for future work.

3   Results

3.1   Complete Enumeration of Non-isomorphic 2-Species CRNs

We applied our CRN enumeration technique to count how many non-isomorphic CRNs there exist with specified numbers of species and reactions. When considering only 2 species, we are able to provide a complete construction, covering all possible numbers of reactions (Fig. 4). Overall, we find that there are 536,884,871 non-isomorphic CRNs with 2 species.

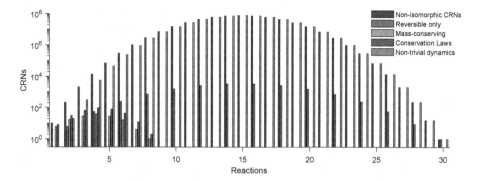

Fig. 4. Enumeration of non-isomorphic CRNs with 2 species. All non-isomorphic 2-species CRNs were enumerated (blue bars), and then filtered according to four criteria (separately): CRNs with only reversible reactions (red bars), mass-conserving CRNs (orange bars), CRNs with conservation laws (purple bars) and CRNs with non-trivial dynamics (green bars). (Color figure online)

The practical utility of non-isomorphic CRN enumeration is that it enables testing of properties of CRNs against a smaller set. Compared to the naive approach of writing all possible reactions among the N and picking all combinations of size M, checking only the non-isomorphic subset amounts to a computational saving of at least $N!$. We also considered four simple filters of properties over the 2-species CRNs. Due to the tree structure of the approach (Fig. 2), applying filters as early as possible is preferable.

The first filter we checked was to restrict the enumeration to CRNs that have only reversible reactions. To achieve this, we modified our approach (Fig. 2) to skip part B (enumeration of directed graphs from undirected graphs), and immediately constructed undirected unassigned CS-graphs. Accordingly, each edge can be viewed as a reversible reaction. This enabled us to rapidly compute all reversible-only 2-species CRNs, of which there were exponentially fewer examples, yet following a similar Gaussian-like distribution over M (Fig. 4).

The second filter we applied was to identify *mass-conserving* CRNs, using the defining feature that there exists a vector $v \in \mathbb{R}^N_{>0}$ (all entries strictly positive) such that $v.\Gamma_k = 0$ for all k, where Γ is the stoichiometry matrix. For example, the CRN $A \to B$ is mass-conserving. It's stoichiometry matrix is $\Gamma = [-1, 1]^\top$, and so $v = [1, 1]$ can satisfy the property. To test for the existence of such a v in general, we used a Fourier-Motzkin algorithm to identify invariants $v \in \mathbb{R}^N$ [10]. As such invariants may include zero entries, we do an additional check to see whether all species participate in an invariant. E.g. strictly positive v can be constructed from the set of invariants. Rather than applying the filter directly to the complete set of non-isomorphic CRNs, we can obtain a computational saving by first removing the naught complex $\emptyset$ from $\mathcal{C}$, since any reaction involving $\emptyset$ would not be mass-conserving. In total, there were only 138 mass-conserving CRNs with 2 species, the largest of which had 8 reactions.

$$A \to B \qquad A + B \to 2A \qquad 2A \to A + B \qquad 2A \to 2B$$
$$B \to A \qquad A + B \to 2B \qquad 2B \to A + B \qquad 2B \to 2A$$

This CRN simply includes all reactions that preserve the total molecule count. However, they are not the only reactions that are mass-conserving on their own. For example, the CRN $2A \to B$ is also mass-conserving, though now B has equivalent mass to 2 copies of A (e.g. this is simply homo-dimerisation). Accordingly, the counting of mass-conserving CRNs is not trivial.

We next identified CRNs for which there exists any conservation law, e.g. there exists a vector v such that $v.\Gamma_k = 0$ for all k, but in contrast to strictly mass-conserving CRNs, now v can include zero entries, as not all species need to participate in a conservation law for one to exist. Single-reaction examples include $A \to B$ ($A + B$ is conserved), $A \to A + B$ (A is conserved) and $A \to 2B$ ($2A + B$ is conserved). As before, we used the Fourier-Motzkin algorithm, but this time simply as a filter applied to the same enumeration approach for the full non-isomorphic set. We found 330 such CRNs (Fig. 4). As for the mass-conserving CRNs, there were no CRNs with more than 8 reactions, though this time an additional CRN was found:

$$\emptyset \to B \qquad \emptyset \to 2B \qquad B \to 2B \qquad A \to A + B$$
$$B \to \emptyset \qquad 2B \to \emptyset \qquad 2B \to B \quad A + B \to A$$

Notably, this CRN includes $\emptyset$ complexes, but these only appear in reactions not interacting with the species A. Instead, the species A only participates by catalysing the production and degradation of B, and is not produced or consumed in these reactions. As such, A is conserved in this CRN.

Finally, we considered "dynamically non-trivial" CRNs [3], which can give rise to positive equilibria, periodic orbits, and other "interesting" properties. Dynamically trivial CRNs, in contrast, have no limit sets. e.g. trajectories grow unbounded in phase space. To enumerate dynamically non-trivial CRNs, we use the definition in [3], that a CRN $\mathcal{N}$ is dynamically trivial if there exists a vector $q > 0$ in $\mathrm{Im}\,\Gamma^\top$. [14] Accordingly, we find the reduced row echelon form of Γ and ask whether any row contains only non-negative entries (though not all zero). As done in [3], we take the set of non-isomorphic CRNs, and then check each CRN. Applying this filter to the 2-species CRNs reveals a considerably smaller number of CRNs with non-trivial dynamics than the full set, when there are few reactions (Fig. 4). As the number of reactions increases, the fraction of CRNs that are dynamically non-trivial tends to 1.

3.2 Enumeration of Non-isomorphic CRNs with More Than 2 Species

Owing to the combinatorial nature of CRNs, simply extending to 3 species leads to an exponential increase in the number of possible CRNs. Using our enumeration method, we found that there are 1,244,363,180 bimolecular 3-species CRNs with $M = 7$ reactions, more than twice the number of all 2-species CRNs (Fig. 5A). Adding another reaction ($M = 8$) increases by a factor of 10 (12,916,870,803) and for $M = 9$ another factor of 10 (117,703,409,335). We enumerated and counted non-isomorphic CRNs with up to 10 species and with a number of reactions that could be evaluated within approximately 2 days of computation. In doing so, we have extended the known number of non-isomorphic CRNs beyond what was previously evaluated in [3], and have tabulated these values in Appendix D (Table 1). We have also evaluated non-isomorphic reversible CRNs with up to 8 species (Fig. 10 and Table 2) and non-isomorphic CRNs with non-trivial dynamics (Fig. 11 and Table 3).

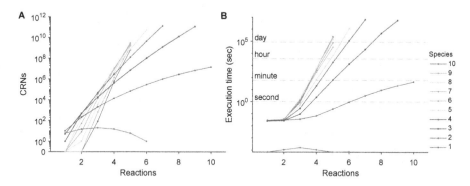

Fig. 5. Counting non-isomorphic CRNs. A. The total number of non-isomorphic CRNs is quantified for up to 10 species and up to 10 reactions. B. Execution time is quantified as if running on a single-core computer, by summing times over a parallel execution on an Intel Xeon Platinum 8168 2.70 GHz machine with 72 cores.

While there is no reason our enumeration method cannot handle more reactions, additional combinatorial complexity leads to longer run times (Figs. 5B, 10B and 11B). We have quantified the execution times for each combination of species and reactions (N, M) by reporting the values as if the calculation was run sequentially on a single-core machine. In practice, we perform calculations in parallel, enumerating each digraph independently and collecting results.

McKay's labelling algorithm is known to have exponential complexity in the worst-case [25] but is well-behaved in practice. Similarly, there is no precise complexity for the canonical construction path method, although it depends on the size of the graph's automorphism groups [21]. In line with this, we found considerable differences in the execution times of each digraph (Fig. 6). The most adversarial digraphs are those which contain the highest number of disconnected sub-digraphs. Examples of such digraphs are those with $2n$ nodes and n edges, resulting in n sub-digraphs. The topology of such digraphs poses little constraints on the topology of the CRNs, and therefore the number of possible CRNs arising from such digraphs is combinatorially larger.

3.3 The Non-isomorphic CRNs Fraction

To gain a more quantitative understanding of the frequency of isomorphisms among sets of CRNs, we computed an *isomorph ratio*, defined simply as the number of non-isomorphic CRNs found, divided by the total number T of CRNs, for a given number of species and reactions. Using

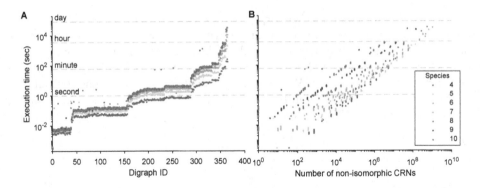

Fig. 6. Execution times vary considerably with input digraph. The execution times of enumerating non-isomorphic CRNs are shown for each of the 365 digraphs of CRNs with 5 reactions. A. The digraphs are sorted by the execution times corresponding to 7 species, which illustrates that the variation in execution time is strongly influenced by digraph structure. B. Execution times are compared against the total number of non-isomorphic CRNs found.

$$T = \binom{L_p(N)(L_p(N) - 1)}{M} - \sum_{k=1}^{s-1} \binom{L_p(k)(L_p(k) - 1)}{M}, \tag{3}$$

where $L_p(N) = \binom{N+p}{p}$ is the number of complex nodes in the CS-graph when there are N species, we computed the isomorph ratios for bimolecular CRNs ($p = 2$) with up to 5 species and 6 reactions (Fig. 7). By considering species relabellings alone, one would naively expect a factor $N!$ saving when considering isomorphisms. As there are $\binom{N+2}{2}$ possible complexes for N species (Eq. 2), there are 6 complexes for 2 species: $\{\emptyset, A, B, 2A, 2B, A + B\}$. As there are $L(L - 1)$ possible (directed) edges connecting L nodes, there are 30 possible reactions for 2 species. Without considering CRN isomorphisms, this would result in $\sum_{r=1}^{30} \binom{30}{r} = 1,073,741,823$ possible CRNs. Whereas, we found that there are 536,884,871 non-isomorphic CRNs with 2 species, which is just more than half of the concretely labelled set. There are $N!$ permutations of N species, and so for most non-isomorphic CRNs, the $N!$ species permutations leads to an $N!$ different CRNs. However, some CRNs are *species-symmetric*, for example $A \rightleftharpoons B$, which means that a species relabelling can sometimes return the exact same CRN. Because such symmetries are automatically resolved in our calculation of the number of non-isomorphic CRNs, but not incorporated into Eq. 3, the isomorph ratio can be less than $1/N!$ (Fig. 7).

3.4   Checking Properties of CRNs with External Tools

As mentioned above, a practical benefit of using non-isomorphic CRN enumeration is that filters can be applied to a stream of CRNs, producing subsets of CRNs satisfying a property of interest. Such a property need not be implemented in the same code base as the CRN enumerator, since results can be piped

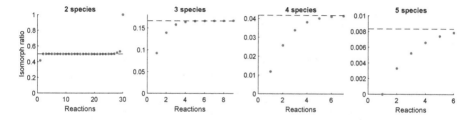

Fig. 7. The isomorph ratio. The ratio of non-isomorphic CRNs to the expected total number of CRNs was computed for different numbers of species and reactions. The expected total number of CRNs was calculated using Eq. 3. The dashed black line indicates the value of $1/N!$, the reciprocal of the number of species relabellings in a CRN with N species.

into external tools. To demonstrate this, we considered the existence of *forward bisimulations* of CRNs [7], using the ERODE tool [8] (Fig. 8). The existence of a forward bisimulation means that a subset of the species can be lumped into a single species, the result being a different CRN with fewer species but with the same behaviour. As such, the analysis of a CRN for which there exists a forward bisimulation can be considered to have been covered by equivalent analysis of CRNs with fewer species. The set of CRNs which are connected, have non-trivial dynamics and are irreducible via forward bisimulation was determined in less than 5 min, despite there being as many as 10^{11} CRNs initially covered by our encoding.

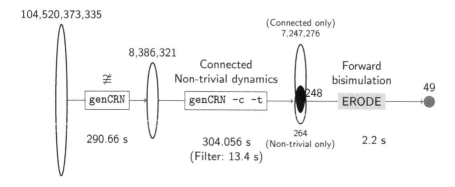

Fig. 8. Identifying non-isomorphic non-trivial connected CRNs with no forward bisimulation. CRNs with 7 species and 4 reactions were enumerated using genCRN, both in the presence and absence of filters for connectedness (-c) and non-trivial dynamics (-t). The resultant 251 CRNs were processed by ERODE, producing 49 CRNs for which no forward bisimulation exists.

4   Discussion

In this paper we have presented a method for the fast enumeration of non-isomorphic Chemical Reaction Networks, which enables *complete* statements on properties of classes of CRNs. The method is based on a novel encoding of CRNs into *Complex-Species graphs*, which are enumerated using established techniques such as the canonical construction path method [6,21], and implemented on top of the NAUTY tool suite [5,22]. We have shown that classes of non-isomorphic CRNs can be further specialised into classes satisfying certain properties of interest, such as mass convervation, non-trivial dynamics, reversible networks and non-lumpability under Forward Bisimulation.

We are the first, to the best of our knowledge, to report that there are precisely 535,852,102 bimolecular 2-species CRNs in total; a surprising number for just two species. The method is highly efficient, and can calculate this count in ≈27 min. It is also highly parallelisable; the same count on a 72 core machine takes ≈ 38seconds. Currently our implementation enumerates unassigned CS-graphs and CRNs in a single step. Deeper parallelisation could be achieved by splitting this step into two, for handling more heavily combinatorial digraphs. We have extended previously known counts for CRNs in excess of 6 species or 6 reactions (see Appendix). Our counts and measurements are reproducible via genCRN, which is available online; comparison with other approaches [3,13] is difficult, because we have not found accompanying tools or time measurements.

Compared to other encodings, Complex-Species graphs are encoded in terms of digraphs and vertex colouring, and as such it has been easier to enumerate them with existing techniques and tools. For comparison, Species-Reaction graphs are expressed using multidigraphs or digraphs with edge labels, in order to express the multiplicity of a species in a reaction. For example, reaction $2A \rightarrow B$ is encoded either by two edges from a species node A to some reaction node R or by an edge with the label 2, to capture the fact that $2A$ is a heterodimer. NAUTY does not support multidigraphs or labelled digraphs natively [5], so CRN encodings relying on these lead to the *production* of isomorphs, which must be stored and filtered out in a subsequent step [3]. Interesting graph-theoretic results exist for complex graphs [15,16] and various directed or undirected bipartite graphs [12,23,24]; once the CRNs have been enumerated (and filtered) as CS-graphs, they can be translated into different representations for further analysis.

In future, it would be interesting to explore more advanced properties of CRNs, such as lumpability, multistability and limit cycles. Our tool allows testing sets of CRNs for overlapping properties, such as CRNs with no conservation laws and lumpable under Forward Bisimulation, or mass-conserving CRNs which are not lumpable under Forward Bisimulation, such as $2A \rightleftharpoons B$. Moreover, it would be interesting to verify properties on the unassigned CS-graphs; we conjecture that trivial dynamics occur when any species, independently from the others, is only assigned to *multiplicity monotonic paths* in an unassigned CS-graph, that is non-cyclic paths where the multiplicity of that species is ever increasing.

Our approach could be beneficial to the study of non-mass-action reaction systems such as Gene Regulatory Networks [18], or reaction-diffusion systems

[19]. For example, special species roles as *fast diffuser, slow diffuser* and *other* could be encoded as ulterior nodes in a CS-graph, connected to species nodes and used as a further enumeration step. Broader applications in computer science might also be possible, to enumerate programs against a set of primitives, unique up to α-conversion. Species nodes might represent variables in the lambda calculus, or channels names in the π-calculus or CCS.

Acknowledgements. We would like to thank Brendan McKay, who extended NAUTY's vertex-colouring algorithm to directed graphs on our request, without which this work would have not been possible. We would also like to thank Andrea Vandin and Mirco Tribastone for providing a command-line version of ERODE and for useful discussions. Finally, we thank an anonymous reviewer for helping us to identify an error in the algorithm used for filtering dynamically trivial networks.

A   Definitions

This section introduces definitions for automorphisms, orbits and the automorphism group for graphs, following [31].

Definition 6 (Permutation). *A permutation of a set S is a total function from S to itself.*

Definition 7 (Cyclic permutation). *A permutation π of the form:*

$$\begin{pmatrix} x & \pi(x) & \pi^2(x) \cdots \pi^{p-2}(x) & \pi^{p-1}(x) \\ \pi(x) & \pi^2(x) & \pi^3(x) \cdots \pi^{p-1}(x) & x \end{pmatrix}$$

is said to be cyclic permutation of period p.

Definition 8 (Disjoint cycle representation). *A disjoint cycle representation of a permutation π on a set S is a composition of cyclic permutations on subsets of S that constitute a partition of S, one cyclic permutation for each subset in the partition.*

Definition 9 (Group). *An algebraic system $<U, \star>$ is a called a group if it has the following properties:*

1. *the operation $\star$ is associative,*
2. *there is an identity element,*
3. *every element of U has an inverse.*

Definition 10 (Permutation group). *A closed non-empty collection P of permutations on a set Y of objects that forms a group under the operation of composition is called a* permutation group. *The combined structure may be denoted $V = [P : V]$. It is often denoted P when the set of Y objects is understood from context.*

Definition 11 (Orbit). *Let* $\mathcal{P} = [P : Y]$ *be a permutation group, and let* $y \in Y$. *The* orbit *of the object* y *under the action of* P *is the set* $\{\pi(y) \mid \pi \in P\}$.

Corollary 1. *Let* $\mathcal{P} = [P : Y]$ *be a permutation group. Then being coorbital is an equivalence relation*

Proof. Identity: by the identity permutation. Commutativity: because each π is invertible. Transitivity: by function composition $\circ$.

B   Proofs

Lemma 1. *Let* $\mathcal{N}$ *be a bimolecular CRN. Then* $[\![\mathcal{N}]\!]_{J_1}^{I_1} \cong [\![\mathcal{N}]\!]_{J_2}^{I_2}$ *holds for any indexing sets* I_1, I_2, J_1, J_2.

Proof. The lemma can be proved by explicitly constructing bijections α and β required by Definition 5. Recall that we indicate with $\{c_i\}_{i \in I}$ and $\{A_j\}_{j \in J}$ respectively the indexed set of the complexes and of the species in $\mathcal{N}$.

Let $\alpha = \{(i_1, i_2) \mid c_{i_1} = c_{i_2}$ for $i_1 \in I_1, i_2 \in I_2\}$ and $\beta = \{(j_1, j_2) \mid A_{j_1} = A_{j_2}$ for $j_1 \in J_1, j_2 \in J_2\}$. These functions are well-defined because the indexing sets all target the same CRN $\mathcal{N}$. It is also easy to show that they are bijections.

The lemma is proved by verifying that α and β satisfy Condition 1 to 3 of Definition 5:

1. $V_1 \alpha \beta = V_2$ because α and β are bijections over the indexed sets;
2. $E_1 \alpha \beta = \{(j_1, i_1) \mid A_{j_1} \in c_{i_1}\} \alpha \beta \cup \{(i_1, i_1') \mid c_{i_1} \to c_{i_1'} \in \mathcal{R}\} \alpha \beta$ by Definition 4
   $= \{(j_2, i_2) \mid A_{j_2} \in c_{i_2}\} \cup \{(i_2, i_2') \mid c_{i_2} \to c_{i_2'} \in \mathcal{R}\}$ by def. of α, β.
   $= E_2$
   which proves the case.
3. Let i_1 be such that $\sigma_1(i_1) = \emptyset$. By Definition 4, $c_{i_1} = \emptyset$, and since $\alpha(i_1) = i_2$ implies that $c_{i_1} = c_{i_2}$, then $c_2 = \emptyset$ as well. Therefore $\sigma(i_2) = \emptyset$ holds by Definition 4, which implies $\sigma(i_1)\alpha = \sigma_2(i_2)$. The proof for the remaining cases (monomers, homodimers and heterodimers) is similar.   □

Lemma 2. *Let* $\mathcal{N}_1$ *and* $\mathcal{N}_2$ *be bimolecular CRNs. If* $\mathcal{N}_1 \cong \mathcal{N}_2$, *then* $[\![\mathcal{N}_1]\!]_{J_1}^{I_1} \cong [\![\mathcal{N}_2]\!]_{J_2}^{I_2}$ *for any indexing sets* I_1, I_2, J_1 *and* J_2.

Proof. By definition of CRN isomorphism (Definition 2), there exists a permutation π over the species of $\mathcal{N}_1$ such that $\mathcal{N}_1 \pi = \mathcal{N}_2$. Notice that by Lemma 1 we can deduce $[\![\mathcal{N}_1 \pi]\!]_{J_1}^{I_1} \cong [\![\mathcal{N}_2]\!]_{J_2}^{I_2}$ for any indexing sets I_1, I_2, J_1 and J_2. The proof of this lemma is similar to Lemma 1, by defining $\alpha = \{(i_1, i_2) \mid c_{i_1} \pi = c_{i_2}$ for $i_1 \in I_1, i_2 \in I_2\}$ and $\beta = \{(j_1, j_2) \mid A_{j_1} \pi = A_{j_2}$ for $j_1 \in J_1, j_2 \in J_2\}$.   □

Lemma 3. *Let* $\mathcal{N}_1$ *and* $\mathcal{N}_2$ *be bimolecular CRNs with indexing sets respectively* I_1, J_1 *and* I_2, J_2. *If* $[\![\mathcal{N}_1]\!]_{J_1}^{I_1} \cong [\![\mathcal{N}_2]\!]_{J_2}^{I_2}$, *then* $\mathcal{N}_1 \cong \mathcal{N}_2$.

Proof. Let $[\![\mathcal{N}_1]\!]_{J_1}^{I_1} = \langle V_1, E_1, \sigma_1, \rho_1 \rangle$ and $[\![\mathcal{N}_2]\!]_{J_2}^{I_2} = \langle V_2, E_2, \sigma_2, \rho_2 \rangle$, such that $[\![\mathcal{N}_1]\!]_{J_1}^{I_1} \cong [\![\mathcal{N}_2]\!]_{J_2}^{I_2}$. By hypothesis, $[\![\mathcal{N}_1]\!]_{J_1}^{I_1} \cong [\![\mathcal{N}_2]\!]_{J_2}^{I_2}$ implies the existence of bijections α and β that satisfy conditions 1 to 3 in Definition 5. Let us define the following permutation of $\mathcal{S}$:

$$\pi = \{(A_{j_1}, A_{j_2}) \mid j_1\beta = j_2\} \circ \pi_I$$

where π_I is the identity function over $\mathcal{S}$. Since β and π_I are bijections, then π is also a bijection; since its domain and range are $\mathcal{S}$, π is a well-defined permutation.

Let $c_{i_1} \to c_{i'_1}$ be a reaction in $\mathcal{N}_1$. By Definition 4 E_1 contains the edge (i_1, i'_1). Since $[\![\mathcal{N}_1]\!]_{J_1}^{I_1}$ and $[\![\mathcal{N}_1]\!]_{J_1}^{I_2}$ are isomorphic by hypothesis, it follows by definition that $E_2 = E_1\alpha\beta$, therefore the edge $(i_1, i'_1)\alpha = (i_2, i'_2)$ also exists in E_2 for $i_2, i'_2 \in I_2$. Because of this, the reaction $c_{i_2} \to c_{i'_2}$ exists in $\mathcal{N}_2$; moreover, by Condition 3 of Definition 5, the complexes have the same stoichiometry.

Similarly, let (j_1, i_1) be an edge in E_1 such that A_{j_1} occurs in c_{i_1}. By Condition 2 of Definition 5, E_2 contains the edge $(j_1, i_1)\alpha\beta = (j_1\alpha, i_1\beta) = (i_2, j_2)$, which means that A_{j_2} occurs in c_{i_2}. By definition of π, $A_{j_1}\pi = A_{j_1}\beta = A_{j_2}$; since c_{i_1} and c_{i_2} also have the same multiplicity by Condition 3 of Definition 5, this implies that $c_{i_1}\pi = c_{i_2}$. A similar line of reasoning shows that $c_{i'_1}\pi = c_{i'_2}$. Therefore $(c_{i_1} \to c_{i'_1})\pi = c_{i_2} \to c_{i'_2}$. Generalising this result to all reactions in $\mathcal{N}_1$, we obtain $\mathcal{N}_1\pi = \mathcal{N}_2$, which concludes the proof. $\square$

Lemma 4. *The number of p-CRNs (reactions have up to p reactants/products) with up to N species grows as $\mathcal{O}(2^{N^{2p}})$.*

Proof. Following Eq. 3, the total number of p-CRNs with up to N and specifically M reactions is given by $\binom{L_p(N)(L_p(N)-1)}{M}$. Given that

$$L(N)(L(N) - 1) = \frac{(N + p)\ldots(N + 1)}{p!} \cdot \frac{(N + p)\ldots(N + 1) - p!}{p!} = \mathcal{O}(N^{2p}),$$

we can use the fact that $\sum_{i=1}^{k} \binom{n}{k} = 2^n$ to characterise the total number of bimolecular CRNs as $\mathcal{O}(2^{N^{2p}})$. $\square$

C Complex-Multiplicity-Species Graph

Section 2.1 has shown how to encode bimolecular CRNs as vertex-coloured digraphs. It is natural to wonder whether this encoding extends to more than bimolecularity. Unfortunately CS-graph cannot encode higher molecularities than 2, however we propose in this section a more general encoding of CRNs called the *Complex-Multiplicity-Species graph* (or CMS graph).

We begin by showing that CS-graphs cannot encode trimolecular reactions. Consider in fact the reaction $2A + B \to A + 2B$. If we added a new color "$2\square + \square$" and connect two node species A and B to it, there would be no way to tell which of the two species is actually the homodimer and which one is the monomer.

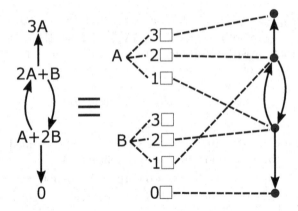

Fig. 9. Complex-Multiplicity-Species graph encoding of a CRN.

To overcome this issue, we propose Complex-Multiplicity-Species graphs, which extend CS-graphs with *multiplicity nodes*, that is distinct coloured nodes that point out the multiplicity of a species in a reaction. If m is the molecularity of interest, then there are $m + 1$ kinds of multiplicity nodes: naught, $\square$, $2\square$, $3\square$ and so on. Each species node is connected to m multiplicity nodes, signifying for example $A, 2A, 3A$ etc. Naught is a separate multiplicity node that cannot be connected to any species. In turn, multiplicity nodes are connected to complex nodes to represent the original CRN's reaction. Figure 9 show an example of a CMS graph; notice that no confusion is possible between complexes $2A + B$ and $A + 2B$.

We believe that CMS graphs are a general encoding of CRNs with any molecularity, but we leave a formal definition and proofs for future work.

D   Counts of Non-isomorphic CRNs

In this appendix, we tabulate the numbers of non-isomorphic CRNs found using genCRN. The tables can be compared against values reported at https://reaction-networks.net/networks/, at the time of writing, which were evaluated using the method in [3]. In each case, we report values for *genuine* CRNs, those which use all N species.

D.1   No Filters

Here, we consider the total number of non-isomorphic CRNs for N species and M reactions. The results are graphically depicted in Fig. 5, but tabulated below (Table 1).

Table 1. Genuine non-isomorphic CRNs. The number of non-isomorphic CRNs is shown for different numbers of species and reactions. Coloured in blue are those counts not available at https://reaction-networks.net/networks/ at time of writing.

Species	Reactions						
	1	2	3	4	5	6	7
1	6	15	20	15	6	1	0
2	10	210	2,024	13,740	71,338	297,114	1,018,264
3	5	495	17,890	414,015	7,262,666	103,511,272	1,244,363,180
4	1	451	47,323	2,900,934	128,328,834	4,518,901,463	133,379,120,523
5	0	204	55,682	7,894,798	763,695,711	56,929,248,832	
6	0	54	35,678	10,704,289	2,069,783,947		
7	0	8	13,964	8,386,321	3,041,467,242		
8	0	1	3,594	4,182,295	2,715,774,734		
9	0	0	639	1,417,784	1,595,551,325		
10	0	0	83	618,885	653,346,685		

D.2   Reversible CRNs

To generate reversible CRNs, we generate undirected graphs of a suitable size and feed these into genCRNin the same way as for general CRNs with irreversible reactions. Reported below are counts for M reversible reactions. As such, the CRNs found have $2M$ unidirectional reactions.

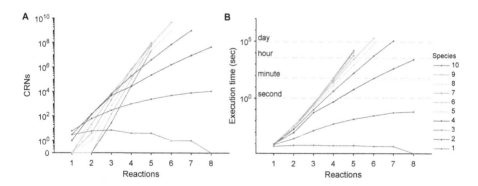

Fig. 10. Counts and execution times for enumeration of genuine non-isomorphic reversible CRNs.

Table 2. Genuine non-isomorphic reversible CRNs. The number of non-isomorphic CRNs with only reversible reactions is shown for different numbers of species and reactions. Coloured entries correspond to comparisons with the counts available at https://reaction-networks.net/networks/ at time of writing. Blue indicates values not available, and red indicates values that differ.

Species	1	2	3	4	5	Reactions 6	7	8
1	3	6	7	4	4	1	1	0
2	6	60	296	989	2,516	4,997	8,241	11,271
3	3	138	4,788	26,988	230,595	1,589,808	9,161,056	45,107,712
4	1	134	6,354	187,005	4,048,219	69,982,180	1,011,965,511	
5	0	65	7,677	513,036	24,186,053	888,323,405		
6	0	21	5,178	709,212	66,152,034	4,674,311,477		
7	0	4	2,188	572,058	98,576,689			
8	0	1	648	298,030	89,754,652			

D.3　Non-trivial Dynamics

As described in the main text, one can test whether a CRN has non-trivial dynamics. To apply this filter to the enumerated non-isomorphic CRNs, one can use the -t flag for genCRN.

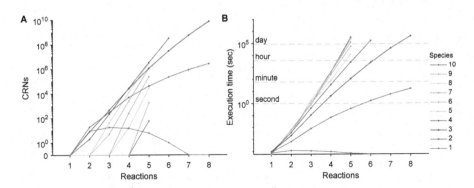

Fig. 11. Counts and execution times for enumeration of genuine non-isomorphic CRNs with non-trivial dynamics.

We found that our counts differ with those reported at https://reaction-networks.net/networks/ (using the method in [3]) for CRNs with at least 4 species and 4 reactions (Table 3). In each of the 5 counts identified as different, the values we report are lower than those reported previously, though within 1.5% relative error. By analysing the sets of CRNs produced in the 6 species and 4 reactions class, we have determined that we incorrectly classify at least one CRN as trivial, and tracked this to numerical discrepancies in our use

Table 3. Genuine non-isomorphic CRNs with non-trivial dynamics. The number of non-isomorphic CRNs with non-trivial dynamics is shown for different numbers of species and reactions. Coloured entries correspond to comparisons with the counts available at https://reaction-networks.net/networks/ at time of writing. Blue indicates values not available, and red indicates values that differ.

Species	1	2	3	4	5	6	7	8
					Reactions			
1	0	9	18	15	6	1	0	0
2	0	19	304	5,016	41,500	221,728	871,330	2,700,277
3	0	8	464	25,272	1,125,465	30,806,874	563,453,020	7,675,100,687
4	0	2	223	27,925	3,276,425	322,473,604		
5	0	0	41	12,310	2,805,266			
6	0	0	5	2,604	1,114,937			
7	0	0	0	264	237,064			
8	0	0	0	17	28,163			
9	0	0	0	0	1,795			
10	0	0	0	0	60			

of Fourier-Motzkin elimination. Further work would be required to refine the numerical procedure used here, in order to improve confidence in our counts of dynamically non-trivial and non-isomorphic CRNs. For instance, implementing Fourier-Motzkin elimination with explicit rational numbers (one integer variable each for the numerator and denominator), could avoid the emergence of values close to zero.

References

1. Angeli, D.: A tutorial on chemical reaction network dynamics. Eur. J. Control **15**(3–4), 398–406 (2009)
2. Angeli, D., De Leenheer, P., Sontag, E.: A Petri Net approach to persistence analysis in chemical reaction networks. In: Queinnec, I., Tarbouriech, S., Garcia, G., Niculescu, S.I. (eds.) Biology and Control Theory: Current Challenges, pp. 181–216. Springer, Heidelberg (2007). https://doi.org/10.1007/978-3-540-71988-5_9
3. Banaji, M.: Counting chemical reaction networks with NAUTY. arXiv e-prints arXiv:1705.10820, May 2017
4. Bayramov, S.K.: New theoretical schemes of the simplest chemical oscillators. Biochem. (Mosc.) **70**(12), 1377–1384 (2005)
5. Brendan, D., McKay, A.P.: Nauty and Traces User's Guide (2013)
6. Brinkmann, G.: Isomorphism rejection in structure generation programs. DIMACS Ser. Discret. Math. Theor. Comput. Sci. **51**(3), 25–38 (2000)
7. Cardelli, L., Tribastone, M., Vandin, A., Tschaikowski, M.: Forward and backward bisimulations for chemical reaction networks. In: CONCUR 2015 (2015)
8. Cardelli, L., Tribastone, M., Tschaikowski, M., Vandin, A.: ERODE: a tool for the evaluation and reduction of ordinary differential equations. In: Legay, A., Margaria, T. (eds.) TACAS 2017. LNCS, vol. 10206, pp. 310–328. Springer, Heidelberg (2017). https://doi.org/10.1007/978-3-662-54580-5_19

9. Chen, Y.J., et al.: Programmable chemical controllers made from DNA. Nat. Nanotechnol. **8**(10), 755 (2013)

10. Colom, J.M., Silva, M.: Convex geometry and semiflows in P/T nets. A comparative study of algorithms for computation of minimal p-semiflows. In: Rozenberg, G. (ed.) ICATPN 1989. LNCS, vol. 483, pp. 79–112. Springer, Heidelberg (1991). https://doi.org/10.1007/3-540-53863-1_22

11. Craciun, G., Feinberg, M.: Multiple equilibria in complex chemical reaction networks: I. The injectivity property. SIAM J. Appl. Math. **65**(5), 1526–1546 (2005)

12. Craciun, G., Feinberg, M.: Multiple equilibria in complex chemical reaction networks: II. The species-reaction graph. SIAM J. Appl. Math. **66**(4), 1321–1338 (2006)

13. Deckard, A.C., Bergmann, F.T., Sauro, H.M.: Enumeration and Online Library of Mass-Action Reaction Networks. arXiv e-prints arXiv:0901.3067, January 2009

14. Farkas, J.: Uber die theorie der einfachen ungeichungen. J. Reine Angew. Math. **124**, 1–24 (1902)

15. Feinberg, M.: Chemical reaction network structure and the stability of complex isothermal reactors-I. The deficiency zero and deficiency one theorems. Chem. Eng. Sci. **42**(10), 2229–2268 (1987)

16. Feinberg, M.: Chemical reaction network structure and the stability of complex isothermal reactors-II. multiple steady states for networks of deficiency one. Chem. Eng. Sci. **43**(1), 1–25 (1988)

17. Hucka, M., et al.: The systems biology markup language (SBML): a medium for representation and exchange of biochemical network models. Bioinformatics **19**(4), 524–531 (2003)

18. Karlebach, G., Shamir, R.: Modelling and analysis of gene regulatory networks. Nat. Rev. Mol. Cell Biol. **9**(10), 770 (2008)

19. Kondo, S., Miura, T.: Reaction-diffusion model as a framework for understanding biological pattern formation. Science **329**(5999), 1616–1620 (2010)

20. McKay, B.D.: Isomorph-free exhaustive generation. J. Algorithms **26**(2), 306–324 (1998)

21. McKay, B.D.: Isomorph-free exhaustive generation. J. Algorithms **26**(2), 306–324 (1998)

22. McKay, B.D., Piperno, A.: Practical graph isomorphism, II. CoRR abs/1301.1493 (2013)

23. Mincheva, M., Roussel, M.R.: A graph-theoretic method for detecting potential turing bifurcations. J. Chem. Phys. **125**(20), 204102 (2006)

24. Mincheva, M., Roussel, M.R.: Graph-theoretic methods for the analysis of chemical and biochemical networks. I. Multistability and oscillations in ordinary differential equation models. J. Math. Biol. **55**(1), 61–86 (2007)

25. Miyazaki, T.: The complexity of McKay's canonical labeling algorithm. In: Groups and Computation, Proceedings of a DIMACS Workshop, New Brunswick, New Jersey, USA, 7–10 June 1995, pp. 239–256 (1995)

26. Murphy, N., Petersen, R., Phillips, A., Yordanov, B., Dalchau, N.: Synthesizing and tuning stochastic chemical reaction networks with specified behaviours. J. R. Soc. Interface **15**(145), 20180283 (2018)

27. Oishi, K., Klavins, E.: Biomolecular implementation of linear I/O systems. IET Syst. Biol. **5**(4), 252–260 (2011)

28. Pedersen, M., Phillips, A.: Towards programming languages for genetic engineering of living cells. J. R. Soc. Interface **6**(suppl-4), S437–S450 (2009)

29. Pólya, G., Read, R.: Combinatorial Enumeration of Groups, Graphs, and Chemical Compounds. Springer, New York (1987). https://doi.org/10.1007/978-1-4612-4664-0
30. Read, R.C., Corneil, D.G.: The graph isomorphism disease. J. Graph Theory **1**(4), 339–363 (1977)
31. Rosen, K.H.: Handbook of Discrete and Combinatorial Mathematics, 2nd edn. Chapman & Hall/CRC, Boca Raton (2010)
32. Shinar, G., Feinberg, M.: Structural sources of robustness in biochemical reaction networks. Science **327**(5971), 1389–1391 (2010)
33. Soloveichik, D., Seelig, G., Winfree, E.: DNA as a universal substrate for chemical kinetics. Proc. Natl. Acad. Sci. **107**(12), 5393–5398 (2010)
34. Srinivas, N., Parkin, J., Seelig, G., Winfree, E., Soloveichik, D.: Enzyme-free nucleic acid dynamical systems. Science **358**(eaal2052), 2052 (2017)
35. Wilhelm, T.: The smallest chemical reaction system with bistability. BMC Syst. Biol. **3**(1), 90 (2009)

A Large-Scale Assessment of Exact Model Reduction in the BioModels Repository

Isabel Cristina Pérez-Verona[1]([envelope]), Mirco Tribastone[1], and Andrea Vandin[2]

[1] IMT School for Advanced Studies Lucca, Lucca, Italy
isabel.perez@imtlucca.it
[2] DTU Technical University of Denmark, Lyngby, Denmark

Abstract. Chemical reaction networks are a popular formalism for modeling biological processes which supports both a deterministic and a stochastic interpretation based on ordinary differential equations and continuous-time Markov chains, respectively. In most cases, these models do not enjoy analytical solution, thus typically requiring expensive computational methods based on numerical solvers or stochastic simulations. Exact model reduction techniques can be used as an aid to lower the analysis cost by providing reduced networks that preserve the dynamics of interest to the modeler. We hereby consider a family of techniques for both deterministic and stochastic networks which are based on equivalence relations over the species in the network, leading to a coarse graining which provides the exact aggregate time-course evolution for each equivalence class. We present a large-scale empirical assessment on the BioModels repository by measuring their compression capability over 667 models. Through a number of selected case studies, we also show their ability in yielding physically interpretable reductions that can reveal dynamical patterns of the bio-molecular processes under consideration.

Keywords: Model reduction · Biological systems · Equivalence relations

1 Introduction

Computational models in systems biology combine biochemical and physiological knowledge to inform highly detailed mechanistic models of biological networks such as signaling pathways, protein-protein interaction networks, and genetic cascades. Mathematical models which equip such interaction networks with kinetic information generally lead to a dynamical-system representation in terms of a formal chemical reaction network (CRN), with two main interpretations based on ordinary differential equations (ODEs) and continuous-time Markov chains (CTMCs), respectively. In either case the model tracks the time-course evolution of all biochemical species in the network. In the ODE interpretation each species is associated with a variable of a system of (typically

© Springer Nature Switzerland AG 2019
L. Bortolussi and G. Sanguinetti (Eds.): CMSB 2019, LNBI 11773, pp. 248–265, 2019.
https://doi.org/10.1007/978-3-030-31304-3_13

nonlinear) ODEs, which are analyzed from an initial condition that represents the initial concentration of each species [51]. In the CTMC interpretation [27], species are tracked discretely and each state is a vector of molecular counts, one component for each species. It is well known that these two representations can be formally related to each other under appropriate conditions, with the ODEs being the thermodynamical limit when the number of molecules in the CRN is large enough [33].

Often it is useful to consider both interpretations—one would take the CTMC semantics as the ground-truth representation and the ODE as an approximation that estimates the first-order moments. Unfortunately, in both cases the analysis can be expensive due to the lack of analytical solutions in general. Indeed, the modeler is typically left with computational approaches such as the numerical integration of ODEs (e.g., [1]) or stochastic simulation [27]. This is a major motivating issue for several lines of research aiming at easing the computational cost of the analysis, including efficient simulation methods (e.g., [26]), approximation methods for stochastic chemical kinetics (e.g., [44]), and simplification techniques for multi-scale biochemical CRNs (e.g., [43]) and rule-based models [23–25].

A complementary approach that can be seen as a generic pre-analysis step consists in the use of an *exact model reduction* algorithm which, given an input CRN, produces a smaller CRN (i.e., consisting of fewer species and reactions) that preserves the output dynamics of interest to the modeler (e.g., [38,49]). This would lead to a coarse-grained CRN which still allows the full observation of the time evolution of some original species (e.g., the phosphorylated forms of downstream molecular complexes in a signaling pathway) while collapsing the behavior of other species into macro-variables. Such an approach may bring about two main advantages. First, being a CRN-to-CRN transformation, the coarse-grained CRN can still be subjected to other techniques to reduce the complexity of the analysis, including *approximate* model reduction methods. Second, the very collapse of several species into one may carry a physical interpretation that increases our understanding of the biology. The latter point appears to be of scientific relevance regardless of the CRN reduction ratio. Therefore, two suitable indicators of the relevance of exact model reduction techniques in practice are the effectiveness and the intelligibility of the reductions.

This paper presents a large-scale assessment on biological models in the literature for recent reduction techniques for CRNs, supporting the ODE and the CTMC semantics [7,8,10,11,13]. The techniques share two main unifying ideas:

(i) Identifying criteria on the species and reactions of a CRN inducing a suitable *species equivalence*, i.e., a partition of the species such that an exactly reduced CRN can be written having a macro-species per partition block.
(ii) Developing an algorithm for computing the largest species equivalence using partition refinement [41], based on iterative refinements of a given initial partition of species (with which, for instance, one can isolate the observable species to be preserved in the reduction).

The definitions of the species equivalences differ according to the underlying semantics to which they are applicable, the assumptions made on the input

CRN, and the kind of reduction that they yield. Specifically, forward equivalence (FE) and backward equivalence (BE) apply to CRNs with ODE semantics based on mass-action kinetics and identify reduced models where each macro-species preserves the sums of original variables belonging to a block [10]; while with FE the time-course of one species cannot be recovered, BE aggregates species that have the same solutions at all time points. Forward differential equivalence (FDE) and backward differential equivalence (BDE) are generalizations that can be applied to CRNs where the underlying ODEs have nonlinearities beyond polynomials such as rational expressions in Hill kinetics [7]. Finally syntactic Markovian bisimulation (SMB) is the species equivalence for stochastic CRNs [11]. It identifies a partition of species which induces a coarse graining of the underlying CTMC in terms of ordinary lumpability [5], aggregating CTMC states that have equal sums of molecular counts across each partition block of species.

Assisted by ERODE [9], a publicly available software tool that implements the aforementioned species equivalences, we carry out an assessment of the BioModels database [37], a well-known repository of quantitative models of biochemical systems.[1] Our goal is to answer the following three evaluation questions:

Q1. *How restrictive are the assumptions required by the species equivalences?* We answer this question by detailing how we translated the BioModels descriptions, available in the SBML format, into the input format of ERODE.

Q2. *What is the effectiveness of exact model reduction by species equivalence?* We measure effectiveness as the percentage of models that can be aggregated, as well as the compression ratio provided by the largest species equivalence that preserves the observables specified in the original model.

Q3. *What is the physical interpretation of the reductions?* For this question, we present a more detailed discussion of a selected number of models.

2   Background

In order to make the paper self-contained, in this section we briefly overview the main results regarding the species equivalences used in our assessment. We refer to the original papers for the details and further examples, while unifying tutorial-like presentations are given in [48,50].

Chemical Reaction Networks. First, we fix the notation and terminology for reaction networks. A CRN is a pair $(\mathcal{S}, \mathcal{R})$ consisting of a finite set of species $\mathcal{S}$ and a finite set of reactions $\mathcal{R}$, where each reaction is in the form $\rho \xrightarrow{f} \pi$ consisting of: a multiset of species ρ, with the multiplicity of species S denoted by ρ_S, that represents the *reactants*; a multiset of species π (the *products*); and the *propensity function* $f : \mathbb{R}^{\mathcal{S}} \to \mathbb{R}_{\geq 0}$. Roughly speaking, it gives the rate at

---

[1] The models are available at https://sysma.imtlucca.it/tools/erode/cmsb2019/.

which the reaction fires based on the current system state; the *net stoichiometry* $\pi - \rho$ gives the state update upon the reaction firing.[2]

Example 1. Let us use a CRN $(\mathcal{S}_E, \mathcal{R}_E)$ with species S_1, S_2, S_3, S_4, S_5, and reactions $S_1 \xrightarrow{2} S_5,\ S_1 \xrightarrow{1} 2S_3,\ S_3 + S_5 \xrightarrow{3} S_3, S_2 \xrightarrow{2} S_3,\ S_2 \xrightarrow{1} 2S_5,\ S_4 + S_5 \xrightarrow{3} S_3$.

According to the deterministic semantics of CRNs [51], a CRN is associated with an ODE system which tracks the time course of the vector of concentrations of the species at time t, $X(t) = (X_S(t))_{S \in \mathcal{S}}$, as follows:

$$\frac{dX_S(t)}{dt} = \sum_{(\rho \xrightarrow{f} \pi) \in \mathcal{R}} (\pi_S - \rho_S) \cdot f(X(t)).$$

In a deterministic *mass-action CRN*, each reaction is associated with a kinetic parameter $\lambda > 0$, and the propensity function, denoted by f_λ, is given by $f_\lambda(x) = \lambda \cdot \prod_{S \in \mathcal{S}} x_S^{\rho_S}$, where ρ is the multiset of reactants. The CRN $(\mathcal{S}_E, \mathcal{R}_E)$ is a mass-action CRN. For example, the ODEs for S_1 and S_2 are:

$$\frac{dX_1(t)}{dt} = -3 \cdot X_1(t)) \qquad \frac{dX_2(t)}{dt} = -3 \cdot X_2(t)$$

According to the stochastic semantics of CRNs [27], a CRN is represented as a Markov population process, a CTMC where each state is a vector $n = (n_S)_{S \in \mathcal{S}}$ of nonnegative integers that tracks the molecular counts of each species. The *initial state* is a vector representing the initial (integer) populations of each species. A transition between any two states n and $n + \pi - \rho$ occurs according to an exponential distribution with parameter $f(n)$ for each reaction $\rho \xrightarrow{f} \pi$. The CTMC underlying a CRN for an initial state consists of all states and transitions generated by applying exhaustively the reactions on all generated states, starting from the initial one. An *elementary* mass-action CRN has reactions in the form $\rho \xrightarrow{f_\lambda} \pi$ where $|\rho| \leq 2$ (i.e., at most two molecules can interact), $\lambda > 0$ is the kinetic parameter, and $f_\lambda(n) = \lambda \cdot \prod_{S \in \mathcal{S}} \binom{n_S}{\rho_S}$, where n is the source state. The CRN in Example 1 is elementary.

Forward and Backward Equivalence (FE and BE). FE and BE are two reduction techniques for deterministic mass-action CRNs given as equivalence relations on species which can be efficiently checked by using only structural conditions on the reactions [10]. For $\chi \in \{FE, BE\}$, both notions can be expressed as:

Given a CRN $(\mathcal{S}, \mathcal{R})$, a partition $\mathcal{H}$ of species is χ if and only if for any two blocks $H, H' \in \mathcal{H}$ and any two species $S_i, S_j \in H$ it holds

$$\mathbf{c}_\chi(S_i, \eta, H') = \mathbf{c}_\chi(S_j, \eta, H') \quad \forall \eta.\ \exists (S_k + \eta \xrightarrow{\lambda} \pi) \in \mathcal{R} \text{ for } S_k \in \{S_i, S_j\}$$

---

[2] As usual, the $+$ and $-$ operators denote multiset union and difference, respectively, while the multiplicity of a species denotes its stoichiometric coefficient.

$\mathcal{H}_f = \{\{S_1, S_2\}, \{S_3, S_4\}, \{S_5\}\}$

$\mathcal{S}_f = \{S_{1,2}, S_{3,4}, S_5\}$

$\mathcal{R}_f = \{S_{1,2} \xrightarrow{1} S_5,\ S_{1,2} \xrightarrow{0.5} 2S_{3,4}, S_{3,4}+S_5 \xrightarrow{3} S_{3,4},$

$\quad S_{1,2} \xrightarrow{1} S_{3,4},\ S_{1,2} \xrightarrow{0.5} 2S_5\}$

(a) FE-reduction

$\mathcal{H}_b = \{\{S_1, S_2\}, \{S_3\}, \{S_4\}, \{S_5\}\}$

$\mathcal{S}_b = \{S_{1,2}, S_3, S_4, S_5\}$

$\mathcal{R}_b = \{S_{1,2} \xrightarrow{1} S_5,\ S_{1,2} \xrightarrow{0.5} 2S_3,\ S_3+S_5 \xrightarrow{3} S_3,$

$\quad S_{1,2} \xrightarrow{1} S_3,\ S_{1,2} \xrightarrow{0.5} 2S_5,\ S_4+S_5 \xrightarrow{3} S_3\}$

(b) BE-reduction

Fig. 1. Coarsest FE/BE, and FE/BE-reductions of $(\mathcal{S}_E, \mathcal{R}_E)$ from Example 1.

where $\mathbf{c}_\chi$ maps a species (S_i, S_j), a multiset of reagent partners (η) and a block (H') into a real number computed by inspecting once the reactions [10].

Figure 1 shows FE partition $\mathcal{H}_f$ and BE partition $\mathcal{H}_b$, as well as their respective reduced CRNs, for the running example (We observe that $\mathcal{H}_b$ is a refinement of $\mathcal{H}_f$, but in general, FE and BE are not comparable [6,8]). FE relates species such that it is possible to rewrite the ODEs underlying the CRN in terms of sums of the variables in each block. Each macro-species in the FE-reduced CRN represents the sum of the corresponding species in the original CRN.For example, in Fig. 1(a) species $S_{1,2}$ and $S_{3,4}$ can be used to study the concentration of the sums of original variables $S_1 + S_2$ and $S_3 + S_4$, respectively.

BE relates species that have same ODE solution at any point in time (which therefore must have same initial condition). In the BE-reduced CRN in Fig. 1(b), $S_{1,2}$ represents the sum of original species $S_1 + S_2$. However, BE ensures that S_1 and S_2 have same ODE solution at all times. Therefore, we can recover each individual solution of by halving that of $S_{1,2}$.

Forward and Backward Differential Equivalence (FDE and BDE). FDE and BDE are generalizations of FE and BE, respectively, for deterministic CRNs beyond mass-action [7,13]. FDE and BDE capture the same dynamical properties of FE and BE, and collapse to them for mass-action deterministic CRNs. The greater generality of FDE/BDE comes at the cost of a more computationally expensive implementation based on encodings in satisfiability modulo theory (SMT) formulas. For instance, the following formula $\psi^{\mathcal{H}_b}$ encodes the check whether partition $\mathcal{H}_b$ is a BDE:

$$\psi^{\mathcal{H}_b} := (X_1 = X_2) \implies (-3 \cdot X_1 = -3 \cdot X_2)$$

which checks that if all variables in same block are equal (the premise) then they must evolve in the same way, i.e. their derivative should evaluate to the same value (the conclusion). The formula has two free real variables, X_1 and X_2, corresponding to S_1 and S_2. By using an SMT solver, e.g., Z3 [19], we can check if $\mathcal{H}_b$ is a BDE by checking for the satisfiability of $\neg\psi^{\mathcal{H}_b}$. If there exists an assignment for X_1 and X_2 that makes $\neg\psi^{\mathcal{H}_b}$ true, then $\mathcal{H}_b$ is not a BDE. This is not the case, and hence it is a BDE (as expected from it being a BE).

Syntactic Markovian Bisimulation (SMB). SMB is a reduction technique for stochastic mass-action elementary CRNs [11]. It is given as an equivalence on species, in the same spirit of FE and BE. Indeed, SMB can be seen as an instantiation of FE to the stochastic semantics of CRNs. We discuss this through our running example. The partition $\mathcal{H}_s = \{\{S_1\}, \{S_2\}, \{S_3, S_4\}, \{S_5\}\}$ is an SMB for the CRN $(\mathcal{S}_E, \mathcal{R}_E)$ from Example 1. The very same notion of FE/BE-reduced CRN applies to SMB as well. The $\mathcal{H}_s$-reduction of $(\mathcal{S}_E, \mathcal{R}_E)$ has species $\mathcal{S}_S = \{S_1, S_2, S_{3,4}, S_5\}$ and reactions $\mathcal{R}_S = \{S_1 \xrightarrow{2} c_5,\ S_1 \xrightarrow{1} 2S_{3,4}, S_{3,4} + S_5 \xrightarrow{3} S_{3,4}, S_2 \xrightarrow{2} S_{3,4},\ S_2 \xrightarrow{1} 2S_5\}$. A state of a CTMC of $(\mathcal{S}_E, \mathcal{R}_E)$ is a vector of size $|\mathcal{S}_E|$ counting the population of each original species, while a state of a CTMC of $(\mathcal{S}_S, \mathcal{R}_S)$ is a vector of size $|\mathcal{S}_S|$ counting the cumulative population of each block of $\mathcal{H}_s$. The CTMCs of $(\mathcal{S}_S, \mathcal{R}_S)$ are reductions in terms of CTMC ordinary lumpability [5] of the ones obtained from $(\mathcal{S}_E, \mathcal{R}_E)$. All states of the original CTMC containing same number of $\mathcal{H}_s$-equivalent species get collapsed in the same macro-state in the reduced CTMC. Therefore, similarly to FE, SMB allows to obtain a coarse-grained version of the original CRN which allows to reason in terms of sums of variables. For example, the states $S_1 + 2S_3$, $S_1 + S_3 + S_4$, and $S_1 + 2S_4$ form an ordinary lumpable partition of a CTMC of the original CRN, and therefore get collapsed in the state $S_1 + 2S_{3,4}$ for the reduced CRN.

We note that $\mathcal{H}_s$ is a refinement of $\mathcal{H}_f$. Indeed, it has been shown that SMB implies FE, but not vice versa [11]. This will be confirmed in Sect. 3.4.

Partition Refinement. Each equivalence is supported by a partition refinement algorithm which refines an initial partition of species (splitting its blocks) until a fixed point. The initial partition can be chosen, e.g., to isolate species that must not be aggregated because they are observables of interest to the modeler. The examples shown in this section are largest refinement of the singleton partition where all species are in a block. Other initial partitions will be used in Sect. 3.

3   Experimental Set-Up

3.1   Overview of the BioModels Repository

The BioModels Database is a repository of computational models of biological processes [37]. It hosts dynamical quantitative models described in peer-reviewed scientific literature as well as models generated automatically from pathway resources such as KEGG [32], BioCarta [40], MetaCyc [14], PID [42] and SABIO-RK [52]. BioModels covers a wide range of models from several biological categories such as biochemical reaction systems, kinetic models, metabolic networks, steady-state models and signaling pathways. Models are available in the Systems Biology Markup Language (SBML) [30], a well-known machine-readable format based on XML for representing quantitative models of biological systems.

The BioModels repository is divided into two sections: the *curated branch* and the *non-curated branch*. The former contains models that have been manually checked and their components annotated using unambiguous identifiers [31] that refer to external biological databases [17,22,46] or ontologies (such as Gene Ontology [2], SBO [18] or ChEBI [20]). Models are curated following

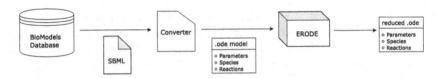

Fig. 2. Workflow overview. Models were downloaded from the BioModels repository in the SBML format. We implemented a tool to translate the SBML description into the CRN-like input (.ode format) of ERODE. The output of ERODE is a reduced CRN with reactions involving *macro-species*, each representing the sum within an equivalence class of original species. We manually inspected the ERODE output to provide a physical interpretation of the obtained equivalences.

the Minimum Information Required in the Annotation of Models guidelines (MIRIAM) [35]. Models that are not MIRIAM-compliant are stored in the non-curated branch, which also contains non-kinetic models such as flux balance analysis models. A more detailed description of BioModels is available at [16].

3.2   Model Conversion

We developed a prototype for translating SBML models into ERODE's format, using the workflow in Fig. 2. SBML files are read using the *jsbml* library (version 1.2) [21,36]. Here we briefly explain the main phases of the conversion process.

The CRN input format of ERODE contains lists of parameters (to be used in kinetic rates), of species (with corresponding initial conditions), and of reactions. This is followed by a list of commands for analysis, reduction, and export.

The following SBML snippet, from *BIOMD0000000030*, specifies a parameter

```
<parameter id="k1" metaid="metaid_0000019" name="k1" value="0.02"/>
```

This is translated into k1 = 0.02 within the parameters list (delimited by begin parameters/end parameters) of the ERODE description.

The next SBML snippet, adapted from the same model by removing the annotation tag containing links to external databases, defines the species M:

```
<species compartment="cell" id="M" initialConcentration="800"
         metaid="metaid_0000005" name="MAPK"/>
```

It describes the compartment in which the species is located, the initial concentration and an identifier. We translate this into M = 800 within ERODE's species declaration section (delimited by begin init/end init).

Instead, the conversion of the reactions is less straightforward, particularly to recognize mass-action models to which the specialized FE, BE, and SMB can be applied. Indeed, SBML allows the direct specification of mass-action reactions by means of appropriate SBO labels in the kineticLaw tag (other labels identify different kinetics such as Michaelis-Menten and Hill). However, we encountered cases of reactions that, although not tagged with mass-action labels, were clearly so upon inspection of the reactions. One such example is given in Fig. 3. It shows

```
1   <reaction id="reaction_0000001" metaid="metaid_0000046"
2           name="binding MAPKK on Tyr site of MAPK">
3     <listOfReactants>
4       <speciesReference metaid="_063184" species="M"/>
5       <speciesReference metaid="_063196" species="MAPKK"/>
6     </listOfReactants>
7     <listOfProducts>
8       <speciesReference metaid="_063208" species="M_MAPKK_Y"/>
9     </listOfProducts>
10    <kineticLaw metaid="_063220">
11      <math xmlns="http://www.w3.org/1998/Math/MathML">
12        <apply>
13          <times/>
14            <ci>cell</ci>
15            <apply>
16              <minus/>
17                <apply>
18                  <times/><ci>k1</ci><ci>M</ci><ci>MAPKK</ci>
19                </apply>
20                <apply>
21                  <times/><ci>k_1</ci><ci>M_MAPKK_Y</ci>
22                </apply>
23            </apply>
24        </apply>
25      </math>
26    </kineticLaw>
27  </reaction>
```

Fig. 3. Sample SBML reaction adapted from *BIOMD0000000030*

the specification of a reaction containing a list of reactants, products (as well as modifiers, not used in this reaction, to model, e.g., catalysts or intermediates in the reaction). The reaction has an optional attribute reversible, by default set to true, indicating if the reaction is reversible. We inferred the forward and reverse rate functions as the left and right operand, respectively, of the topmost minus MathML tag (Line 16). This leads to the two following ERODE irreversible reactions (as ERODE does not support reversible reactions):

```
M + MAPKK -> M_MAPKK_Y, arbitrary cell * k1 * M * MAPKK
M_MAPKK_Y -> M + MAPKK, arbitrary cell * k_1 * M_MAPKK_Y
```

Here, the left- and right-hand sides of the reactions are taken from the SMBL lists (and modifiers are added in both sides if present), whereas the arbitrary keyword denotes a reaction with a generic non-mass-action propensity function. However, one can notice that these two reactions are actually equivalent to mass-action reactions with kinetic parameters cell * k1 and cell * k_1, respectively. We manually detected such occurrences of non-tagged mass-action reactions and translated into ERODE mass-action ones. In this example we get:

```
M + MAPKK -> M_MAPKK_Y,  cell *  k1
M_MAPKK_Y -> M + MAPKK,  cell * k_1
```

ERODE can export the ODEs underlying a model as a Matlab function. Likewise, in BioModels all models come with an encoding as Matlab functions. We tested our converter over a large random selection of BioModels files by checking that their Matlab functions and those exported by ERODE corresponded.

3.3   Repository Preprocessing

In our experiments we used the BioModels repository snapshot 26 July 2017. It consists of 640 models in the curated branch (from id *BIOMD0000000001* to *BIOMD0000000640*) and 1000 models in the non-curated branch (with ids ranging from *MODEL0072364382* to *MODEL9811206584*).

We performed a preprocessing step to filter out models that could not be used for the analysis (cf. evaluation question **Q1** in Sect. 1). In the non-curated branch only 491 models are kinetic models described as ODE systems, while the others are described in formalisms, such as logical or flux balance analysis models, that are outside the scope of applicability of species equivalences.

Overall, we could process 448 models from the curated branch and 219 from the non-curated one, for an overall sanitized dataset of 667 models. Of these, 43 were recognized as mass-action CRNs (as detailed in Sect. 3.2); all of them were found to be elementary mass-action CRNs, hence analyzable by SMB. The most frequent reasons for discarding a model were (within parenthesis we give the frequency in the curated branch, which we assume to be more stable):

- syntactic limitations in our converter prototype, including the lack of support for models without explicit reactions where the dynamics is given by rate rules over a set of parameters, e.g., as in *BIOMD0000000020* (114);
- models with unsupported propensity functions such as tanh and exp (31);
- models with species with *Assignment Rules*, used to model features such as delayed equations and hybrid systems, not supported by ERODE (47).

3.4   Reduction Results

Here we report the summary of the reduction results. Non mass-action models were analyzed using FDE and BDE, while for mass-action ones we used FE, BE, and SMB. In a preliminary analysis we considered the maximal equivalences for all cases, computed by starting the partition-refinement algorithms with the initial singleton partition with a single block containing all species in the CRN. However, in 32 cases we found that the maximal FDE/FE collapsed *all* species and reactions. This is because these CRNs are *closed* and *mass-preserving*, meaning that the concentrations (represented by the ODE solutions for each species) just flow among the species, but the *total* cumulative concentration is constant. Therefore these systems can be self-consistently written as a single-equation ODE with zero derivative (and initial concentration equal to that total cumulative concentration). We dismissed such partitions as degenerate/uninteresting. Instead, for these cases we built more meaningful (ad-hoc) initial partitions to be used in the partition-refinement algorithm: we isolated variables of interest to the modeler, as evinced from the related scientific publication.

For each equivalence we computed the reduced CRN, recording the resulting number of species and reactions as a measure of the effectiveness of the exact reduction techniques (cf. **Q2** in Sect. 1). Figure 4 counts the models that could be reduced by at least one technique, regardless of the reduction ratio. For the non

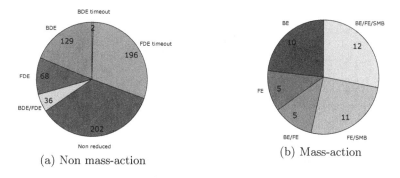

(a) Non mass-action

(b) Mass-action

Fig. 4. Reduction results.

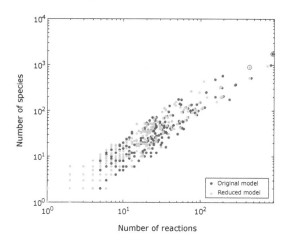

Number of reactions

Fig. 5. Comparison among original and reduced species and reactions (log scale). (Color figure online)

mass-action models (Fig. 4a), 233 models (37%) could be reduced. In particular, only 36 models could be reduced by both FDE and BDE, proving that they are not comparable. Several models (196, 31%) could not be analyzed due to the excessive computational cost of FDE, while only 2 due to BDE (we used a time-out of 8 hours). This is consistent with the more (and more complex) SMT checks required by FDE with respect to BDE [7].

All the mass-action models (Fig. 4b) could be reduced by at least one equivalence relation. Ten models (23%) could be reduced with BE and 5 (12%) with FE only. Twelve models (28%) could be reduced with the three methods, while 11 (25%) could be reduced with SMB and FE, and 5 (12%) with FE and BE. The presence of models that could be reduced only by FE and not SMB shows that FE does not imply SMB, while the converse is true, as discussed.

Figure 5 shows a scatter plot to summarize the reduction ratio for each model using the species equivalence that yielded the best reduction.

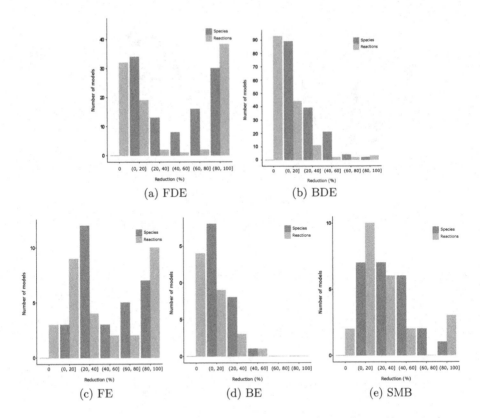

Fig. 6. Reduction ratios for the species and reactions for each species equivalence. (Color figure online)

MODEL3632127506, the largest model processed, denoted with blue circled dots in the figure, was reduced from 872 species and 1750 reactions to 436 species and 900 reactions, with a reduction of about 50% in the number of species and reactions. Overall, the average compression ratio is 36% for the species and 26% for the reactions.

The average reduction ratio in the number of species and reactions varies with each method: BDE (23% for species, 8% for reactions), FDE (50%, 48%), BE (19%, 8%), FE (51%, 47%), SMB (35%, 29%). Figure 6 illustrates the reductions obtained. For each species equivalence, we group the models in 5 histogram bins (0%–20%, ..., 80%–100%) in two series showing the reduction ratio of the species (red) and the reactions (blue). It is possible to observe cases with models showing no reductions in the number of reactions. This can be due to an equivalence among species with no dynamical role in the network, as they can be interpreted as distinct auxiliary species that are used to model zero-order reactions, such as I in reaction $I \rightarrow I + A$, a purely catalytic species C in a reaction like $A + C \rightarrow B + C$, or $SINK$ in a degradation reaction such as $A \rightarrow SINK$. In the first two cases, these species are associated with zero-derivative variable, while in the last case the variable for $SINK$ does not appear in any ODE in the system.

(a)   (b)

Fig. 7. (a) Mechanisms for the initial interaction of M and Mpp with MAPKK and MKP from [39]. Phosphorylation of M starts with the binding of MAPKK in either of terminus (T or Y) or M. Dephosphorylation occurs when MKP binds to an active molecule of M, in this case Mpp. (b) Reduced mechanism. BE equates the molecular complexes up to their phosphorylated residue.

(a)   (b)

Fig. 8. (a) Adaptation of the SPOC dynamical model from [15]. The SPB compartment is depicted in the yellow-circle background. Reactions crossing the compartment boundary represent the intrinsic Tem1 (blue rectangle) GTPase-cycle and reversible SPB association in terminal T. (b) Reduced mechanism where both FE and SMB equate all Tem1 molecules up to their GTP (green)- or GDP (red)-bound state (indicated by the green/red ellipsis). (Color figure online)

4 Case Studies

We hereby report selected case studies to highlight the physical interpretability of the reductions (cf. **Q3** in Sect. 1).

BE Example: MAPK Double Phosphorylation. Multisite phosphorylation is a well-studied model in computational systems biology [29,47]. The double (de)phospho-rylation model depicted in Fig. 7 reflects the changes in the phosphorylated state of MAPK in *BIOMOD0000000030*. MAPK cascades are evolutionary conserved and consist of several (usually 3) levels, where the activated kinase at each level phosphorylates the kinase at the next level down the cascade. MAPK (M) is a molecule with two residues: tyrosine (Y) and threonine (T), thus requires double phosphorylation from a MAPK Kinase to become active (Mpp), and double dephosphorylation from a MAPK phosphathase to return to its original inactive state. This dynamics is represented in a model with 18 species and 32 reactions. BE equates the MAPK complexes regardless of their binding with MAPK or MKP, yielding a reduced CRN with 16 species and 28 reactions.

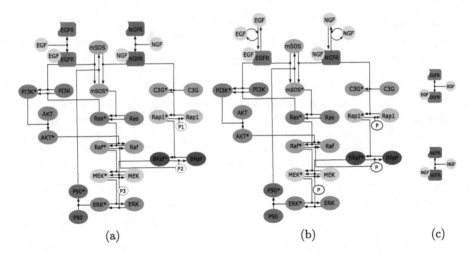

Fig. 9. (a) Adaptation of the signaling network in [4]. The activation of the molecular SOS by either of the receptors triggers the Ras cascade, concluding in ERK activation. EGF can also use the left branch involving PI3K to modulate Erk activity through Raf1 downregulation, and NGF can upregulate Mek using the right branch containing Rap1. P1, P2 and P3 represent unregulated phosphatases. Molecular components in (a) with the same color are grouped together in the same FDE equivalence class. (b) BDE reduction. (c) FDE reduction. (Color figure online)

FE Example: SPOC. Model *BIOMOD0000000705* is a CRN of the Spindle Position Checkpoint (SPOC) [15]. SPOC intervenes in the process of cell division by verifying all requirements to pass to the next phase in the cell cycle. In particular, it prevents the separation of the duplicated chromosomes until each chromosome is properly attached to the spindle apparatus. The most upstream event of the pathway involves GTPase Tem1. Tem1 binds to the yeast centrosomes (called spindle pole bodies, SPBs) via GAP-dependent and GAP-independent sites (Fig. 8a). The intrinsic GTPase switching cycle of Tem1 is modeled as a reversible first-order reaction that converts $\text{Tem}_1^{\text{GTP}}$ into $\text{Tem}_1^{\text{GDP}}$ and vice versa. The model consists of 24 species and 71 reactions. FE equates the two forms of the GPTase Tem1 (Fig. 8), moreover this equivalence extends to all Tem1 molecular complexes, yielding a reduced model with 16 species and 36 reactions . In this example, the largest SMB yields the same reduction.

BDE/FDE Example: Signaling Cascade. Model *BIOMOD0000000033* is a signaling pathway concluding in ERK activation [4]. Its most upstream event (Fig. 9) starts with the binding of EGF and NGF to their respective receptors (EGFR, NGFR). Once bound, both receptors can activate molecular SOS and trigger the Ras cascade. Here, molecular components are modeled representing the species active and inactive state, i.e mSOS* and mSOS, yielding a model with 32 species and 26 reactions. For BDE, the free EGF and the free receptor EGFR are aggregated, simplifying the process of EGF binding to EGFR. Similarly, this

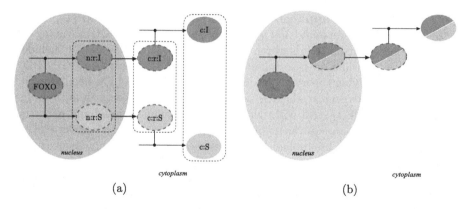

(a)   (b)

Fig. 10. (a) Adaptation of the FOXO-dependent IsnR and Sod2 synthesis mechanism in [45]. Species labels are $x{:}y{:}z$, where x is the species compartment, y indicates binding with molecular RNA, and z is the first letter of the name of the protein, e.g., n:r:I encodes the nuclear RNA-bound IsnR, c:S encodes cytoplasmic Sod2. RNA-bound molecules are rounded by a dotted circle. SMB equivalences are represented by a dotted rectangle. (b) SMB/FE reduced mechanism.

occurs for NFG and NFGR. Finally, phosphatases P1, P2, and P3, whose role is purely catalytic, are aggregated (in a macro-species denoted by P). The BDE reduction has 27 species and 26 reactions (Fig. 9b). Instead, FDE collapses the active and inactive form of those species. Moreover, the dynamics of the active and inactive species sum up to zero if aggregated. As above, the phosphatases P1, P2, and P3 are aggregated in the same class. This results in the FDE reduction in Fig. 9c, with 18 species and 4 reactions.

SMB Example: Proteins with Same Synthesis Mechanism. As observed, our methods can help detecting symmetries among molecular components. We show this in *BIOMOD0000000705*, a FOXO-dependent synthesis mechanism involving IsnR and Sod2. Forkhead Box-type O (FOXO) is a family of transcription factors responsible for various biological processes including apoptosis, cell metabolism, differentiation, and drug resistance [34]. The model has 56 species and 135 reactions describing processes such as FOXO-dependent and basal transcription, export, translation, and degradation of RNA and proteins. The kinetic parameters for FOXO-dependent protein synthesis (Fig. 10a) for both IsnR and Sod2 are assumed to be equal. This gives an SMB reduction with 36 species and 110 reactions where IsnR and Sod2 molecules are aggregated in each step of the protein synthesis mechanism (Fig. 10b). FE leads to the same reduction.

5 Concluding Remarks

The empirical assessment of exact model reduction on the BioModels repository has provided a number of findings along the main evaluation questions **Q1–Q3** introduced in Sect. 1, which can be summarized as follows.

Q1. Assumptions for applicability of model reductions. In the preprocessing phase (Sect. 3.3), we found 300 models not supported by ERODE. Among the reasons for incompatibility it is worth commenting on the models which included exponential expressions in rate functions. This is not accepted by FDE/BDE because the underlying theory is not decidable. A workaround has been sketched in [10,13] and builds on a systematic technique which transforms an initial value problem for an ODE system with derivatives containing rational and exponential expressions into an equivalent problem with polynomial derivatives [28], to which BE and FE can be applied. In future work we plan to implement such a transformation in order to extend the range of applicability of species equivalences. Instead, the limitation of SMB to elementary CRNs did not turn out to be practically impeding for the analysis of the BioModels repository, since all the CRNs were in this form; it is however theoretically interesting to extend the theory to non-elementary mass-action kinetics.

Q2. Effectiveness of the reductions. Overall, we found exact model reductions effective in terms of both the number of cases in which a CRN could be reduced by at least one technique (40%) and the overall compression ratio achieved on average (36% for number of species and 26% for the number of reactions). Unfortunately, the analysis of FDE on a rather appreciable number of models (196) was not conclusive due to timeouts, because of the relative complexity of the SMT checks that are required. This challenges the practical applicability of FDE to realistic case studies (BDE, on the other hand, timed out only twice in our tests whereas BE, FE, and SMB are supported by minimization algorithms that enjoy polynomial time and space complexity), prompting alternative approaches to computing FDE, for example by parallelizing the computations.

Q3. Physical interpretability. In the selected case studies herein presented, the exact model reductions have revealed that symmetries in certain signalling pathways carry over to equivalences at the level of the underlying quantitative semantics. Given their moderate size, the considered models would be computationally tractable even without reduction. However, the equivalences can be used as an aid in developing more complex models where such symmetries are present in some components. In addition, we remark that exact model reduction can still be useful when the complexity is due to the many repetitions that are required (e.g., for sensitivity analysis or for simulation with tight confidence intervals) or for particularly difficult analyses such as parametric inference [44].

Future Work. This empirical study suggests potential benefits in the application of exact model reduction techniques in biological models from the literature. This motivates the development of our ERODE translator into a more mature tool to be further integrated with BioModels/SBML. The availability of ready-to-use model conversions in a simple CRN format such as ERODE's might stimulate similar assessments with other model reduction techniques (e.g., [3,12]).

In this paper we focused on reducing models with parameterizations given as in the respective original publications. If we wish to draw more general conclusions about the relevance of the reductions and the presence of certain symmetrical patterns in signaling pathways, it becomes important to test their *robustness* with respect to the model parameters. Theoretically, this does not seem to be particularly difficult, at least for CRNs with deterministic semantics. For example, model parameters could be interpreted as further variables in the SMT formulas used for checking FDE and BDE. Such an extension is currently not implemented in ERODE and is subject to the aforementioned caveats about the scalability of SMT-based reduction techniques, hence left for future work.

Acknowledgement. The authors are grateful to Andreas Dräguer (Institut für Informatik Zentrum für Bioinformatik Tübingen) for his support with JSBML.

References

1. Ascher, U.M., Petzold, L.R.: Computer Methods for Ordinary Differential Equations and Differential-Algebraic Equations. SIAM (1988)
2. Ashburner, M., et al.: Gene ontology: tool for the unification of biology. Nat. Genet. **25**(1), 25 (2000)
3. Boreale, M.: Algebra, coalgebra, and minimization in polynomial differential equations. In: Esparza, J., Murawski, A.S. (eds.) FoSSaCS 2017. LNCS, vol. 10203, pp. 71–87. Springer, Heidelberg (2017). https://doi.org/10.1007/978-3-662-54458-7_5
4. Brown, K.S., et al.: The statistical mechanics of complex signaling networks: nerve growth factor signaling. Phys. Biol. **1**(3), 184 (2004)
5. Buchholz, P.: Exact and ordinary lumpability in finite Markov chains. J. Appl. Probab. **31**(1), 59–75 (1994)
6. Cardelli, L., Tribastone, M., Tschaikowski, M., Vandin, A.: Efficient syntax-driven lumping of differential equations. In: Chechik, M., Raskin, J.-F. (eds.) TACAS 2016. LNCS, vol. 9636, pp. 93–111. Springer, Heidelberg (2016). https://doi.org/10.1007/978-3-662-49674-9_6
7. Cardelli, L., Tribastone, M., Tschaikowski, M., Vandin, A.: Symbolic computation of differential equivalences. In: POPL, pp. 137–150 (2016). https://doi.org/10.1145/2837614.2837649
8. Cardelli, L., Tribastone, M., Tschaikowski, M., Vandin, A.: Forward and backward bisimulations for chemical reaction networks. In: 26th International Conference on Concurrency Theory, CONCUR, pp. 226–239 (2015). https://doi.org/10.4230/LIPIcs.CONCUR.2015.226
9. Cardelli, L., Tribastone, M., Tschaikowski, M., Vandin, A.: ERODE: a tool for the evaluation and reduction of ordinary differential equations. In: Legay, A., Margaria, T. (eds.) TACAS 2017. LNCS, vol. 10206, pp. 310–328. Springer, Heidelberg (2017). https://doi.org/10.1007/978-3-662-54580-5_19
10. Cardelli, L., Tribastone, M., Tschaikowski, M., Vandin, A.: Maximal aggregation of polynomial dynamical systems. Proc. Nat. Acad. Sci. **114**(38), 10029–10034 (2017)
11. Cardelli, L., Tribastone, M., Tschaikowski, M., Vandin, A.: Syntactic Markovian bisimulation for chemical reaction networks. In: Aceto, L., Bacci, G., Bacci, G., Ingólfsdóttir, A., Legay, A., Mardare, R. (eds.) Models, Algorithms, Logics and Tools. LNCS, vol. 10460, pp. 466–483. Springer, Cham (2017). https://doi.org/10.1007/978-3-319-63121-9_23
12. Cardelli, L., Tribastone, M., Tschaikowski, M., Vandin, A.: Guaranteed error bounds on approximate model abstractions through reachability analysis. In: 15th International Conference on Quantitative Evaluation of Systems (QEST) (2018)

13. Cardelli, L., Tribastone, M., Tschaikowski, M., Vandin, A.: Symbolic computation of differential equivalences. Theor. Comput. Sci. **777**, 132–154 (2019)
14. Caspi, R., et al.: The MetaCyc database of metabolic pathways and enzymes and the BioCyc collection of pathway/genome databases. Nucleic Acids Res. **42**(D1), D459–D471 (2013)
15. Caydasi, A.K., Lohel, M., Grünert, G., Dittrich, P., Pereira, G., Ibrahim, B.: A dynamical model of the spindle position checkpoint. Mol. Syst. Biol. **8**(1), 582 (2012)
16. Chelliah, V., Laibe, C., Novère, N.L.: Biomodels database: a repository of mathematical models of biological processes. In: Dubitzky, W., Wolkenhauer, O., Cho, K.H., Yokota, H. (eds.) Encyclopedia of Systems Biology, pp. 134–138. Springer, New York (2013). https://doi.org/10.1007/978-1-4419-9863-7
17. Consortium, U.: UniProt: a hub for protein information. Nucleic Acids Res. **43**(D1), D204–D212 (2014)
18. Courtot, M., et al.: Controlled vocabularies and semantics in systems biology. Mol. Syst. Biol. **7**(1), 543 (2011)
19. de Moura, L., Bjørner, N.: Z3: an efficient SMT solver. In: Ramakrishnan, C.R., Rehof, J. (eds.) TACAS 2008. LNCS, vol. 4963, pp. 337–340. Springer, Heidelberg (2008). https://doi.org/10.1007/978-3-540-78800-3_24
20. Degtyarenko, K., et al.: ChEBI: a database and ontology for chemical entities of biological interest. Nucleic Acids Res. **36**(Suppl. 1), D344–D350 (2007)
21. Dräger, A., et al.: JSBML: a flexible Java library for working with SBML. Bioinformatics **27**(15), 2167–2168 (2011). https://doi.org/10.1093/bioinformatics/btr361
22. Federhen, S.: The NCBI taxonomy database. Nucleic Acids Res. **40**(D1), D136–D143 (2011)
23. Feret, J., Henzinger, T., Koeppl, H., Petrov, T.: Lumpability abstractions of rule-based systems. Theor. Comput. Sci. **431**, 137–164 (2012)
24. Feret, J., Danos, V., Krivine, J., Harmer, R., Fontana, W.: Internal coarse-graining of molecular systems. Proc. Nat. Acad. Sci. **106**(16), 6453–6458 (2009). https://doi.org/10.1073/pnas.0809908106
25. Ganguly, A., Petrov, T., Koeppl, H.: Markov chain aggregation and its applications to combinatorial reaction networks. J. Math. Biol. **69**(3), 767–797 (2014). https://doi.org/10.1007/s00285-013-0738-7
26. Gillespie, D.T.: Stochastic simulation of chemical kinetics. Annu. Rev. Phys. Chem. **58**(1), 35–55 (2007)
27. Gillespie, D.: Exact stochastic simulation of coupled chemical reactions. J. Phys. Chem. **81**(25), 2340–2361 (1977)
28. Gu, C.: QLMOR: a projection-based nonlinear model order reduction approach using quadratic-linear representation of nonlinear systems. IEEE Trans. Comput. Aided Des. Integr. Circuits Syst. **30**(9), 1307–1320 (2011). https://doi.org/10.1109/TCAD.2011.2142184
29. Gunawardena, J.: Multisite protein phosphorylation makes a good threshold but can be a poor switch. Proc. Nat. Acad. Sci. U.S.A. **102**(41), 14617–14622 (2005). https://doi.org/10.1073/pnas.0507322102
30. Hucka, M., et al.: The systems biology markup language (SBML): a medium for representation and exchange of biochemical network models. Bioinformatics **19**(4), 524–531 (2003)
31. Juty, N., Le Novere, N., Laibe, C.: Identifiers.org and MIRIAM registry: community resources to provide persistent identification. Nucleic Acids Res. **40**(D1), D580–D586 (2011)

32. Kanehisa, M., Goto, S.: KEGG: kyoto encyclopedia of genes and genomes. Nucleic Acids Res. **28**(1), 27–30 (2000)
33. Kurtz, T.G.: The relationship between stochastic and deterministic models for chemical reactions. J. Chem. Phys. **57**(7), 2976–2978 (1972)
34. Lam, E.W.F., Brosens, J.J., Gomes, A.R., Koo, C.Y.: Forkhead box proteins: tuning forks for transcriptional harmony. Nat. Rev. Cancer **13**, 482 EP (2013)
35. Le Novère, N., et al.: Minimum information requested in the annotation of biochemical models (MIRIAM). Nat. Biotechnol. **23**(12), 1509 (2005)
36. Le Novère, N., et al.: JSBML 1.0: providing a smorgasbord of options to encode systems biology models. Bioinformatics **31**(20), 3383–3386 (2015). https://doi.org/10.1093/bioinformatics/btv341
37. Li, C., et al.: BioModels database: an enhanced, curated and annotated resource for published quantitative kinetic models. BMC Syst. Biol. **4**, 92 (2010)
38. Li, G., Rabitz, H.: A general analysis of exact lumping in chemical kinetics. Chem. Eng. Sci. **44**(6), 1413–1430 (1989). https://doi.org/10.1016/0009-2509(89)85014-6
39. Markevich, N.I., Hoek, J.B., Kholodenko, B.N.: Signaling switches and bistability arising from multisite phosphorylation in protein kinase cascades. J. Cell Biol. **164**(3), 353–359 (2004)
40. Nishimura, D.: BioCarta. Biotech Soft. Internet Rep.: Comput. Softw. J. Scient **2**(3), 117–120 (2001)
41. Paige, R., Tarjan, R.: Three partition refinement algorithms. SIAM J. Comput. **16**(6), 973–989 (1987). https://doi.org/10.1137/0216062
42. Pratt, D., et al.: NDEx, the network data exchange. Cell Syst. **1**(4), 302–305 (2015)
43. Radulescu, O., Gorban, A.N., Zinovyev, A., Noel, V.: Reduction of dynamical biochemical reactions networks in computational biology. Front. Genet. **3**(131) (2012). https://doi.org/10.3389/fgene.2012.00131
44. Schnoerr, D., Sanguinetti, G., Grima, R.: Approximation and inference methods for stochastic biochemical kinetics – a tutorial review. J. Phys. A: Math. Theor. **50**(9), 093001 (2017)
45. Smith, G.R., Shanley, D.P.: Modelling the response of FOXO transcription factors to multiple post-translational modifications made by ageing-related signalling pathways. PLoS ONE **5**(6), e11092 (2010)
46. Stoesser, G., et al.: The embl nucleotide sequence database. Nucleic Acids Res. **30**(1), 21–26 (2002)
47. Thomson, M., Gunawardena, J.: Unlimited multistability in multisite phosphorylation systems. Nature **460**(7252), 274–277 (2009). https://doi.org/10.1038/nature08102
48. Tribastone, M., Vandin, A.: Speeding up stochastic and deterministic simulation by aggregation: an advanced tutorial. In: 2018 Winter Simulation Conference, WSC 2018, Gothenburg, Sweden, 9–12 December 2018, pp. 336–350 (2018). https://doi.org/10.1109/WSC.2018.8632364
49. Turanyi, T., Tomlin, A.S.: Analysis of Kinetic Reaction Mechanisms. Springer, Heidelberg (2014). https://doi.org/10.1007/978-3-662-44562-4
50. Vandin, A., Tribastone, M.: Quantitative abstractions for collective adaptive systems. In: SFM 2016, pp. 202–232. Bertinoro Summer School (2016). https://doi.org/10.1007/978-3-319-34096-8_7
51. Voit, E.O.: Biochemical systems theory: a review. ISRN Biomath. **2013**, 53 (2013). https://doi.org/10.1155/2013/897658
52. Wittig, U., et al.: SABIO-RK–database for biochemical reaction kinetics. Nucleic Acids Res. **40**(D1), D790–D796 (2011)

Computing Difference Abstractions of Metabolic Networks Under Kinetic Constraints

Emilie Allart[1,2(✉)], Joachim Niehren[1,3], and Cristian Versari[1,2]

[1] BioComputing Team, CRIStAL Lab, Lille, France
[2] Université de Lille, Lille, France
emilie.allart@univ-lille.fr
[3] Inria, Lille, France

Abstract. Algorithms based on abstract interpretation were proposed recently for predicting changes of reaction networks with partial kinetic information. Their prediction precision, however, depends heavily on which heuristics are applied in order to add linear consequences of the steady state equations of the metabolic network. In this paper we ask the question whether such heuristics can be avoided while obtaining the highest possible precision. This leads us to the first algorithm for computing the difference abstractions of a linear equation system exactly without any approximation. This algorithm relies on the usage of elementary flux modes in a nontrivial manner, first-order definitions of the abstractions, and finite domain constraint solving.

Keywords: Gene knockout prediction · Reaction networks · Constraints · Systems biology · Synthetic biology · Metabolism

1 Introduction

Flux balance analysis [16,17] can be used to predict the effect of influx changes of metabolic networks at steady state. Such predictions can be based on reasoning with linear equations systems that describe the rates of the reactions in a steady state of the metabolic network, by using Gaussian elimination, elementary flux modes (EFMs) [13], or optimisation methods [6,14]. Most importantly, precise quantitative kinetic information is not required in contrast to classical mathematical analysis methods for reaction networks [2,7,9,11]. In fact, even when the kinetic functions associated to chemical reactions are known, the values of rate constants are most often missing, since it is difficult to measure them experimentally in the precise state of the regulation of the metabolic network at the time point of interest.

Recently, abstract interpretation [3,5,8] has been exploited to design novel algorithms [4,12,15] that can use partial kinetic information beneficially for predicting changes of metabolic networks. They can in particular exploit the

© Springer Nature Switzerland AG 2019
L. Bortolussi and G. Sanguinetti (Eds.): CMSB 2019, LNBI 11773, pp. 266–285, 2019.
https://doi.org/10.1007/978-3-030-31304-3_14

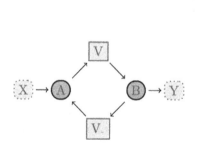

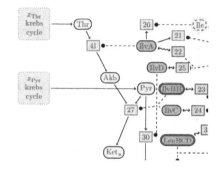

Fig. 1. A toy metabolic network with a simple cycle.

Fig. 2. A glimpse of the formal model from [4] of leucine production in *B. subtilis*.

knowledge about the enzymes and inhibitors. Similarly to flux balance analysis, the linear equations describing steady states are used, but in addition to them, kinetic constraints are inferred from the partial kinetic information of inhibitors and enzymes.

The steady state equations and the kinetic constraints enable gene knockout predictions based on abstract interpretation (in the finite relational structure Δ_6), based on the linear equations from the metabolic network and the constraints on its regulatory control. The unknown kinetic parameters are abstracted away, by interpretation over some finite relational structure, that contains a finite number of abstract differences rather than concrete differences in $\mathbb{R}_+^2$. Eventually, the prediction algorithm will apply a finite domain constraint solver for Δ_6 that we implemented in Minizinc [18] to enumerate all the changes that may or must lead to the target change.

The prediction quality of abstract interpretation approaches heavily depends on heuristics that find and add linear equations entailed by the steady state equations before constraint solving over Δ_6. This is necessary to enable global reasoning, since local reasoning alone is not able to deal with cycles in metabolic part of the network as we will illustrate in Sect. 2. On the other hand, it is impossible to add the infinity of all entailed linear equations before abstract interpretation. Therefore, these algorithms can at best approximate the abstraction of differences of solution set of linear equations. Whether this abstraction can be computed exactly is a long standing open problem, as well as how to measure the quality of approximation heuristics.

In this paper, we present the first exact algorithm that can compute the Δ_6 difference abstraction of the solution set of a linear equation system without any overapproximation. We apply it to the prediction of leucine overproduction, a benchmark task that is best studied with abstract interpretation. In this case, we need to deal with kinetic constraints in addition that are naturally interpreted over Δ_6. It turns out that a new heuristic that we also propose in the present paper does indeed compute well the difference abstraction at this benchmark task from systems biology, although being inexact in the general case. The main advantage of this heuristic is that it outperforms the exact algorithm dramatically in computation time: only 5 min are needed for the knockout prediction rather than 5 h with the always exact algorithm.

2 Qualitative Reasoning on Metabolic Reaction Networks

The application of abstract interpretation to the analysis of metabolic reaction networks is based on an intuitive qualitative reasoning. Its aim is to predict how a living organism or its environment should be changed in order to maximize the production of some metabolite of interest, without exact knowledge of the quantitative parameters of the system. A change of the organism is represented for example by any modification of its genome, such as a gene knockout. A change of the environment is typically represented by the modification of the culture medium, which results in an increase or decrease of some inflows. Since changes can be arbitrarily combined to obtain or improve the wanted results, the problem that we tackle is highly combinatorial.

Examples of formal metabolic networks on which abstract interpretation can be applied are shown in Figs. 1 and 2. Figure 1 shows a toy metabolic network with a simple cycle that will be used in the following to introduce the key ideas of the reasoning. In this network, metabolites are displayed in yellow rounded boxes, while reactions are in gray squared boxes, in the tradition of Petri nets. Reactions with dotted contour are inflows or outflows of the system.

Figure 2 shows a glimpse of a bigger metabolic reaction network with regulation, where the regulatory part is represented by the enzymes in blue rounded boxes. The full network of Fig. 1 models a part of the metabolism of the gram positive bacterium *B. subtilis*. Our benchmark application – taken from [4] – is the overproduction of one of the metabolites of this network, the branched chain amino acid Leucine (Leu). This amino acid is a precursor of surfactin, a non ribosomal peptide with several applications in food and pharmaceutical industry.

Let us now reconsider the toy model with a simple cycle in Fig. 1. This network is composed of two chemical species A and B. The species A is continuously produced by an inflow at a fixed rate X, and is transformed into B by the reaction with rate V. The inverse reaction with rate V_- transforms B back into A. The species B has an outflow with rate Y. All reactions but the inflow are controlled internally by the system. The outflow rate Y in particular is determined by the concentration of B which in turn depends on the rates V and V_-.

The only possible change in this toy network is the increase or decrease of the inflow X. In order to illustrate the reasoning method, we set the increase of Y as our final target. As usually done in flux balance analysis [16,17], we consider the system at the steady state, that gives us the following linear system of equations:

$$\exists V \exists V_-.\quad V = X + V_- \quad \wedge \quad V = Y + V_- \tag{1}$$

The existential quantifiers for V and V_- allow us to hide the internal behaviour of the network, so to project to inflows and outflows. While the consequences of steady state equations may be difficult to interpret without using Gauss' algorithm, in this particular case it is easy to see that by subtracting the first equation in (1) from the second we obtain an important relation:

$$X = Y \qquad (2)$$

Equation (2) tells us that the only way for Y to increase is that X increases too. We formalize now this intuition by means of abstract interpretation, this time applied to *concrete differences* in $\mathbb{R}^2_+$. Concrete differences capture the essence of a change in the spirit of [15]: a change of the value of X for instance can be thought of as a pair of positive reals $(r^{\text{before}}, r^{\text{after}})$ representing the value of X at steady state respectively *before* and *after* the modification of the environment. We need to consider positive reals since the rates of irreversible reactions are positive. We call the above concrete differences an *increase* if $r^{\text{before}} < r^{\text{after}}$, a *decrease* if $r^{\text{before}} > r^{\text{after}}$ and a no-change if $r^{\text{before}} = r^{\text{after}}$. This intuition motivates the usage of abstract values in $\Delta_3 = \{\triangle, \triangledown, \approx\}$ where $\triangle = (0,1)$ represents an increase, $\triangledown = (1,0)$ a decrease, and $\approx = (0,0)$ a no-change. The canonical mapping of concrete differences in $\mathbb{R}^2_+$ to the abstract differences in Δ_3 can be seen as a homomorphism between the relational structures $\mathbb{R}^2_+$ to Δ_3. This algebraic view of abstractions as homomorphisms enables various generalizations. An example is the abstraction from $\mathbb{R}^2_+$ to Δ_6 – as considered for gene knockout prediction [4,12] – which refines each of the three $\mathbb{R}^2_+$ equivalence classes produced by Δ_3 into two parts in Δ_6.

Any abstraction of concrete differences enables some form of *abstract qualitative reasoning* [10] based on operations of the relational structure of abstract differences, that can be used for change prediction in systems biology. Let us illustrate how to reason with Δ_3. As a first example, assume that we know for some reason that X and V_- both increase, that is $X = \triangle$ and $V_- = \triangle$. Then we can use the first equation in (1) to deduce for sure V will also increase, since $\triangle +^{\Delta_3} \triangle = \triangle$. The full table defining the summation operator $+^{\Delta_3}$ on abstract difference is given in Fig. 3.

The above qualitative reasoning method, however, is quite weak when relying only on the steady state equations in (1). The main reason is that all reasoning steps are local, so that they overlook global properties of the network that are arising for example with metabolic cycles. To see this, suppose that we want to predict which environmental change may lead to an increase of Y. We can use the second equation in (1) to infer some constraints on the values of V and V_-: if $Y = \triangle$, we can for example infer that V cannot be $\triangledown$ if V_- is $\triangle$. In fact, $\triangle + \triangle$ can never be equal to $\triangledown$. However, $V = \triangle, V_- = \triangledown$ is a partial solution that seems *compatible* with an increase of Y. This illustrates that we cannot infer any constraint on X with this kind of local reasoning.

Equation $X = Y$ in Eq. (2), in contrast, expresses a global property of the network that immediately implies that the only value for X compatible with an increase of Y is $\triangle$. In other words, when reasoning with equations over abstract differences, Eq. (2) is *no longer an implicit consequence* of the system (1). Therefore, to be taken into account, such equations must be explicitly included in the system *before* applying the abstract reasoning. Unfortunately, the number of entailed linear equations is infinite in general. For instance, in our small network all the equations of the family $nV + mX = nV_- + (m+1)Y$ for any two naturals n, m are consequences of (1). So, instead of trying to infer the set of *all* the consequences of our system, we may try to compute a "good" subset of it, by including only the consequences that more heavily constrain the variables. The prediction quality of the existing approaches heavily depends on the heuristics chosen for adding entailed linear consequences.

One evident advantage of Eq. (2) is its small number of variables (there are only two: X and Y). If we consider an individual linear equation, the intuition is that removing a variable increases the constraining power on the remaining variables. So, we propose as a heuristics the inclusion of all linear consequences involving a minimal subset of variables. This idea is at the core of the first result that we present in this paper: a heuristic algorithm that enriches the steady state equations of the metabolic network with minimal support consequences before applying abstract interpretation. While this algorithm has been internally used for some time to increase the precision of the abstract interpretation, some key questions have been always open about it:

1. is the set of minimal support equations *complete*? that is, does it represent all the deducible constraints on the abstract system?
2. are these constraints sufficient to compute the *exact* set of abstract solutions?
3. if this is not the case, is there a method to compute them?

The definitive answers to these questions is the second main contribution of this paper. We show in particular that the above heuristic *does not* cover all the entailed constraints on the abstract system, i.e. it does not allow to compute the exact abstraction of a linear system in general. Intuitively, this happens because the approach takes for granted that the abstract reasoning is based on the linear system computed at the steady state, that is on the matrix equation $AX = 0$ where A is the stoichiometry matrix associated to the metabolic network, and $\mathbf{X}$ is the set of metabolic flows representing the unknowns of the linear system. However, it is easy to notice that as soon as concrete differences are introduced in the reasoning, there is not only one, but actually *two* linear systems to consider: one *before* the environmental change, and one *after* it, that is one system for each value of the pairs representing concrete changes. Informally, we should therefore consider a bigger matrix equation including somehow both $A\mathbf{X}^{\text{before}} = 0$ and $A\mathbf{X}^{\text{after}} = 0$.

This idea is the starting point of development of the main contribution of the present paper: a method for the exact computation of the abstraction of a linear system, that we call the *exact* algorithm. This method not only provides us with the counterexamples to the exactness of the heuristic based on minimal support consequences, but gives us also an exact measure of its goodness as well as of the goodness of all the other heuristics used to improve our abstract analysis.Remarkably, both the heuristic and the exact algorithm have their root in the rewriting of a linear system in terms of its EFMs. The key difference between the two methods lies in the choice of the linear system (i.e. the matrix equation) initially used to compute the EFMs.

3 Preliminaries

Set Notation. We start with usual notation for sets. Let $\mathbb{N}$ be the set of natural numbers and $\mathbb{R}_+$ the set of positive real numbers, both including 0. For any set A and $n \in \mathbb{N}$, the set of n-tuples of elements in A is denoted by A^n. The i-th projection function on n-tuples of elements in A, where $1 \leq i \leq n$ is the function $\pi_i : A^n \to A$ such that $\pi_i(a_1, \ldots, a_n) = a_i$ for all $a_1, \ldots, a_n \in A$. If A is finite the number of elements of A is denote by $|A|$.

Σ- Algebras and Σ- Structures. We next recall the notions of Σ-algebras, Σ-structures, and homomorphism between Σ-structures. Let $\Sigma = \cup_{n \geq 0} F^{(n)} \uplus C$ be a ranked signature. The elements of $f \in F^{(n)}$ are called the n-ary function symbols of Σ and the elements in $c \in C$ its constants.

Definition 1. *A Σ-algebra $S = (dom(S), .^S)$ consists of a set $dom(S)$ and an interpretation $.^S$ such that $c^S \in dom(S)$ for all $c \in C$, and $f^S : dom(S)^n \to dom(S)$ for all $f \in F^{(n)}$ and $n \in \mathbb{N}$.*

We next reinterpret n-ary function symbols of Σ as $n{+}1$-ary relation symbols, so that we can reuse the same signature Σ for defining Σ-structures.

Definition 2. *A Σ-structure $\Delta = (dom(\Delta), .^\Delta)$ consists of a set $dom(\Delta)$ and an interpretation $.^\Delta$ such that $c^\Delta \in dom(\Delta)$ for all $c \in C$ and $f^\Delta \subseteq dom(\Delta)^{n+1}$ for all $f \in F^{(n)}$ and $n \in \mathbb{N}$.*

In this manner, any Σ-algebra is also a Σ-structure since any n-ary function is an $n{+}1$-ary relation. Note also that symbols in $F^{(0)}$ are interpreted as monadic relations in Σ-structures, i.e., as subsets of the domain, in contrast to constants in C that are interpreted as elements of the domain.

Definition 3. *A homomorphism between two Σ-structures S and Δ is a function $h : dom(S) \to dom(\Delta)$ such that for $c \in C$, $n \in \mathbb{N}$, $f \in F^{(n)}$, and $s_1, \ldots, s_{n+1} \in dom(S)$:*

1. $h(c^S) = c^\Delta$, and
2. if $(s_1, \ldots, s_{n+1}) \in f^S$ then $(h(s_1), \ldots, h(s_{n+1})) \in f^\Delta$.

If we consider $n{+}1$-ary relations as n-ary set valued functions, the second condition can be rewritten equivalently as $h(f^S(s_1, \ldots, s_n)) \subseteq f^\Delta(h(s_1), \ldots, h(s_n))$. For Σ-algebras, this condition is equivalent to $h(f^S(s_1, \ldots, s_n)) = f^\Delta(h(s_1), \ldots, h(s_n))$.

4 Σ-Abstractions

Throughout the paper we will use the signature $\Sigma = F^{(2)} \uplus C$ with two binary function symbols in $F^{(2)} = \{+, *\}$ and two constants $C = \{0, 1\}$. We will only consider Σ-algebras S in which $+^S$ and $*^S$ are associative and commutative, with neutral element 0^S and 1^S respectively.

Example 4. The set of positive real numbers $\mathbb{R}_+$ can be turned into a Σ-algebra with domain $\mathbb{R}_+$, by interpreting $+$ as the addition of positive real numbers $+^{\mathbb{R}+}$, $*$ as the multiplication of positive real numbers $*^{\mathbb{R}+}$, and interpreting the constants by themselves $0^{\mathbb{R}+} = 0$ and $1^{\mathbb{R}+} = 1$. We will deliberately confuse the set $\mathbb{R}_+$ with the Σ-algebra $(\mathbb{R}_+, .^{\mathbb{R}+})$.

Example 5. The set of Booleans $\mathbb{B} = \{0, 1\} \subseteq \mathbb{R}_+$ can be turned into a Σ-algebra with domain $\mathbb{B}$ by interpreting $+^{\mathbb{B}} = \vee^{\mathbb{B}}$ as disjunction, $*^{\mathbb{B}} = \wedge^{\mathbb{B}}$ as conjunction, and the constants by themselves $0^{\mathbb{B}} = 0$ and $1^{B} = 1$. We will deliberatly confuse the set $\mathbb{B}$ with the Σ-algebra $(\mathbb{B}, .^{\mathbb{B}})$.

We can abstract positive real numbers into booleans by defining a function $h_{\mathbb{B}} : \mathbb{R}_+ \to \mathbb{B}$ such that $h_{\mathbb{B}}(0) = 0$ and $h_{\mathbb{B}}(r) = 1$ for all $r \in \mathbb{R}_+ \setminus \{0\}$.

Lemma 6. *The function $h_{\mathbb{B}} : \mathbb{R}_+ \to \mathbb{B}$ is a homomorphism between Σ-algebras.*

The homomorphism $h_{\mathbb{B}}$ is the prime example of what we will call a Σ-abstraction.

Definition 7. *A Σ-abstraction is a homomorphism between Σ-structures S and Δ such that $dom(\Delta) \subseteq dom(S)$.*

5   Abstracting Concrete Differences

Concrete differences are pairs of positive real numbers such as $(r^{\text{before}}, r^{\text{after}}) \in \mathbb{R}_+^2$ in the example Sect. 2. We show how to abstract concrete differences into abstract differences in some finite Σ-structure.

The Tuple Σ-Algebra S^n. For any Σ-algebra S and natural number $n \in \mathbb{N}$ we define the Σ-algebra of n-tuples $S^n = (dom(S)^n, .^{S^n})$ such that for all $s_1, \ldots, s_n, s_1', \ldots, s_n' \in dom(S)$ and $\odot \in F^{(2)}$:

$$(s_1, \ldots, s_n) \odot^{S^n} (s_1', \ldots, s_n') = (s_1 \odot^S s_1', \ldots, s_n \odot^S s_n')$$

The constants $c \in C$ are interpreted as $c^{S^n} = (c^S, \ldots, c^S)$. Note that if 0^S is the neutral element of $+^S$, then 0^{S^n} is the also the neutral element of $+^{S^n}$. In analogy, if 1^S is the neutral element of $*^S$ then 1^{S^n} is also the neutral element of $*^{S^n}$. Furthermore, the associativity and commutativity of $+^{S^n}$ and $*^{S^n}$ inherit from $+^S$ and $*^S$ respectively.

Note that we deliberately confuse the set $\mathbb{R}_+^2$ with the Σ-algebra $(\mathbb{R}_+^2, .^{\mathbb{R}_+^2})$ with our notation. Given this, it follows from the above, that the algebra $\mathbb{R}_+^2$ has the neutral element $(0, 0)$ for $+^{\mathbb{R}_+}$ and the neutral element $(1, 1)$ for $*^{\mathbb{R}_+}$, and that these operations are associative and commutative.

For any function $h : A \to B$ and $n \in \mathbb{N}$ we define the function $h^n : A^n \to B^n$ such that $h^n(a_1, \ldots, a_n) = (h(a_1), \ldots, h(a_n))$ for all $a_1, \ldots, a_n \in A$.

Lemma 8. *If h is Σ-abstraction from S to Δ then h^n is a Σ-abstraction from S^n to Δ^n.*

δ	δ'	$\delta +^{\Delta_3} \delta'$	$\delta *^{\Delta_3} \delta'$
$\triangle$	$\triangle$	$\{\triangle\}$	$\{\triangle\}$
$\triangle$	$\triangledown$	$\{\triangle,\approxeq,\triangledown\}$	$\{\triangle,\approxeq,\triangledown\}$
$\triangle$	$\approxeq$	$\{\triangle\}$	$\{\triangle,\approxeq\}$

δ	δ'	$\delta +^{\Delta_3} \delta'$	$\delta *^{\Delta_3} \delta'$
$\approxeq$	$\approxeq$	$\{\approxeq\}$	$\{\approxeq\}$
$\triangledown$	$\triangledown$	$\{\triangledown\}$	$\{\triangledown\}$
$\triangledown$	$\approxeq$	$\{\triangledown\}$	$\{\triangledown,\approxeq\}$

c	c^{Δ_3}
0	$\approxeq$
1	$\approxeq$

Fig. 3. Interpretation of Σ-structure Δ_3.

δ	δ'	$\delta +^{\Delta_6} \delta'$	$\delta *^{\Delta_6} \delta'$
$\uparrow$	$\uparrow$	$\{\uparrow \}$	$\{\uparrow \}$
$\uparrow$	$\downarrow$	$\{\uparrow ,\sim,\downarrow\}$	$\{\uparrow ,\sim,\downarrow\}$
$\uparrow$	$\sim$	$\{\uparrow \}$	$\{\uparrow \}$
$\uparrow$	$\Uparrow$	$\{\uparrow \}$	$\{\Uparrow\}$
$\uparrow$	$\Downarrow$	$\{\uparrow ,\downarrow,\sim\}$	$\{\Downarrow\}$
$\uparrow$	$\approx$	$\{\uparrow \}$	$\{\approx\}$
$\Uparrow$	$\downarrow$	$\{\uparrow ,\sim,\downarrow\}$	$\{\Uparrow\}$

δ	δ'	$\delta +^{\Delta_6} \delta'$	$\delta *^{\Delta_6} \delta'$
$\Uparrow$	$\sim$	$\{\uparrow \}$	$\{\Uparrow\}$
$\Uparrow$	$\Uparrow$	$\{\Uparrow\}$	$\{\Uparrow\}$
$\Uparrow$	$\Downarrow$	$\{\uparrow ,\sim,\downarrow\}$	$\{\approx\}$
$\Uparrow$	$\approx$	$\{\Uparrow\}$	$\{\approx\}$
$\sim$	$\sim$	$\{\sim\}$	$\{\sim\}$
$\sim$	$\approx$	$\{\sim\}$	$\{\approx\}$
$\sim$	$\downarrow$	$\{\downarrow\}$	$\{\downarrow\}$

δ	δ'	$\delta +^{\Delta_6} \delta'$	$\delta *^{\Delta_6} \delta'$
$\sim$	$\Downarrow$	$\{\downarrow\}$	$\{\Downarrow\}$
$\approx$	$\approx$	$\{\approx\}$	$\{\approx\}$
$\approx$	$\downarrow$	$\{\downarrow\}$	$\{\Downarrow\}$
$\approx$	$\Downarrow$	$\{\Downarrow\}$	$\{\Downarrow\}$
$\downarrow$	$\downarrow$	$\{\downarrow\}$	$\{\downarrow\}$
$\downarrow$	$\Downarrow$	$\{\downarrow\}$	$\{\Downarrow\}$
$\Downarrow$	$\Downarrow$	$\{\Downarrow\}$	$\{\Downarrow\}$

c	c^{Δ_6}
0	$\approx$
1	$\sim$

Fig. 4. Interpretation of Σ-structure Δ_6.

Abstractions of Concrete Differences. A generic manner to abstract concrete differences in $\mathbb{R}_+^2$ is to start with a finite set $\Delta \subseteq \mathbb{R}_+^2$ of so called *abstract differences*, and some function $h : \mathbb{R}_+^2 \to \Delta$ that says how to abstract any concrete differences to some abstract difference. The function h defines a partition of $\mathbb{R}_+^2$ into the equivalences classes of concrete differences that are mapped to the same abstract difference.

Given such a function h, there is a unique manner to define an interpretation $.^{\Delta}$ such that $(\Delta,.^{\Delta})$ becomes Σ-structure and h a Σ-abstraction. For any constant $c \in C$ we have to define $c^{\Delta} = h(c^{\mathbb{R}_+^2})$ and for any function symbol $\odot \in F^{(2)}$ we have to define a ternary relation $\odot^{\Delta}$, which seen as set-valued function $\odot^{\Delta} : \Delta \times \Delta \to 2^{\Delta}$ must satisfy for all abstract values $\delta_1, \delta_2 \in \Delta$:

$$\delta_1 \odot^{\Delta} \delta_2 = \{h(r_1 \odot^{\mathbb{R}_+} r_2, r_1' \odot^{\mathbb{R}_+} r_2') \mid h(r_1, r_1') = \delta_1,\ h(r_2, r_2') = \delta_2\}$$

Lemma 9. $h : \mathbb{R}_+^2 \to \Delta$ *is a* Σ-*abstraction.*

The Σ-Structure Δ_3. Our next objective is to recall the abstraction of concrete differences into the finite Σ-structure with domain $\Delta_3 = \{\triangle, \triangledown, \approxeq\}$ that is well-known from qualitative reasoning (see e.g. [10]). For this we start with the function $h_{\Delta_3}(r, r') \in \Delta_3$ such that for any all $r, r' \in \mathbb{R}_+$:

$$h_{\Delta_3}(r, r') = \begin{cases} \triangledown = (1,0) & \text{if } r > r' \\ \triangle = (0,1) & \text{if } r < r' \\ \approxeq = (0,0) & \text{if } r = r' \end{cases}$$

The relations $+^{\Delta_3}$ and $*^{\Delta_3}$ are is the symmetric closure of the relation in Fig. 1. Furthermore, $h_{\Delta_3} : \mathbb{R}_+^2 \to \Delta_3$ is a Σ-abstraction by Lemma 9.

The Σ-Structure Δ_6. We next recall the abstraction of concrete differences to the finite Σ-structure with domain $\Delta_6 = \{\uparrow, \downarrow, \sim, \Uparrow, \Downarrow, \approx\}$ that was introduced for gene knockout prediction in [15]. For defining this Σ-structure, we start with the function $h_{\Delta_6} : \mathbb{R}_+^2 \to \Delta_6$ such that for any two numbers $r, r' \in \mathbb{R}_+$:

$$h_{\Delta_6}(r,r') = \begin{cases} \uparrow \ = (1,2) & \text{if } 0 \neq r < r' \\ \downarrow = (2,1) & \text{if } r > r' \neq 0 \\ \sim = (1,1) & \text{if } r = r' \neq 0 \end{cases} \qquad h_{\Delta_6}(r,r') = \begin{cases} \Uparrow = (0,2) & \text{if } 0 = r < r' \\ \Downarrow = (2,0) & \text{if } r > r' = 0 \\ \approx = (0,0) & \text{if } r = r' = 0 \end{cases}$$

The relations $+^{\Delta_6}$ and $*^{\Delta_6}$ are the symmetric closure of the relations in Fig. 4. By Lemma 9, $h_{\Delta_6} : \mathbb{R}_+^2 \to \Delta_6$ is a Σ-abstraction.

6   First-Order Logic

We first recall the standard first-order logic and then show how to enhance it with n-tuples without increasing the expressiveness.

We fix a set of variables $\mathcal{V}$ (for instance $\mathcal{V} = \mathbb{N}$). The variables in $\mathcal{V}$ will be ranged over by x and y. The set of first-order expressions $e \in \mathcal{E}_\Sigma$ and first-order formulas $\phi \in \mathcal{F}_\Sigma$ are constructed according to the abstract syntax in Fig. 5 from the symbols in the signature Σ, the variables in $\mathcal{V}$, the first-order connectives, and the equality symbol $\doteq$. As shortcuts, we define the formula $true =_{\text{def}} 1 \doteq 1$ and for any sequence of formulas $\phi_1, \ldots, \phi_n$ we define $\wedge_{i=1}^n \phi_i$ as $\phi_1 \wedge \ldots \wedge \phi_n$ which is equal to $true$ if $n = 0$. We define formulas $e \not\doteq 0$ by $\neg e \doteq 0$.

The semantics of a formula $\phi \in \mathcal{F}_\Sigma$ is a truth value, which depends on the Σ-structures S of interpretation and on a variable assignment $\alpha : \mathcal{V} \to dom(S)$. Any Σ-expressions $e \in \mathcal{E}_\Sigma$ denotes a subset of values in $dom(S)$, which will be singleton in case that S was a Σ-algebra. The semantic of equations $e \doteq e'$ is, as expected when interpreted over Σ-algebras S: the unique values of e and e' in S must be equal. However, we will also need to interpret equations $e \doteq e'$ over Σ-structures. This is why, any expression e denotes a subset of the Σ-structure, not just a single element. We can then interpret equality as nondisjointness, i.e., $e \doteq e'$ holds in a Σ-structure S if e and e' are interpreted as nondisjoint subsets of $dom(S)$.

A variable assignment into a Σ-structure S is a partial function $\alpha : V \to dom(S)$ for some subset $V \subseteq \mathcal{V}$. Let S be a Σ-structure and α a variable assignment to S. Any Σ-expression e with $\mathcal{V}(e) \subseteq V$ can be interpreted as an element of $dom(S)$ and any Σ-formula $\phi \in \mathcal{F}_\Sigma$ with $\mathcal{V}(\phi) \subseteq V$ as a Boolean value. The set of solutions of a formula $\phi \in \mathcal{F}_\Sigma$ over a Σ-structure S with respect to some set of variables $V \supseteq \mathcal{V}(\phi)$ is defined by:

$$sol_V^S(\phi) = \{\alpha : V \to dom(S) \mid [\![\phi]\!]^{S,\alpha} = 1\}$$

If $V = \mathcal{V}(\phi)$ then we omit the index V, i.e., $sol^S(\phi) = sol_V^S(\phi)$.

We next extend the first-order logic to n-tuples where the parameter n is fixed. In applications, we will use the case $n = 2$, that is the first-order logic with pairs. Back and forth compilers from first-order logic with and without tuples will be convenient later on.

The syntax of first-order logic with n-tuples is given in Fig. 6. The expressions $o \in \mathcal{O}_\Sigma^n$ are like the expression $e \in \mathcal{E}_\Sigma$ except that variables x are now replaced

First-order expressions and formulas:

$$e \in \mathcal{E}_\Sigma \ ::= \ x \ | \ c \ | \ e \odot e' \qquad \text{where } \odot \in F^{(2)}, c \in C$$
$$\phi \in \mathcal{F}_\Sigma \ ::= \ e \dot{=} e \ | \ \exists x.\phi \ | \ \phi \wedge \phi \ | \ \neg\phi \text{ where } x \in \mathcal{V}$$

Interpretation of expressions as sets of elements $[\![e]\!]^{S,\alpha} \subseteq dom(S)$, where S is a Σ-structures and $\alpha : V \to dom(S)$ where V contains all free variables.

$$[\![c]\!]^{S,\alpha} = c^S \qquad [\![x]\!]^{S,\alpha} = \{\alpha(x)\} \qquad [\![e \odot e']\!]^{S,\alpha} = \cup\{s \odot^S s' \ | \ s \in [\![e]\!]^{S,\alpha}, s' \in [\![e']\!]^{S,\alpha}\}$$

Interpretation of formulas as truth values $[\![\phi]\!]^{S,\alpha} \in \mathbb{B}$:

$$[\![e \dot{=} e']\!]^{S,\alpha} = \begin{cases} 1 \text{ if } [\![e]\!]^{S,\alpha} \cap [\![e']\!]^{S,\alpha} \neq \emptyset \\ 0 \text{ else} \end{cases} \qquad [\![\phi \wedge \phi']\!]^{S,\alpha} = [\![\phi]\!]^{S,\alpha} \wedge^\mathbb{B} [\![\phi']\!]^{S,\alpha}$$

$$[\![\neg\phi]\!]^{S,\alpha} = \neg^\mathbb{B}([\![\phi]\!]^{S,\alpha}) \qquad [\![\exists x.\phi]\!]^{S,\alpha} = \begin{cases} 1 \text{ if exists } s \in dom(S). \ [\![\phi]\!]^{S,\alpha[x/s]} = 1 \\ 0 \text{ else} \end{cases}$$

Fig. 5. Syntax and semantics of expressions and formulas of first-order logic.

$$o \in \mathcal{O}_\Sigma^n \ ::= \ \dot{\pi}_i(x) \ | \ c \ | \ o \odot o \qquad \text{where } \odot \in F^{(2)}, c \in C, 1 \leq i \leq n$$
$$\psi \in \mathcal{F}_\Sigma^n \ ::= \ o \dot{=} o' \ | \ \exists x.\psi \ | \ \psi \wedge \psi \ | \ \neg\psi \text{ where } x \in \mathcal{V}$$

Fig. 6. Σ-expressions and Σ-formulas of the first-order logic with n-tuples.

by projection expressions $\dot{\pi}_i(x)$ where $1 \leq i \leq n$. The reason is that any variable does now denote an n-tuple of values, rather than a single value (while the interpretation of constants and function symbols remain unchanged). The only change in the semantics is that variables assignment β do now map to n-tuples of values of the domain, and that $[\![\dot{\pi}_i(x)]\!]^{S,\beta} = \{\pi_i(\beta(x))\}$. The set of solutions of a formula $\psi \in \mathcal{F}_\Sigma^n$ over a Σ-structure S is defined as follows:

$$n\text{-}sol^S(\psi) = \{\beta : \mathcal{V}(\psi) \to dom(S)^n \ | \ [\![\psi]\!]^{S,\beta} = 1\}$$

We next show how to express any first-order formulas in $\mathcal{F}_\Sigma$, interpreted over a tuple algebra S^n, by some formulas in $\mathcal{F}_\Sigma^n$, interpreted over S. In a first step, we convert first-order expression in $e \in \mathcal{E}_\Sigma$ – that we will interpret over the Σ-algebra S^n – to n projected expressions $\Pi_i(e) \in \mathcal{O}_\Sigma^n$ where $1 \leq i \leq n$. For all operators $\odot \in F^{(2)}$ and constants $c \in C$ we define:

$$\Pi_i(e \odot e') =_{\text{def}} \Pi_i(e) \odot \Pi_i(e') \qquad \Pi_i(x) =_{\text{def}} \dot{\pi}_i(x) \qquad \Pi_i(c) =_{\text{def}} c$$

In the second step, we convert any formula $\phi \in \mathcal{F}_\Sigma$ without tuples – that will be interpreted in the tuple algebra S^n – to some formula $\langle\phi\rangle^n \in \mathcal{F}_\Sigma^n$ with tuples.

$$\langle e \dot{=} e'\rangle^n =_{\text{def}} \wedge_{i=1}^n \Pi_i(e) \dot{=} \Pi_i(e') \qquad \langle\phi \wedge \phi'\rangle^n =_{\text{def}} \langle\phi\rangle^n \wedge \langle\phi'\rangle^n$$
$$\langle\neg\phi\rangle^n =_{\text{def}} \neg\langle\phi\rangle^n \qquad \langle\exists x.\phi\rangle^n =_{\text{def}} \exists x.\langle\phi\rangle^n$$

Proposition 10. *For any* $\phi \in \mathcal{F}_\Sigma$, Σ-*structure* S, *and* $n \geq 1$: $sol^{S^n}(\phi) = n\text{-}sol^S(\langle\phi\rangle^n)$.

Example 11. Let $\mathbf{3} =_{\text{def}} 1 + 1 + 1$ and $\mathbf{4} =_{\text{def}} 1 + 1 + 1 + 1$. The formula $\phi \in \mathcal{F}_\Sigma$ equal to:

$$\mathbf{3} * x + \mathbf{4} * y \dot{=} 0$$

then has the same solutions over $\mathbb{R}_+^2$ than the formula $\langle \phi \rangle^2 \in \mathcal{F}_\Sigma^2$ over $\mathbb{R}_+$ below:

$$\mathbf{3} * \tilde{\pi}_1(x) + \mathbf{4} * \tilde{\pi}_1(y) \dot{=} 0 \wedge \mathbf{3} * \tilde{\pi}_2(x) + \mathbf{4} * \tilde{\pi}_2(y) \dot{=} 0$$

We next show how to rewrite any first-order formulas with tuples $\psi \in \mathcal{F}_\Sigma^n$ into some first-order formula $\tilde{\nu}(\psi) \in \mathcal{F}_\Sigma$ without tuples. The idea is to replace all projections $\pi_i(x)$ by new variables $\nu_i(x)$. For this, we first fix n generators of fresh variables $\nu_1, \ldots, \nu_n : \mathcal{V} \to \mathcal{V}$. Second, we map any expression $o \in \mathcal{O}_\Sigma^n$ with projections to some expressions $\tilde{\nu}(o) \in \mathcal{E}_\Sigma$ without new variables:

$$\tilde{\nu}(\tilde{\pi}_i(x)) =_{\text{def}} \nu_i(x), \qquad \tilde{\nu}(c) =_{\text{def}} c, \qquad \tilde{\nu}(o_1 \odot o_2) =_{\text{def}} \tilde{\nu}(o_1) \odot \tilde{\nu}(o_2).$$

Third, we map any formula $\psi \in \mathcal{F}_\Sigma^n$ with projections to some formula $\tilde{\nu}(\psi) \in \mathcal{F}_\Sigma$ with fresh variables:

$$\tilde{\nu}(o = o') =_{\text{def}} \tilde{\nu}(o) = \tilde{\nu}(o') \quad \tilde{\nu}(\neg\psi) =_{\text{def}} \neg\tilde{\nu}(\psi)$$
$$\tilde{\nu}(\psi \wedge \psi') =_{\text{def}} \tilde{\nu}(\psi) \wedge \tilde{\nu}(\psi') \quad \tilde{\nu}(\exists x.\psi) =_{\text{def}} \exists \nu_1(x) \ldots \exists \nu_n(x). \ \tilde{\nu}(\psi)$$

Given an variable assignment $\beta : V \to dom(S)^n$ with $V \subseteq \mathcal{V}$, we define $\nu(\beta) : \uplus_{i=1}^n \nu_i(V) \to dom(S)$ such that for all $x \in V$:

$$\nu(\beta)(\nu_i(x)) = \pi_i(\beta(x)))$$

Function ν is a bijection with range $\{\alpha \mid \alpha : \uplus_{i=1}^n \nu_i(V) \to dom(S)\}$. The inverse of this function satisfies $\nu^{-1}(\alpha)(x) = (\alpha(\nu_1(x)), \ldots, \alpha(\nu_n(x)))$ for all α in the range and all $x \in V$.

Proposition 12. *For any* $\psi \in \mathcal{F}_\Sigma^n$, *$\Sigma$-structure S, and $n \geq 1$: $n\text{-}sol^S(\psi) = \nu^{-1}(sol^S(\tilde{\nu}(\psi)))$.*

Proposition 13. *For any subset R of variable assignments of type $V \to dom(S)$ where $V \subseteq \mathcal{V}$, $n \geq 1$, and Σ-abstraction $h : S \to \Delta$: $\nu^{-1}(h \circ R) = h^n \circ \nu^{-1}(R)$.*

7 Difference Abstraction

We next show how to recast the notions of difference abstractions from [4,12,15] by applying our notion of Σ-abstractions to the Σ-algebra $\mathbb{R}_+^2$.

Let S be a Σ-algebra and $V \subseteq \mathcal{V}$ a subset of variables. For any two variable assignments $\alpha, \alpha' : V \to dom(S)$, we define an assignment of variables to pairs of elements in the domain of the structure $\textit{diff}(\alpha, \alpha') : V \to dom(S)^2$ – that we call the differences of α and α' – such that for all variables $x \in V$, $\textit{diff}(\alpha, \alpha')(x) =$

$(\alpha(x), \alpha'(x))$. For any subset R of variable assignments of type $V \to dom(S)$ we define the *set of differences of assignments in R* by:

$$diff(R) = \{ diff(\alpha, \alpha') \mid \alpha, \alpha' \in R \}$$

Furthermore, for any Σ-abstraction $h : S^2 \to \Delta$ and subset R' of difference abstractions of type $V \to dom(S)^2$ we define the application of the abstraction h to R' by $h \circ R' =_{\mathrm{def}} \{ h \circ \beta \mid \beta \in R' \}$

Definition 14. *For any Σ-abstraction $h : S^2 \to \Delta$ and formula $\phi \in \mathcal{F}_\Sigma$ we define the difference abstraction of the S-solution set of ϕ by: $sol^S(\phi)^\Delta = h \circ diff(sol^S(\phi)))$.*

The original definition of $sol(\phi)^{\Delta_6}$ in [15] did not make explicit the roles of $diff$ and $h_{\Delta_6} : \mathbb{R}_+^2 \to \Delta_6$. Having done so, we can now see that the difference abstraction of the $\mathbb{R}_+$-solution sets of a formula is the $\mathbb{R}_+^2$-solution set of the same formula.

Lemma 15. *For any formula $\phi \in \mathcal{F}_\Sigma$ and Σ-structure S: $diff(sol^S(\phi)) = sol^{S^2}(\phi)$.*

As an immediate consequence, we have for any Σ-abstraction $h : S^2 \to \Delta$ that $sol(\phi)^\Delta = h_\Delta \circ sol^{S^2}(\phi)$. Our next objective is to show that we can overapproximate the set $sol(\phi)^\Delta$ by $sol^\Delta(\phi)$ (Corollary 19). In order to show this, let $h' : S' \to \Delta$ be a Σ-abstraction and α be a variable assignment into $dom(S')$:

Lemma 16. *For any expression $e \in \mathcal{E}_\Sigma$ with $V(e) \subseteq dom(\alpha)$: $h'(\llbracket e \rrbracket^{S',\alpha}) \subseteq \llbracket e \rrbracket^{\Delta, h' \circ \alpha}$.*

Proposition 17. *For any positive formula $\phi \in \mathcal{F}_\Sigma$ with $V(\phi) \subseteq dom(\alpha)$: $\llbracket \phi \rrbracket^{S',\alpha} \leq \llbracket \phi \rrbracket^{\Delta, h' \circ \alpha}$.*

Theorem 18. *For any positive formula $\phi \in \mathcal{F}_\Sigma$: $h' \circ sol^{S'}(\phi) \subseteq sol^\Delta(\phi)$.*

Corollary 19. *For any Σ-abstraction $h : S^2 \to \Delta$ and positive first-order formula $\phi \in \mathcal{F}_\Sigma$:*

$$sol^S(\phi)^\Delta = h \circ diff(sol^S(\phi)) \subseteq sol^\Delta(\phi)$$

This is an obvious consequence from Theorem 18 and Proposition 15. If Δ is finite then the set $sol^\Delta(\phi)$ is finite. If furthermore ϕ is a conjunctive formula, we can therefore compute the set $sol^\Delta(\phi)$ by a finite domain constraint solver (such as e.g. Minizinc [18]). In contrast, it remains unclear how to compute the finite set $h \circ diff(sol^S(\phi))$ for infinite structures S. The problem is open, even if ϕ is a system of homogenous linear equations and $S = \mathbb{R}_+$, so that the infinite set $sol^S(\phi)$ has a finite solved form by a triangular matrix. This is the core of the objective that we tackle in the remainder of the present paper.

8   Objective

We formalize the full algorithmic problem that we will solve in this paper and illustrate its relevance to our benchmark application to systems biology.

Once having fixed the parameter $\Delta \in \{\Delta_3, \Delta_6\}$ the algorithmic problem has three inputs:

Linear system over $\mathbb{R}_+$: a first-order formula $\phi \in \mathcal{F}_\Sigma$ that represents a linear equation system. (This formula typically captures the steady state equations of the model.)

Constraint over Δ: a first-order formula $\phi' \in \mathcal{F}_{\Sigma \cup \Delta}$ where the signature Σ is extended with additional constants of Δ that will be interpreted by themselves. (This formula typically expresses the partial kinetic knowledge on the reactions in the model and the change target of the prediction task (e.g. overproduction of some metabolites).)

Set of observable variables: a finite subset of variables $V \subseteq \mathcal{V}(\phi) \cup \mathcal{V}(\phi')$. (This set typically contains the control variables such as inflows and gene knockouts as well as the target variables, but not the variables for the rate of the internal metabolic reactions. Since the number of solutions may be of cardinality $|\Delta|^{|V|}$, it is essential to choose V as small as possible.)

The algorithmic output that has to be produced is the Δ-abstraction of differences of $\mathbb{R}_+$-solutions of ϕ, but constrained to the solutions of ϕ' over the structure Δ, and projected to the variables of V. In other words, the algorithm will compute the following finite set where $V' = \mathcal{V}(\phi) \cup \mathcal{V}(\phi')$:

$$\{\beta_{|V} \mid \beta \in sol_{V'}^{\mathbb{R}_+}(\phi)^\Delta \cap sol_{V'}^\Delta(\phi')\}$$

The only restriction on the inputs that we will impose is that the first formula ϕ must represent a homogeneous system of linear equations in $\mathcal{F}_\Sigma$. For instance, the linear equation $x - 2y = 0$ is captured by the equation $x \dot{=} y + y$ in $\mathcal{F}_\Sigma$ where we cannot use the minus operator. See Sect. 9 for the general definition. The constraint $\phi' \in \mathcal{F}_{\Sigma \cup \Delta}$ in contrast may be arbitrary, including nonlinear equations and universal quantifiers but must be interpreted abstractly over Δ, while the linear equation system is valid over $\mathbb{R}_+$. Note however, that any universal quantifiers in ϕ' can be expressed by a simple conjunction, given that the interpretation domain Δ is finite.

In Fig. 7 we illustrate how the inputs will be instantiated for our benchmark application of leucine overproduction (a glimpse of the reaction network was given in Fig. 2). In this case, we choose the parameter $\Delta = \Delta_6$. The observable variables in V stand for the rates of the inflows Threonine (x_{Thr}), Akb (x_{Akb}), etc, the rate of the target outflow Leucine (y_{Leu}), and the possible gene knockouts. The system of linear equations ϕ contains the steady state equations for the metabolic reactions in the network. These require that all metabolites must be produced and consumed at the same rate. For instance, Pyruvate is produced by the inflow of Threonine at rate x_{Thr} and consumed by reactions 27 and 30 at rates r_{27} and r_{30} respectively. The yields the linear equation (Thr) of Fig. 7.

Observable variables	Linear equations in $\mathcal{F}_\Sigma$	Δ_6 constraints in $\mathcal{F}_{\Sigma \cup \Delta_6}$
$V = \{\, x_{Thr},$ $x_{Akb},$ $y_{Leu},$ $\ldots\}$	$\phi =$ $x_{Thr} \dot{=} r_{27} + r_{30}$ (Thr) $\wedge\ r_{41} \dot{=} r_{27}$ (Akb) $\wedge\ r_{45} \dot{=} r_{35} + y_{Leu}$ (Leu) $\wedge \ldots$	$\phi' =$ $y_{Leu} \dot{=} \uparrow$ (target) $\wedge\ r_{27} \dot{=} Pyr * Akb * IlvD$ $\qquad * IlvBH * IlvC$ (27) $\wedge \ldots$

Fig. 7. Inputs of our algorithm on the benchmark example from of leucine overproduction.

Species *Akb* is produced by reaction 41 and consumed by 27, leading to the steady state equation (*Akb*). Leucine is produced by reaction 45 and consumed by its outflow, leading to equation (*Leu*). The constraint $\phi \in \mathcal{F}_{\Sigma \cup \Delta_6}$ contains the overproduction target $y_{Leu} = \uparrow$ in (*target*) and the kinetic constraints for all reactions, of which we show only constraint (27) for reaction 27. The kinetic constraints must be interpreted abstractly over Δ_6 according the formal semantics of the modeling language [15] rather than concretely over $\mathbb{R}_+$. Therefore, the meaning of the constraints is purely qualitative and not at all quantitative. For instance, the constraint (27) states (beside others) that rate of reaction 27 increases if either of the concentrations of the reactants *Pyr* and *Akb* or of the enzymes *IlvD*, *IlvBH* or *IlvC* increase. Nothing is said about quantities of these increases.

9   Exact Algorithms

We now present an exact solution of the problem presented in the previous section. Our approach is to characterize the abstraction of the solution set of a linear equation system as the solution set of some first-order formula over the abstract domain. We consider the abstractions $h_\mathbb{B}$, h_{Δ_3}, and h_{Δ_6} in this order.

Characterizing $\mathbb{B}$-Abstractions. We now present a result from [1] that shows that the boolean abstraction of the $\mathbb{R}_+$-solution set of a mixed linear system can be computed exactly. The development of this result was motivated by the needs of the present paper, but given that it is of independent interest and nontrivial to prove, we decided to present it independently.

Any natural numbers n can be described by the expression $\mathbf{n} =_{def} \sum_{i=1}^{n} 1$ in $\mathcal{E}_\Sigma$. This permits to define *linear equations* as equations in $\mathcal{F}_\Sigma$ that have the form:

$$\mathbf{n}_1 * x_1 + \ldots \mathbf{n}_m * x_m \dot{=} \mathbf{n}_{m+1} * x_{m+1} + \ldots + \mathbf{n}_p * x_p \qquad (3)$$

where $m, p, n_1, \ldots, n_p \in \mathbb{N}$ and $x_1, \ldots, x_p \in \mathcal{V}$.

Mixed Linear Systems. A *product-zero-equation* in $\mathcal{F}_\Sigma$ is an equation of the form $x * y \dot{=} 0$ where $x, y \in \mathcal{V}$. A *mixed linear system* is a conjunctive formula in $\mathcal{F}_\Sigma$ of the form $\exists z.\ L \wedge E$ where L is a conjunction of linear equations and E a conjunction of product-zero-equations.

Elementary Flux Modes. The support of a variable assignment $\alpha : V \to \mathbb{R}$ with $V \subseteq \mathcal{V}$ is $supp(\alpha) = \{x \in dom(\alpha) \mid \alpha(x) \neq 0\}$. Given a linear system ϕ, the

EFMs of ϕ are the minimal support solutions of ϕ over $\mathbb{R}_+$. The $\mathbb{R}$-EFMs of ϕ are the minimal support solutions of ϕ over $\mathbb{R}$. Note that the interpretation of $\mathbb{R}$ is natural for the steady-state equations of metabolic networks with reversible reactions, while the reactions of our networks are always irreversible.

Theorem 20 [1]. *Let ϕ be a mixed linear system. We can compute in at most exponential time some conjunctive formula ϕ' with existential quantifiers such that $h_{\mathbb{B}} \circ sol^{\mathbb{R}_+}(\phi) = sol^{\mathbb{B}}(\phi')$.*

Proof Sketch. There are quite some insights behind this theorem that we can only sketch here. First, any linear equation L system can be rewritten in the form $A\mathbf{y} \dot{=} 0$ where A is an integer matrix and $\mathbf{y}$ a sequence of pairwise distinct variables such that $V(\mathbf{y}) = V(L)$. Let P be a positive integer matrix whose columns contain all the EFMs of A. These can be computed in at most exponential time [19]. Then $sol^{\mathbb{R}_+}(A\mathbf{y} \dot{=} 0)$ is equal to $sol^{\mathbb{R}_+}(\exists \mathbf{x}.\ P\mathbf{x} \dot{=} \mathbf{y})$ given that any solution of $A\mathbf{y} \dot{=} 0$ over $\mathbb{R}_+$ can be obtained from some linear combination of the EFMs of A. Second, a formula $\phi \in \mathcal{F}_\Sigma$ is called $h_{\mathbb{B}}$-exact if $sol^{\mathbb{B}}(\phi) = h_{\mathbb{B}} \circ sol^{\mathbb{R}_+}(\phi)$. Unfortunately, not every linear systems is $h_{\mathbb{B}}$-exact. However, the formula ϕ'' equal to $\exists \mathbf{x}.P\mathbf{x} \dot{=} \mathbf{y}$ can be shown to be $h_{\mathbb{B}}$-exact, roughly since matrix P contains only positive integers. Third, it was noticed that any conjunction of product-zero equations is $h_{\mathbb{B}}$-exact as well. Fourth, for any system of product-zero equations E and any sequence of variables $\mathbf{z}$, the formula ϕ' equal to $\exists \mathbf{z}.\phi'' \wedge E$ can be shown to be $h_{\mathbb{B}}$-exact. Finally, any mixed linear systems ϕ can be brought into the form of ϕ' by computing the EFMs of the matrix A of the linear subsystem of ϕ in exponential time. □

Exact Algorithm. In order to compute the $h_{\mathbb{B}}$-abstraction of a mixed linear system ϕ, we first compute ϕ' along the lines of the sketch of the proof ideas of Theorem 20. Second, given that ϕ' is a conjunctive formula, we compute $sol^{\mathbb{B}}(\phi')$ by finite domain constraint programming.

Characterizing Δ_3-Abstractions. We present a characterization of Δ_3-abstractions of linear equation systems and show that it provides an exact algorithm solving the objective in the case of Δ_3. We first decompose h_{Δ_3} into $h_{\mathbb{B}}$ and the binary relation, that is defined by the following first-order formula in the logic with pairs $\mathcal{F}_\Sigma^2$:

$$proj_G(x,y) =_{\text{def}} \dot{\pi}_1(x) + \dot{\pi}_2(y) = \dot{\pi}_2(x) + \dot{\pi}_1(y) \wedge \dot{\pi}_1(y) * \dot{\pi}_2(y) = 0$$

We are mainly interested in interpreting this formula over $\mathbb{R}_+^2$.

Lemma 21. *The relation $proj_G{}^{\mathbb{R}_+^2}$ is a function satisfying $h_{\Delta_3} = h_{\mathbb{B}}^2 \circ proj_G{}^{\mathbb{R}_+^2}$.*

We next define applications of $proj_G$ in FO-logic. For any sequence of variables $\mathbf{y}$ and FO-formula $\phi(\mathbf{y}) \in \mathcal{F}_\Sigma^2$ with $\mathcal{V}(\phi(\mathbf{y})) \subseteq \{\mathbf{y}\}$ we define a formula $proj_G(\phi(\mathbf{y})) \in \mathcal{F}_\Sigma^2$ describing the application of $proj_G$ to the solutions of $\phi(\mathbf{y})$ by $\exists \mathbf{z}.\ \phi(\mathbf{z}) \wedge proj_G(\mathbf{z},\mathbf{y})$ where $\phi(\mathbf{z})$ is obtained from $\phi(\mathbf{y})$ by replacing the variables in $\mathbf{y}$ by arbitrarily but fixed fresh variables $\mathbf{z}$.

Lemma 22. $proj_G^{\mathbb{R}^2_+} \circ 2\text{-}sol^{\mathbb{R}+}(\phi(\boldsymbol{y})) = 2\text{-}sol^{\mathbb{R}+}(proj_G(\phi(\boldsymbol{y})))$.

Theorem 23. *For any linear formula $L \in \mathcal{F}_{\Sigma}$ we can compute in at most exponential time a positive conjunctive formula with existential quantifiers $\phi \in \mathcal{F}_{\Sigma}$ such that:*

$$h_{\Delta_3} \circ \textit{diff}(sol^{\mathbb{R}+}(L)) = \nu^{-1}(sol^{\mathbb{B}}(\phi))$$

Proof. Let $L(\mathbf{y})$ be a linear system with $\mathcal{V}(L(\mathbf{y})) = \mathcal{V}(\mathbf{y})$ where $\mathbf{y}$ is a sequence of distinct variables. The time for computing ϕ is dominated by the time for computing the elementary modes, which can be done in at most exponential time.

$$
\begin{aligned}
& \left| h_{\Delta_3} \circ \textit{diff}(sol^{\mathbb{R}+}(L(\mathbf{y}))) \right. \\
\textit{Proposition 15} \quad &= h_{\Delta_3} \circ sol^{\mathbb{R}^2_+}(L(\mathbf{y})) \\
\textit{Pair FO Proposition 10} \quad &= h_{\Delta_3} \circ 2\text{-}sol^{\mathbb{R}+}(L_2(\mathbf{y})) \text{ with } L_2(\mathbf{y}) = \langle L(\mathbf{y}) \rangle^2 \\
\textit{Decomposition Lemma 21} \quad &= h_{\mathbb{B}}^2 \circ proj_G^{\mathbb{R}^2_+} \circ 2\text{-}sol^{\mathbb{R}+}(L_2(\mathbf{y}))) \\
\textit{FO-Definition Lemma 22} \quad &= h_{\mathbb{B}}^2 \circ 2\text{-}sol^{\mathbb{R}+}(proj_G(L_2(\mathbf{y}))) \\
\textit{Proposition 12} \quad &= h_{\mathbb{B}}^2 \circ \nu^{-1}(sol^{\mathbb{R}+}(\tilde{\nu}(proj_G(L_2(\mathbf{y}))))) \\
\textit{Proposition 13} \quad &= \nu^{-1}(h_{\mathbb{B}} \circ sol^{\mathbb{R}+}(\tilde{\nu}(proj_G(L_2(\mathbf{y}))))) \\
\textit{Definition of } proj_G(L_2(y)) \quad &= \nu^{-1}(h_{\mathbb{B}} \circ sol^{\mathbb{R}+}(\tilde{\nu}(\exists \mathbf{z}.\ L_2(\mathbf{z}) \wedge proj_G(\mathbf{z}, \mathbf{y})))) \\
\textit{Mixed linear systems Theorem 20} \quad &= \nu^{-1}(sol^{\mathbb{B}}(\phi)) \\
& \quad \text{where } \phi \text{ is an equivalent conjunctive formula for the} \\
& \quad \text{mixed linear sytem } \tilde{\nu}(\exists \mathbf{z}.\ L_2(\mathbf{z}) \wedge proj_G(\mathbf{z}, \mathbf{y}))
\end{aligned}
$$

$\square$

Note that $sol^{\mathbb{B}}(\phi)$ can be computed by finite domain constraint programming. This yields an exact algorithm for computing the Δ_3-abstraction of a system of linear equations, which is a special case of our general objective without kinetic constraints.

For adding a treatment of kinetic constraints, we consider the union $\mathbb{B} \cup \Delta_3$ as a relational structure providing the values and functions of both structures $\mathbb{B}$ and Δ_3. The signature of this mixed structure consists of the function symbols in $\{+^{\mathbb{B}}, *^{\mathbb{B}}, +^{\Delta_3}, *^{\Delta_3}\}$ and the constants in the set $\mathbb{B} \cup \Delta_3$, all of which are interpreted by themselves in the mixed structure $\mathbb{B} \cup \Delta_3$. The set of first-order formulas over the mixed signature is denoted by $\mathcal{F}_{\mathbb{B} \cup \Delta_3}$. For any $\alpha : V \to dom(S)$, we can define its restriction $\alpha_{|V'} : V' \to dom(S)$ such that for all $y \in V' \subseteq V$, $\alpha_{|V'}(y) = \alpha(y)$.

Proposition 24. *For any formulas $\phi \in \mathcal{F}_{\Sigma}$ and $\phi' \in \mathcal{F}_{\Sigma \cup \Delta_3}$ and sets of variables $V \subseteq V' = \mathcal{V}(\phi) \cup \mathcal{V}(\phi')$ we can compute in linear time a formula $\phi_M \in \mathcal{F}_{\mathbb{B} \cup \Delta_3}$ over the mixed signature such that: $sol^{\mathbb{B} \cup \Delta_3}(\phi_M) = \{\beta_{|V} \mid \beta \in \nu^{-1}(sol_{V'}^{\mathbb{B}}(\phi)) \cap sol_{V'}^{\Delta_3}(\phi')\}$.*

The set $sol^{\mathbb{B} \cup \Delta_3}(\phi_M)$ can be computed by a finite domain constraint programming, since $\mathbb{B} \cup \Delta_3$ is a finite structure. By combining Theorem 23 and Proposition 24 we obtain an algorithm for solving the general problem of Sect. 8 in the cases of Δ_3.

Characterizing Δ_6-***Abstractions.*** The case of Δ_6 is considerably more envolved that the case of Δ_3, even though following the same general approach. For this, we consider the abstraction h_{Δ_6} as an element of the algebra of total functions on $\mathbb{R}_+^2$, that we denote as $\mathbb{R}_+^2 \to \mathbb{R}_+^2$. The following lemma shows that h_{Δ_6} is the sum of h_{Δ_3} and $h_{\mathbb{B}}^2$ in this Σ-algebra.

Lemma 25. $h_{\Delta_6} = h_{\mathbb{B}}^2 + h_{\Delta_3}$ *where* $+ = +^{\mathbb{R}_+^2 \to \mathbb{R}_+^2}$.

Let $idproj_G^{\mathbb{R}_+^2} : \mathbb{R}_+^2 \to (\mathbb{R}_+^2)^2$ such that for any $p \in \mathbb{R}_+^2$ $idproj_G^{\mathbb{R}_+^2}(p) = (p, proj_G^{\mathbb{R}_+^2}(p))$. Furthermore, we define for any two functions $g : A \to B \times C$ and $f : B \times C \to D$ the pseudo composition $f \bullet g : A \to D$ such that for all $a \in A$: $(f \bullet g)(a) = f(\pi_1(g(a)), \pi_2(g(a)))$. The Σ-abstraction $h_{\mathbb{B}}^2 : \mathbb{R}_+^2 \to \mathbb{B}^2$ allows us to define $(h_{\mathbb{B}}^2)^2 : (\mathbb{R}_+^2)^2 \to (\mathbb{B}^2)^2$.

Lemma 26 Decomposition. $h_{\Delta_6} = +^{\mathbb{R}_+^2} \bullet (h_{\mathbb{B}}^2)^2 \circ idproj_G^{\mathbb{R}_+^2}$.

We can now define the ternary relation $idproj_G^{\mathbb{R}_+^2}$ in the first-order logic with pairs by $idproj_G : \mathcal{V} \times \mathcal{V}^2 \to \mathcal{F}_\Sigma^2$ such that for all $x, y_1, y_2 \in \mathcal{V}$:

$$idproj_G(x, y_1, y_2) =_{\text{def}} \langle x = y_1 \rangle^2 \wedge proj_G(x, y_2)$$

We next define applications of $idproj_G$ in FO-logic. For any sequence of variables $\mathbf{y}$ and FO-formula $\phi(\mathbf{y}) \in \mathcal{F}_\Sigma^2$ with $\mathcal{V}(\phi(\mathbf{y})) \subseteq \{\mathbf{y}\}$ we define a formula $idproj_G(\phi(\overline{\mathbf{y}})) \in \mathcal{F}_\Sigma^2$ for describing the application of $idproj_G$ to the solution set of $\phi(\mathbf{y})$. We let $idproj_G(\phi(\overline{\mathbf{y}})$ be $\exists \mathbf{z}. \phi(\mathbf{z}) \wedge idproj_G(\mathbf{z}, \mathbf{y}^1, \mathbf{y}^2)$ where $\phi(\mathbf{z})$ is obtained from $\phi(\mathbf{y})$ by replacing the variables in $\mathbf{y}$ by arbitrarily but fixed fresh variables $\mathbf{z}$ and by fixing two sequences of fresh variables $\mathbf{y}^1, \mathbf{y}^2 \in \mathcal{V}^m$ such that $\overline{\mathbf{y}} = (\mathbf{y}^1, \mathbf{y}^2)$.

Lemma 27. $idproj_G^{\mathbb{R}_+^2} \circ 2\text{-}sol^{\mathbb{R}+}(\phi(\mathbf{y})) = \{[\mathbf{y}/(\alpha(\mathbf{y}^1), \alpha(\mathbf{y}^2)] | \alpha \in 2\text{-}sol^{\mathbb{R}+}(idproj_G (\phi(\overline{\mathbf{y}})))\}$.

Theorem 28. *For any linear formula* $L(\mathbf{y})$ *with free distinct variable* $\mathbf{y}$ *we can compute in at most exponential time a positive conjunctive formula with existential quantifiers* $\phi' \in \mathcal{F}_\Sigma$ *and sequences of variables* $\mathbf{y}^1, \mathbf{y}^2$ *such that:*

$$h_{\Delta_6} \circ \textit{diff}(sol^{\mathbb{R}+}(L(\mathbf{y}))) = \{[\mathbf{y}/(\beta(\mathbf{y}^1) +^{\mathbb{R}_+^2} \beta(\mathbf{y}^2)] \mid \beta \in \nu^{-1}(sol^{\mathbb{B}}(\phi'))\}$$

Proof. Let $L(\mathbf{y})$ be a linear formula L with free distinct variable $\mathbf{y} \in \mathcal{V}^m$.

$$
\begin{aligned}
& h_{\Delta_6} \circ \textit{diff}(sol^{\mathbb{R}+}(L(\mathbf{y}))) \\
\text{\textit{Proposition 15}} \quad &= h_{\Delta_6} \circ sol^{\mathbb{R}_+^2}(L(\mathbf{y})) \\
\text{\textit{Proposition 10}} \quad &= h_{\Delta_6} \circ 2\text{-}sol^{\mathbb{R}+}(L_2(\mathbf{y})) \text{ with } L_2(\mathbf{y}) = \langle L(\mathbf{y}) \rangle^2 \\
\text{\textit{Decomposition Lemma 26}} \quad &= +^{\mathbb{R}_+^2} \bullet (h_{\mathbb{B}}^2)^2 \circ idproj_G^{\mathbb{R}_+^2} \circ 2\text{-}sol^{\mathbb{R}+}(L_2(\mathbf{y})) \\
\text{\textit{FO-Definition Lemma 27}} \quad &= +^{\mathbb{R}_+^2} \bullet (h_{\mathbb{B}}^2)^2 \circ \{[\mathbf{y}/(\alpha(\mathbf{y}^1), \alpha(\mathbf{y}^2)] \mid \alpha \in 2\text{-}sol^{\mathbb{R}+}(idproj_G(L_2(\overline{\mathbf{y}})))\} \\
&= +^{\mathbb{R}_+^2} \bullet \{[\mathbf{y}/(\beta(\mathbf{y}^1), \beta(\mathbf{y}^2)] \mid \beta \in h_{\mathbb{B}}^2 \circ 2\text{-}sol^{\mathbb{R}+}(idproj_G(L_2(\overline{\mathbf{y}})))\} \\
&= \{[\mathbf{y}/(\beta(\mathbf{y}^1) +^{\mathbb{R}_+^2} \beta(\mathbf{y}^2)] \mid \beta \in h_{\mathbb{B}}^2 \circ 2\text{-}sol^{\mathbb{R}+}(idproj_G(L_2(\overline{\mathbf{y}})))\}
\end{aligned}
$$

Network	Count type	pure abstr. interpr.	min. support consequences	exact
Simple metabolic cycle (Fig. 1)	abstract solutions	19	6	6
Leucine overproduction [4]	knockouts	16	14	14
	abstract solutions	292	228	228
Counterexample	abstract solutions	≥ 10000	4454	4374

Fig. 8. Predictions for the networks analysed in this paper, obtained respectively by pure abstract interpretation, the heuristic based on minimal support consequences and the exact algorithm.

We can now finish the proof by computing the $h_{\mathbb{B}}^2$ abstraction of the above solution set similarly to the case of Δ_3.

$$
\begin{aligned}
& h_{\mathbb{B}}^2 \circ 2\text{-}sol^{\mathbb{R}+}(idproj_G(L_2)(\overline{\mathbf{y}})) \\
\textit{Proposition 12} \quad &= h_{\mathbb{B}}^2 \circ \nu^{-1}(sol^{\mathbb{R}+}(\tilde{\nu}(idproj_G(L_2(\overline{\mathbf{y}}))))) \\
\textit{Proposition 13} \quad &= \nu^{-1}(h_{\mathbb{B}} \circ sol^{\mathbb{R}+}(\tilde{\nu}(idproj_G(L_2(\overline{\mathbf{y}}))))) \\
\textit{Definition of } idproj_G(L_2(\overline{y})) \quad &= \nu^{-1}(h_{\mathbb{B}} \circ sol^{\mathbb{R}+}(\tilde{\nu}(\exists \mathbf{z}.\ L_2(\mathbf{z}) \wedge idproj_G(\mathbf{z}, \overline{\mathbf{y}})))) \\
\textit{Mixed linear systems Theorem 20} \quad &= \nu^{-1}(sol^{\mathbb{B}}(\phi')) \\
& \text{where } \phi' \text{ is conjunctive formula equivalent to the} \\
& \text{mixed linear system } \tilde{\nu}(\exists \mathbf{z}.\ L_2(\mathbf{z}) \wedge idproj_G(\mathbf{z}, \overline{\mathbf{y}}))
\end{aligned}
$$

$\square$

For adding a treatment of kinetic constraints, we consider the union $\mathbb{B} \cup \Delta_6$ as a relational structure providing the values and functions of both structures $\mathbb{B}$ and Δ_6 in analogy to the case of Δ_3. We denote the of first-order formulas over the signature of this mixed structure by $\mathcal{F}_{\mathbb{B}\cup\Delta_6}$.

Proposition 29. *For any formula* $\phi(\mathbf{y}) \in \mathcal{F}_\Sigma$ *with distinct free variables* $\mathbf{y}$, *formula* $\phi' \in \mathcal{F}_{\Sigma\cup\Delta_6}$, *and variable sets* $V \subseteq V' = \mathcal{V}(\mathbf{y}) \cup \mathcal{V}(\phi')$ *we can compute in linear time a formula* $\phi_M \in \mathcal{F}_{\mathbb{B}\cup\Delta_6}$ *such that* $sol^{\mathbb{B}\cup\Delta_6}(\phi_M) = \beta_{|V} \mid \beta \in \{[\mathbf{y}/(\beta'(\mathbf{y}^1) +_+^{\mathbb{R}^2} \beta'(\mathbf{y}^2))] \mid \beta' \in \nu^{-1}(sol_{V'}^{\mathbb{B}}(\phi))\} \cap sol_{V'}^{\Delta_6}(\phi')$.

The set $sol^{\mathbb{B}\cup\Delta_6}(\phi_M)$ can be computed by a finite domain constraint programming, since $\mathbb{B} \cup \Delta_6$ is a finite structure. By combining Theorem 28 and Proposition 29 we obtain an algorithm for solving the general problem of Sect. 8 in the case of Δ_6.

10 Heuristic Algorithm Based on Minimal Support Consequences

The intuition behind the heuristic with minimal support consequences relies on two facts: first, adding consequences to a given linear system L before applying abstract interpretation can improve the precision of the abstraction, as already discussed (2); second, the smaller the number of variables in an equation, the more constraining generally is its abstract interpretation. The heuristics is therefore very simple: before abstracting from $\mathbb{R}_+$ to Δ, the linear system L containing the steady state equations of the system is replaced by a linear system $L_{\min}$ containing all the minimal support $\mathbb{R}_+$-consequences of the equations in L. The linear system $L_{\min}$ can be computed by applying any existing algorithm for the calculation of $\mathbb{R}$-EFMs to the orthogonal complement of $L^\perp$ as follows:

1. From L compute a linear system $L^\perp$ whose solution space – seen as a subspace of the vector space $\mathbb{R}^{\mathcal{V}(L)}$ – is the orthogonal complement of $sol^{\mathbb{R}_+}(L)$. This can be done for example by using a variant of Gauß' triangularization method.
2. From $L^\perp$ compute the $\mathbb{R}$-EFMs $l_1^{\min}, \ldots, l_k^{\min}$ with any known $\mathbb{R}$-EFMs algorithm.
3. Build $L_{\min}$ by using $l_1^{\min}, \ldots, l_k^{\min}$ as the coefficients of the equations of $L_{\min}$.

11   Experimental Results

We experimentally compare three algorithms for overapproximating the objective $\{\beta_{|V} \mid \beta \in sol_{V'}^{\mathbb{R}_+}(\phi)^{\Delta_6} \cap sol_{V'}^{\Delta}(\phi')\}$ given a linear system of equations ϕ, kinetic constraints ϕ', and observable variables V. The first algorithm directly applies pure abstract interpretation to ϕ to compute $sol^{\Delta_6}(\exists \overline{V}.(\phi \wedge \phi'))$ by finite domain programming where $\overline{V} = \mathcal{V}(\phi) \setminus V$, overapproximating the objective by Theorem 18. The second algorithm enriches the linear system ϕ with its minimal support consequences as discussed in Sect. 10 before applying abstract interpretation. The third algorithm is the exact algorithm that can be derived from Theorem 28.

The experimental results are summarized in Fig. 8. The first instance verifies our expectations on the toy metabolic network with a simple cycle in Fig. 1, without kinetic constraints and $V = \{X, Y\}$ as observable variables. The exact algorithm shows that there are 6 abstract solutions, one for each value of Δ_6. The heuristic with minimal support consequences finds exactly these same 6 abstract solutions, while by applying pure abstract interpretation we find 19 abstract solutions (out of 36 possible assignments), thus a large overapproximation.

The second real scale instance treats leucine overproduction on the network from Fig. 2, see Fig. 7 for a discussion of the precise inputs. The heuristic and the exact algorithm produce the same result with 226 abstract solutions, while by pure abstract interpretation 292 abstract solutions are found. Thereby, the two new algorithms both remove the same 2 wrong gene knockout predictions with respect to baseline algorithm by pure abstract interpretation.

However, the heuristics with minimal support consequences is not always exact: we found a counter example given on the right for which it slightly overapproximates the exact solution set.

On the other hand, the heuristic algorithm with *EFM-consequences* is remarkably faster than the exact algorithm – in the benchmark on leucine overproduction, we have 5 min versus 5 h – while still being equally precise in most cases.

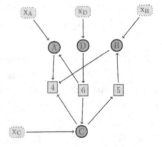

References

1. Allart, E., Niehren, J., Versari, C.: Computing sign abstractions of linear systems. Working paper or preprint, hal-02279942, v1, September 2019. https://hal.archives-ouvertes.fr/hal-02279942
2. Calzone, L., Fages, F., Soliman, S.: BIOCHAM: an environment for modeling biological systems and formalizing experimental knowledge. Bioinformatics **22**(14), 1805–1807 (2006)
3. Cousot, P., Cousot, R.: Systematic design of program analysis frameworks. In: POPL, pp. 269–282 (1979)
4. Coutte, F., Niehren, J., Dhali, D., John, M., Versari, C., Jacques, P.: Modeling leucine's metabolic pathway and knockout prediction improving the production of surfactin, a biosurfactant from bacillus subtilis. Biotechnol. J. **10**(8), 1216–34 (2015)
5. Danos, V., Feret, J., Fontana, W., Harmer, R., Krivine, J.: Abstracting the differential semantics of rule-based models: exact and automated model reduction. In: LICS, pp. 362–381. IEEE Computer Society (2010)
6. Facchetti, G., Altafini, C.: Partial inhibition and bilevel optimization in flux balance analysis. BMC Bioinform. **14**, 344 (2013)
7. Fages, F., Gay, S., Soliman, S.: Inferring reaction systems from ordinary differential equations. Theoret. Comput. Sci. **599**, 64–78 (2015)
8. Fages, F., Soliman, S.: Abstract interpretation and types for systems biology. Theoret. Comput. Sci. **403**(1), 52–70 (2008)
9. Feinberg, M.: Chemical reaction network structure and the stability of complex isothermal reactors. Chem. Eng. Sci. **42**(10), 2229–2268 (1987)
10. Forbus, K.D.: Qualitative reasoning. In: Tucker, A.B. (ed.) The Computer Science and Engineering Handbook, pp. 715–733. CRC Press, Boca Raton (1997)
11. Hoops, S., et al.: Copasi-a complex pathway simulator. Bioinformatics **22**(24), 3067–3074 (2006)
12. John, M., Nebut, M., Niehren, J.: Knockout prediction for reaction networks with partial kinetic information. In: Giacobazzi, R., Berdine, J., Mastroeni, I. (eds.) VMCAI 2013. LNCS, vol. 7737, pp. 355–374. Springer, Heidelberg (2013). https://doi.org/10.1007/978-3-642-35873-9_22
13. Lotz, K., Hartmann, A., Grafahrend-Belau, E., Schreiber, F., Junker, B.H.: Elementary flux modes, flux balance analysis, and their application to plant metabolism. In: Sriram, G. (ed.) Plant Metabolism. MMB, vol. 1083, pp. 231–252. Humana Press, Totowa, NJ (2014). https://doi.org/10.1007/978-1-62703-661-0_14
14. Maranas, C.D., Zomorrodi, A.R.: Flux balance analysis and LP problems. Chap. 3, pp. 53–80. Wiley-Blackwell (2016)
15. Niehren, J., Versari, C., John, M., Coutte, F., Jacques, P.: Predicting changes of reaction networks with partial kinetic information. BioSystems **149**, 113–124 (2016)
16. Orth, J.D., Thiele, I., Palsson, B.O.: What is flux balance analysis? Nat. Biotechnol. **28**(3), 245–248 (2010)
17. Papin, J.A., Stelling, J., Price, N.D., Klamt, S., Schuster, S., Palsson, B.O.: Comparison of network-based pathway analysis methods. Trends Biotechnol. **22**(8), 400–405 (2004)
18. Rendl, A., Guns, T., Stuckey, P.J., Tack, G.: MiniSearch: a solver-independent metasearch language for MiniZinc. In: Pesant, G. (ed.) CP 2015. LNCS, vol. 9255, pp. 376–392. Springer, Cham (2015). https://doi.org/10.1007/978-3-319-23219-5_27
19. Zanghellini, D., Ruckerbauer, D.E., Hanscho, M., Jungreuthmayer, C.: Elementary flux modes in a nutshell: properties, calculation and applications. Biotechnol. J. **8**, 1009–1016 (2013)

Tool Papers

BRE:IN - A Backend for Reasoning About Interaction Networks with Temporal Logic

Judah Goldfeder[1] and Hillel Kugler[2]([✉])

[1] Yeshiva University, New York, USA
ygoldfed@gmail.com
[2] Bar-Ilan University, Ramat-Gan, Israel
hillelk@biu.ac.il

Abstract. We present the BRE:IN tool, a Backend for Reasoning about Interaction Networks. Our tool supports the framework and methodology originally introduced by the RE:IN tool, where an Abstract Boolean Network (ABN) specifies partial information about the network topology, and experimental observations are used to constrain the ABN, allowing to synthesize consistent models, or prove that no consistent model exists. RE:IN has been used successfully to derive mechanistic models of biological systems allowing to gain new insights into cellular decision-making and to make predictions that were validated experimentally. BRE:IN implements translations of experimental observations to temporal logic and captures the semantics of ABNs as transition systems, enabling to use off-the-shelf model checking algorithms. We make our tool and benchmarks publicly available and demonstrate the utility of the tool, providing speed-up gains for some benchmarks, while also enabling extensions of the experimental observations specification language currently supported in RE:IN by using the rich expressive power of temporal logic.

1 Introduction

One of the main challenges in Biology is deciphering and understanding the complex networks of genetic interactions driving cellular behavior and decision-making. Often such genetic *Interaction Networks* are represented in the biological literature as a directed graph, where an edge from one component to another represents a direct genetic interaction, that can have an effect of either activation or repression. Providing formal semantics to such diagrams opens the way to utilizing computational methods for comparing the dynamics of the networks to known experimental measurements, and making new predictions that can be tested experimentally.

Previously, the Reasoning Engine for Interaction Networks (RE:IN), a formal verification based approach and tool [6,17] enabling the synthesis and analysis of *Boolean Networks (BNs)* [11] was developed, where only partial information on the genetic interactions and update rules is explicitly known. The RE:IN

© Springer Nature Switzerland AG 2019
L. Bortolussi and G. Sanguinetti (Eds.): CMSB 2019, LNBI 11773, pp. 289–295, 2019.
https://doi.org/10.1007/978-3-030-31304-3_15

tool is provided with experimental constraints representing measurements of the different components under various experimental conditions, and then it can automatically synthesize BNs that match the partial information known about the network topology and also satisfy all experimental constraints. Here we present BRE:IN, a new backend implementation for RE:IN which is based on using off-the-shelf model checking methods and an encoding of experimental observations via temporal logic [9]. BRE:IN is a command line tool made publicly available with the aim of extending the synthesis capabilities of RE:IN in terms of expressive power and performance.

The main contributions of our new tool are: (1) We encode the experimental observations specification language supported by RE:IN in temporal logic allowing us to specify more expressive temporal properties, or simplify the representation of some properties that can already be represented in RE:IN but in a more complex manner; (2) We capture the semantics of Abstract Boolean Networks (ABNs) as a transition system using the SMV format of the NuSMV model checking tool [3], making explicit the semantic interpretation provided to interaction networks in RE:IN and enabling the application of formal verification methods to the analysis and synthesis of consistent models; and (3) We make the tool and benchmarks publicly available and open source to encourage reproducibility and support further research in scaling up the methods to the challenges provided by current real-world biological models and data.

2   Background

In this section we briefly explain the synthesis-based approach supported by RE:IN [6,17]. The modeling starts with an ABN (Fig. 1(A)), which extends classical BNs [11], a well studied and widely used formalism providing Boolean abstractions of genetic systems. In a BN every gene is represented by a Boolean variable specifying whether the gene is active or inactive. The concept of an ABN allows the representation of models with partial information on the network topology and dynamics.

Rather than allowing the use of any arbitrary update function as allowed in classical BNs, a set of biologically plausible update function templates, which are called *regulation conditions*, is used to define the combined effect of activators and repressors. The regulation conditions satisfy monotonicity, and only consider whether none, some, or all potential activators or repressors are active.

To capture the possible uncertainty in the precise network topology, the ABN formalism allows some interactions to be marked as *optional*. Each optional interaction could be included or excluded from a synthesized *concrete model*. Thus, in terms of network topology, for n optional interactions, an ABN model specifies a set of 2^n concrete models, each corresponding to a unique selection of optional interactions. Additionally, a choice of several possible regulation conditions for each gene is allowed. An ABN is transformed into a concrete model by selecting a subset of the optional interactions to be included and assigning a specific regulation condition for each gene.

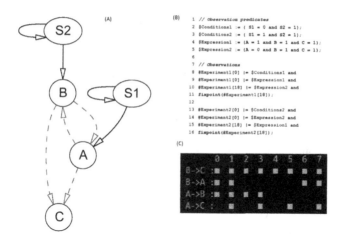

Fig. 1. Toy Model: (A) An Abstract Boolean Network with components S1, S2, A, B, C. Four of the interactions A → B, B → A, A → C and B → C are optional interactions that are depicted with dashed lines in the diagram. (B) Experimental observations are specified on the dynamics of the network using an observation specification language defined in RE:IN. (C) After running synthesis, 8 solutions are found to be consistent with all the experimental constraints, they are shown graphically as columns 0 .. 7 highlighting which optional interactions were chosen in each concrete model.

A set of *experimental observations* that each concrete model needs to be able to satisfy are encoded as predicates over system states (Fig. 1(B)) which limits the possible consistent choices of the optional interactions and regulation conditions. The synthesis problem is, given an ABN and a set of experiments, find a choice of optional interactions and regulation conditions for each gene, guaranteeing that the resulting concrete model is consistent with all experimental observations (Fig. 1(C)), or prove that no consistent models exist. To synthesize consistent models, RE:IN encodes the synthesis problem using the Z3 Satisfiability Modulo Theories (SMT) solver [5] and utilizes a bounded model checking strategy to search for a consistent model or prove that no such model exists. In this work we develop BRE:IN to try and improve the running time of the synthesis methods towards tackling larger networks with more experimental constraints than currently feasible, utilizing model checking algorithms for temporal logic properties.

3   BRE:IN Tool

BRE:IN is a command line tool that serves as an extended backend for reasoning about interaction networks and is publicly available at github [8]. It uses the NuSMV model checker [3] to perform the synthesis on the translated models, whereas the RE:IN tool uses the Z3 SMT solver [5]. BRE:IN allows to use off-the-shelf verification algorithms to analyze RE:IN models and synthesize concrete consistent networks. BRE:IN can also make use of the expressiveness of temporal logic to specify behavioral properties not directly expressible in RE:IN. BRE:IN can encode the models in two different encodings, referred to henceforth as modes. While both modes can encode the same ABNs, they differ in terms of what types of specifications they can handle while performing synthesis. The first mode, time_step, can handle specifications of the same form that RE:IN deals with, namely where values are ascribed to different experiments at specific time steps. The second mode, temporal_logic, supports specifications that can contain more complex dynamics using temporal logics. BRE:IN supports both computation tree logic (CTL) [4] and linear temporal logic (LTL) [14]. In both modes BRE:IN synthesizes and then enumerates all consistent models (up to a specified limit) that satisfy the topology, regulation conditions and observation constraints, see (Fig. 1(C)). Next we describe the functionality of both modes.

time_step mode: This mode is designed to deal with the same set of specifications as in RE:IN. It accepts the same files that RE:IN reads from, namely a file with a .net extension that describes the network and a file with a .spec extension that describes the specification. Like in RE:IN, the observation specification language only allows for assertions at specific time steps. In several cases, BRE:IN running in this mode was able to outperform RE:IN.

temporal_logic mode: This mode can deal with a larger range of specifications than RE:IN. It reads from the same model file as before (.net extension), but instead of reading the specification from .spec files, it deals with one of two file types, .ctlspec extension (for CTL properties) and .ltlspec extension (for LTL properties).

This added expressiveness brings two primary advantages. First off, in some cases, converting the time-step specifications to LTL/CTL can lead to a performance boost. The second advantage is that various specifications not expressible in the more limited time-step format of RE:IN can now be specified. BRE:IN supports arbitrarily complex CTL and LTL specifications, to allow for more nuanced and complex specifications than RE:IN. We have also implemented in the tool a validation component that independently of the synthesis algorithm checks that the synthesized models indeed satisfy all the experimental observations, to detect and avoid errors in our tool implementation.

We have implemented interaction network synthesis algorithms in BRE:IN, and evaluated them on key biological benchmarks studied in RE:IN [6,17], the runtime for our new tool BRE:IN and a command line version of RE:IN is shown in Table. 1. There are some cases in which BRE:IN outperformed the original RE:IN, although in general there are other cases in which RE:IN performed better. The main advantage in our view is the enhanced expressive power of temporal logics that can be added on top of the existing specification language. Temporal logic can be used naturally to specify steady-state behaviour, which is common in biological experiments (e.g. embryonic stem cell model [6]). RE:IN's specification language would require to precisely specify the timepoint in which the steady-state occurs while our extension provides a more robust way to specify such dynamics. Another example of temporal logic's relevance is specifying required ordering between events (e.g. cell cycle model [17]) without having to specify the absolute timing in which the specific events occur. LTL enables specifying explicitly that a certain behaviour must always occur, whereas the existing specification language can only specify that a certain behaviour is possible.

Table 1. Runtime for benchmarks analyzed with our BRE:IN tool and command line version of RE:IN, time shown in seconds running on a standard laptop with Intel i5-4210 CPU at 1.70 GHz and 8.0 GB memory. The benchmarks called ltl1 .. ltl4, ctl1, ctl2 use temporal logic specifications so can only be directly analyzed with BRE:IN.

Benchmark	BRE:IN	RE:IN	Nodes	Semantics
toy	0.5	0.9	5	Sync
pluri1	112.0	130.3	16	sync
pluri2	66.3	72.2	16	sync
pluri3	183.3	588.2	16	sync
pluri4	42.2	19.1	16	sync
myeloid	1.5	2.7	11	async
ltl1	0.3	–	5	sync
ltl2	0.4	–	5	sync
ltl3	6.2	–	16	sync
ltl4	5.7	–	16	sync
ctl1	0.4	–	5	sync
ctl2	0.5	–	11	async

BRE:IN synthesis algorithms can handle Boolean networks with either synchronous or asynchronous semantics (In Table 1 models myeloid, ctl2 are asynchronous, while all the other models are synchronous). In terms of performance, additional work is still required to scale up the synthesis algorithms to tackle even larger interaction networks with more experimental constraints, thus we hope our tool can serve as a reference implementation and encourage additional research in this direction.

4 Related Work

Developing synthesis and verification methods for biological systems is an active area of research [1,2,7,10,12,13,16]. Synthesis methods can automate the process of model development constrained by experimental data and enable rapid construction of predictive models. The inherent complexity of synthesis methods is a major challenge that needs to be addressed to make synthesis algorithms more broadly applicable in biology. Synthesis in BRE:IN does not consider an adversarial environment as is typically the case in reactive synthesis [15], extending the framework in this direction remains a future research direction.

Acknowledgment. The research was partially supported by the Horizon 2020 research and innovation programme for the Bio4Comp project under grant agreement number 732482. This research was also supported by the ISRAEL SCIENCE FOUNDATION (grant No. 190/19).

References

1. Bartocci, E., Lió, P.: Computational modeling, formal analysis, and tools for systems biology. PLoS Comput. Biol. **12**(1), e1004591 (2016)
2. Chabrier, N., Fages, F.: Symbolic model checking of biochemical networks. In: Priami, C. (ed.) CMSB 2003. LNCS, vol. 2602, pp. 149–162. Springer, Heidelberg (2003). https://doi.org/10.1007/3-540-36481-1_13
3. Cimatti, A., Clarke, E., Giunchiglia, F., Roveri, M.: NuSMV: a new symbolic model checker. Int. J. Softw. Tools Technol. Transf. **2**(4), 410–425 (2000)
4. Clarke, E.M., Emerson, E.A.: Design and synthesis of synchronization skeletons using branching time temporal logic. In: Kozen, D. (ed.) Logic of Programs 1981. LNCS, vol. 131, pp. 52–71. Springer, Heidelberg (1982). https://doi.org/10.1007/BFb0025774
5. de Moura, L., Bjørner, N.: Z3: an efficient SMT solver. In: Ramakrishnan, C.R., Rehof, J. (eds.) TACAS 2008. LNCS, vol. 4963, pp. 337–340. Springer, Heidelberg (2008). https://doi.org/10.1007/978-3-540-78800-3_24
6. Dunn, S.-J., Martello, G., Yordanov, B., Emmott, S., Smith, A.G.: Defining an essential transcription factor program for naïve pluripotency. Science **344**(6188), 1156–1160 (2014)
7. Fisman, D., Kugler, H.: Temporal reasoning on incomplete paths. In: Margaria, T., Steffen, B. (eds.) ISoLA 2018. LNCS, vol. 11245, pp. 28–52. Springer, Cham (2018). https://doi.org/10.1007/978-3-030-03421-4_3
8. Goldfeder, J., Kugler, H.: https://github.com/kuglerh/BREIN (2019)
9. Goldfeder, J., Kugler, H.: Temporal logic based synthesis of experimentally constrained interaction networks. In: Chaves, M., Martins, M.A. (eds.) MLCSB 2018. LNCS, vol. 11415, pp. 89–104. Springer, Cham (2019). https://doi.org/10.1007/978-3-030-19432-1_6
10. Guziolowski, C., et al.: Exhaustively characterizing feasible logic models of a signaling network using answer set programming. Bioinformatics **29**(18), 2320–2326 (2013)
11. Kauffman, S.A.: Metabolic stability and epigenesis in randomly constructed genetic nets. J. Theor. Biol. **22**(3), 437–467 (1969)
12. Koksal, A.S., Pu, Y., Srivastava, S., Bodik, R., Fisher, J., Piterman, N.: Synthesis of biological models from mutation experimentss. In: SIGPLAN-SIGACT Symposium on Principles of Programming Languages. ACM (2013)
13. Paoletti, N., Yordanov, B., Hamadi, Y., Wintersteiger, C.M., Kugler, H.: Analyzing and synthesizing genomic logic functions. In: Biere, A., Bloem, R. (eds.) CAV 2014. LNCS, vol. 8559, pp. 343–357. Springer, Cham (2014). https://doi.org/10.1007/978-3-319-08867-9_23
14. Pnueli, A.: The temporal logic of programs. In: Proceedings 18th IEEE Symposium Foundations of Computer Science, pp. 46–57 (1977)
15. Pnueli, A., Rosner, R.: On the synthesis of a reactive module. In: Proceedings 16th ACM Symposium Principles of Program Language, pp. 179–190 (1989)

16. Woodhouse, S., Piterman, N., Wintersteiger, C.M., Göttgens, B., Fisher, J.: SCNS: a graphical tool for reconstructing executable regulatory networks from single-cell genomic data. BMC Syst. Biol. **12**(1), 59 (2018)
17. Yordanov, B., Dunn, S.J., Kugler, H., Smith, A., Martello, G., Emmott, S.: A method to identify and analyze biological programs through automated reasoning. NPJ Syst. Biol. Appl. **2**, 16010 (2016)

The Kappa Simulator Made Interactive

Pierre Boutillier[(✉)]🆔

Harvard Medical School, Boston, USA
Pierre_Boutillier@hms.harvard.edu

Abstract. Like during software development, interactivity is of tremendous help during model development. The more and the earlier feedback come, the more efficiently the target is reached. This is true for human as well as during mechanical model construction. If you try to mechanically learn some parameters for a model by streaming potential values for example, you would better stop as quickly as possible the simulations that behave the worst toward the goal. The Kappa simulator KaSim has been refactored to give the control to the user (human or an other program) during the simulation, allowing to pause, restart, observe, modify, prematurely stop, continue after the original end. Interventions on a simulation that can be offered as well as their consequences on the design of a stochastic simulator of rule-based models are describe here.

Keywords: Interactivity · Rule-based modelling · Stochastic simulation

1 Introduction

Models in systems biology are useful when in silico runs provide more than what can easily be observed in vivo. The outcome of executing a model has to supersede the cost of gathering necessary information to write it. Computing more stuff is not enough. Biologists must be able to get these results in a exploitable way.

Tools have been developed to take advantages of progresses made in formal methods in the area of system-biology [1–3,5] and efforts have been put for accessibility to users [4,6]. Nevertheless, unlike in the area of physical simulation where tremendous effort in term of software development have been put to allow a real dialogue during runs [7], tools to work at the level of quantitative or logical simulation are often provided as a kind of black box that is fed at the beginning of runs with a model, parameters and possibly some queries, and computes outputs to be analysed afterwards without any possibility to interact during executions. Model development and analyses are not linear processes at all though. The first versions of models contain mistakes such as first versions of software contain bugs. Every computational biologist can tell horror stories where tardily reported errors cost a night of wasted computation.

© Springer Nature Switzerland AG 2019
L. Bortolussi and G. Sanguinetti (Eds.): CMSB 2019, LNBI 11773, pp. 296–301, 2019.
https://doi.org/10.1007/978-3-030-31304-3_16

As for software, model development is made more efficient by static analyses returning quickly potential loopholes in models. The other axis to reduce the length of the modelling loop cycle is interactive execution.

This article defines the basis of an interaction language with a stochastic simulator and presents how a simulator can be turned from a monolithic software into an interactive server. It takes the example of KaSim, an interpreter for the Kappa language. It is more critical in Kappa as temporal transformation are described at the level of patterns not fully specified species so the state space is so big that it is even difficult to anticipate what will be interesting to observe.

2   User Interfaces

There are 3 ways to use Kappa. Practical details to use them is on the website https://kappalanguage.org

2.1   Command Line Interface

The historical way is to use the command line interface. In this case, the model is written without interactive feedback in files using an external text editor. The only interactivity comes from a bit of information about the simulation progress printed on the standard output and the fact that Ctrl-c does not kill the simulation but pauses it, allowing the user to fire some interventions (presented later) and resume (or prematurely stop) the simulation.

2.2   Graphical User Interface

The educational way is to use *Kappapp*, a graphical interface available online As it runs in a web browser, its efficiency relies on the one of the javascript interpreter.

The possibility to run simulations at native speed is also provided.

In the Kappapp, the model is editable in a text editor embedded in the app. On the fly, the model is parsed and errors are immediately reported as you type. When the parsing is OK, the model is asynchronously sent to KaSa, the static analyser for Kappa that reports back as soon as the results of its analyses are available, triggering warning or exhibiting invariants of the model. The user can therefore notice unexpected consequences of an input when she has it still in mind.

Once the model seems to match the goal, a simulation can be launched directly in the app. Results are returned and graphically rendered as soon as they are available. During the track of the run, a pause button is there to immediately give back the control, pause and wait for intervention. The intervention can be either some unplanned requests for more feedback in order to get a better understanding of the current state/behavior or an anticipated change in the experimental setting (addition/deletion of a species, shut down/up of a rule, ...) because what was expected has occurred already so you can move on to the next step.

2.3   Programmatic Interface

The scientific way to use Kappa is by running a batch of runs of (variants of) a model.

The first option to do so is non interactive. One can use the command line interface to launch all the runs and read the output files they generated to gather and analyse the result.

The best option is to use Kappy, a python wrapper to drive Kappa simulation.

For example, running a simulation for 80 time units of a dummy model where 100 agents A form dimers reversibly and printing the abundance of monomers every time units looks like:

```
import kappy
client = kappy.KappaStd()

model = ("%agent: A(x[x.A]) "
         "%var: k_on 1e-2 "
         "A(x[.]),A(x[.]) <-> A(x[1]),A(x[1]) @ k_on, 1 "
         "%plot: |A(x[.])| "
         "%init: 100 A()")
client.add_model_string(model)
ast = client.project_parse()
sim_param = kappy.SimulationParameter(pause_condition="[T] > 80",
                                      plot_period=1)
client.simulation_start(sim_param)
results = client.simulation_plot()
client.shutdown()

import matplotlib.pyplot as pyplot
pyplot.plot([ d[0] for d in results['series'] ],
            [d[1] for d in results['series'] ],
            label=results['legend'][1])
pyplot.legend()
pyplot.show()
```

We can then screen the effect of the constant rate k_on on the amount of monomer at equilibrium by launching runs for different values.

3   Software Architecture

Internally, all the modes of interaction relies on the same infrastructure. The elementary binaries are state-full servers that communicate on their standard input/output through a specific asynchronous JSON based protocol.

This architecture is convenient because it is compatible both with unix pipe-based communication between processes and WebWorkers, the framework to mimic threads in javascript. It allows to build both a version purely embedded in a web-browser and a native code efficient one.

Asynchronicity is very important. As there is only 1 sequential communication channel, preventing any communication during the wait for a reply of a computationally heavy request is not an option. Moreover some intermediary result may be of interest for the user and monopolising the communication channel forbids to send them. Therefore, a request that may be non immediate is split in two. One request to launch the computation that just returns "it's ongoing" unless there is an obvious problem and one request to poll the results that also replies immediately either the result once it's available or "come back later".

An ecosystem of server realizing elementary tasks is also an easy way to introduce parallelization. Instead of multi-threading one process to run a batch of simulations, you spawn one scheduler and as many simulator processes as you have simulations to run.

Having a JSON based protocol and documenting it in a "standard" description language (swagger in this case) facilitates the construction of bindings for script languages as it is done with Python. The R language is the obvious next candidate where it would be relevant.

4   Intervention Language

Kappa intervention language offers 3 kinds of actions.

real perturbations $UPDATE variable value; to change the rate constant by which rules are fired and $APPLY integer rule; to apply a rule a given amount of time in the current state.

$ADD integer species; and $DELETE integer pattern; are syntactic sugar available for clarity in the predominant cases where the rule to apply has no left/right hand side.

$RUN; and $STOP; to resume execution or definitely stopping it.

immediate observations that do not interfere with the simulation:

$SNAPSHOT; returns the exact content of the current state of the simulation.

$PRINT algebraic_expression; computes the value of an expression containing the number of matches in the current state of some graph patterns. For example, $PRINT |A(x{p})| / |A()| returns the ratio of agent A whose site x is currently in state p.

continuous observations that make sense only on a time window for which the recording is started or stopped.

This concern the record of effects of rule application on the activity of rules by $DIN file_name boolean; and the record of occurrences of pattern of interest via $TRACK pattern boolean;

5   Tutorial

We consider a model made of proteins C with 2 sites a and b that serve as a catalyst between some proteins A and B by binding independently to an A and to a B to form some complex ACB so that A and B can potentially interact. Said in kappa:

```
%agent :  A( x )
%agent :  B( x )
%agent :  C( a  b )

%var :  k_on  1e−2

A( x [.]) ,  C( a [.])  <−>  A( x [1]) ,  C( a [1])  @  1e−2, 1
B( x [.]) ,  C( b [.])  <−>  B( x [1]) ,  C( b [1])  @  k_on , 1

%plot :  |A( x [1]) ,  C( a [1]  b [2]) ,  B( x [2])|

%init :  100  A() ,  B()
.  −>  C()  @  1e−3
```

The last rule is a way to slowly add some C to see the effect on the amount of ACB at equilibrium.

During the beginning of the simulation, this number increases with the amount of C. But, as we follow the progress live, we start to see a stagnation and even the beginning of a decrease. We stop to check if the number of C really increases thanks to a

$PRINT |C()|;

Nothing special to see. Maybe, it is a stability problem, let's try to increase the stickiness between B and C and restart by

$UPDATE k_on 1e−1; $RUN;

Things are even worse! The number of ABC has decreased and keep doing so. Let's refill in B maybe:

$ADD 50 B();

it helps temporarily but it keeps decreasing again as the simulation continues. We pause and inspect the number of available B for binding:

$PRINT |B(x[.])|;

There are barely none. Where are the Bs? Only one way to see:

$SNAPSHOT;

Here, we visualize the explanation: As there are tons of ACs and BCs, there are no available monomers to bind to an existing dimer. We've just re-rediscovered that too many catalyser ultimately put reactives apart.

6 Conclusion

Learning is an interactive process. Making a tool interactive makes easier (if not simply possible) for potential users (and especially students) to learn how to use it, for researcher to learn new facts using it and for computers to learn thanks to experimenting with it.

The transformation of the Kappa simulator toward this goal involved mainly to understand in which (software and human) ecosystem it evolves so that to sort out the mandatory pipes and faucets to provide. Detailing them could ease the work of adapting some other tools.

Now that simulation runs are interactive, the bottleneck of interactivity is back to the model construction phase. Model construction is interactive in the sens that parsing and static analyses are run automatically after every edit but it is done by redoing all the computation from scratch with the new model. In order to smoothly handle large models, the framework has to become incremental: new outputs for modified inputs have to be computed from the previous output using only the diff between the 2 versions. This is where efforts are put now in the Kappa ecosystem.

References

1. Danos, V., Feret, J., Fontana, W., Harmer, R., Krivine, J.: Rule-based modelling of cellular signalling. In: Caires, L., Vasconcelos, V.T. (eds.) CONCUR 2007. LNCS, vol. 4703, pp. 17–41. Springer, Heidelberg (2007). https://doi.org/10.1007/978-3-540-74407-8_3
2. Fages, F., Soliman, S.: On robustness computation and optimization in BIOCHAM-4. In: Češka, M., Šafránek, D. (eds.) CMSB 2018. LNCS, vol. 11095, pp. 292–299. Springer, Cham (2018). https://doi.org/10.1007/978-3-319-99429-1_18
3. Goldstein, B., Faeder, J.R., Blinov, M.L., Hlavacek, W.S.: BioNetGen: software for rule-based modeling of signal transduction based on the interactions of molecular domains. Bioinformatics **20**(17), 3289–3291 (2004). https://doi.org/10.1093/bioinformatics/bth378
4. Gyori, B.M., Bachman, J.A., Subramanian, K., Muhlich, J.L., Galescu, L., Sorger, P.K.: From word models to executable models of signaling networks using automated assembly. Mol. Syst. Biol. **13**(11), 954 (2017). http://msb.embopress.org/content/13/11/954
5. Naldi, A., Berenguier, D., Fauré, A., Lopez, F., Thieffry, D., Chaouiya, C.: Logical modelling of regulatory networks with ginsim 2.3. Biosystems **97**(2), 134–139 (2009). https://doi.org/10.1016/j.biosystems.2009.04.008. http://www.sciencedirect.com/science/article/pii/S0303264709000665
6. Naldi, A., et al.: The CoLoMoTo interactive notebook: accessible and reproducible computational analyses for qualitative biological networks. Front. Physiol. **9**, 680 (2018). https://doi.org/10.3389/fphys.2018.00680
7. Tek, A., Chavent, M., Baaden, M., Delalande, O., Bourdot, P., Ferey, N.: Advances in human-protein interaction - interactive and immersive molecular simulations. In: Cai, W., Hong, H. (eds.) Protein-Protein Interactions, chapater 2. IntechOpen, Rijeka (2012). https://doi.org/10.5772/36568

Biochemical Reaction Networks with Fuzzy Kinetic Parameters in Snoopy

George Assaf[(⊠)], Monika Heiner, and Fei Liu

Computer Science Institute, Brandenburg Technical University,
Postbox 10 13 44, 03013 Cottbus, Germany
{George.Assaf,monika.heiner}@b-tu.de, feiliu@scut.edu.cn
https://www-dssz.informatik.tu-cottbus.de

Abstract. *Snoopy* is a powerful modelling and simulation tool for various types of Petri nets, which have been applied to a wide range of biochemical reaction networks. We present an extended version of *Snoopy*, now supporting stochastic, continuous and hybrid Petri Nets with fuzzy kinetic parameters. Fuzzy parameters are specifically useful when kinetic parameter values can not be precisely measured or estimated. By running fuzzy simulation we obtain output bands of the variables of interest induced by the effect of the fuzzy kinetic parameters.

Keywords: Fuzzy logic · Fuzzy kinetic parameters ·
Fuzzy continuous · Stochastic and hybrid Petri nets ·
Modelling and simulation

1 Objectives

Modelling of biochemical reaction networks is often hampered by parametric uncertainty. This uncertainty usually comes from unavailable or imprecise parameters due to some environmental factors or lack of exact knowledge. When stochastic methods are not appropriate to deal with such models, analysing them by giving an uncertain band of all outputs of interest might be an alternative. These uncertain bands describe the effect of uncertain kinetic parameters.

To address these issues, quantitative Petri nets, such as continuous Petri nets ($\mathcal{CPN}$) and stochastic Petri nets ($\mathcal{SPN}$), have been extended by fuzzy kinetic parameters yielding fuzzy continuous Petri nets ($\mathcal{FCPN}$) [4] and fuzzy stochastic Petri nets ($\mathcal{FSPN}$) [6], respectively. We are going to translate this idea to hybrid Petri nets ($\mathcal{HPN}$), which complements our family of related Petri nets by fuzzy hybrid Petri nets ($\mathcal{FHPN}$); compare Fig. 1. In all these fuzzy quantitative Petri nets ($\mathcal{FPN}$ for short), a kinetic parameter can either be represented - as usual - as a crisp number or as a fuzzy number, if the parameter cannot be measured or estimated precisely.

Electronic supplementary material The online version of this chapter (https://doi.org/10.1007/978-3-030-31304-3_17) contains supplementary material, which is available to authorized users.

L. Bortolussi and G. Sanguinetti (Eds.): CMSB 2019, LNBI 11773, pp. 302–307, 2019.
https://doi.org/10.1007/978-3-030-31304-3_17

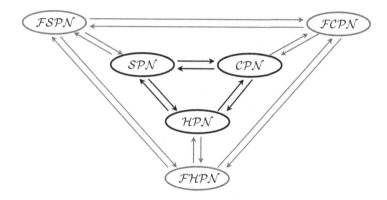

Fig. 1. Export relation between some of *Snoopy*'s Petri net classes. Fuzzy nets differ from their crisp counterparts just by additional pre-defined data types, supporting fuzzy numbers, which can be used as kinetic parameters. The new net classes and their export relation are coloured in red. (Color figure online)

Fuzzy logic uses a degree of belonging defined by a membership function to describe an element, and thus can represent uncertainties in a model. The fundamental concept of fuzzy logic is the fuzzy set [7]. A fuzzy set is defined on a universal set $\mathbb{X}$ by its membership function μ which only takes real values in the closed (unit) interval $[0, 1]$, thus specifying a membership degree for each element of the universal set. In contrast, for traditional (crisp) sets, the membership function only takes values of the set $\{0, 1\}$.

A fuzzy number is a special (convex and normalized) fuzzy set with the universal set $\mathbb{X}$ given by the set of real numbers. Commonly used fuzzy numbers include triangular, trapezoidal and Gaussian fuzzy numbers. *Snoopy* supports so far triangular fuzzy numbers (TFNs). The α-cut of a fuzzy set at a given membership degree $\alpha \in [0, 1]$ (formally called α level), consists of a crisp subset of $\mathbb{X}$, in which each element has a membership degree greater than or equal to the given α level; compare Fig. 2.

2 $\mathcal{FPN}$ Simulation

The general idea for simulation and analysis of $\mathcal{FPN}$ follows Zadeh's extension principle [7], according to which a fuzzy number is represented as a union of its α-cuts. For this purpose, we decompose all fuzzy parameters into α-cuts, typically equally spread over the continuous interval $[0, 1]$. Then we draw samples at each α level and run – depending on the given model class – stochastic/continuous/hybrid simulations for each sample combination, and we obtain the corresponding α-cut for each output of interest. Finally, we compose all the α-cuts and obtain the membership function for each output and how they evolve over time, which reflect the effect of the uncertainties of the kinetic parameters.

To obtain the output bands of the time series data, we could simply print all simulation traces together into one plot. This would require to keep all simulation traces. Instead, to reduce the memory load, we determine for each output variable the minimum and maximum values of the traces as they evolve over time. The whole procedure is sketched in Algorithm 1; see also [5] for more details.

Algorithm 1. $\mathcal{FPN}$ simulation algorithm.

Data: $\mathcal{FPN}$ with M variables (places, species) and K fuzzy kinetic parameters, J - number of α levels.

Result: Output bands & membership functions of the M variables over time.

1 **for** *each* α *level* $\alpha_j, j = 0, 1, \ldots J - 1$ **do**
2   **for** *each fuzzy kinetic parameter* **do**
3   Obtain its α-cut;
4   Sampling: discretise the α-cut and obtain crisp values;
5   **end**
6   **for** *each combination of values for the K fuzzy kinetic parameters* **do**
7   Run stochastic/continuous/hybrid simulation;
8   **end**
9 **end**
10 **for** *each variable* $Y_m, m = 1, 2, \ldots M$ **do**
11   Determine minimum & maximum output values over time to obtain its output band;
12   Compose all the α-cuts of Y_m to obtain its membership function over time;
13 **end**

Discretising each α-cut of the fuzzy number(s) independently into crisp values may produce redundant samples over all levels. This causes unnecessary simulation runs. To address this issue, a more efficient discretising method needs

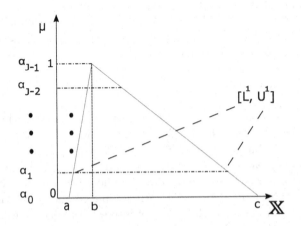

Fig. 2. A triangular fuzzy number (TFN) defined by the triple (a, b, c), with $0 \leq a \leq b \leq c$, where a can be read as the pessimistic value, b as the most possible value, and c as the optimistic value; and its α-cuts, each defining an α level.

to be designed to eliminate redundant samples. *Snoopy* supports two sampling strategies.

- **Basic Sampling** neglects the problem of redundant samples. This strategy discretises each α level with the same number of samples, except for $\alpha = 1$, and samples are equally spread over each α level, compare Fig. 3a. In this case, the total number of simulation runs is given by $N^K \times J + 1$, with N being the number of samples per level.
- **Reduced Sampling** takes redundant samples into consideration by reusing the samples at $\alpha = 0$ for all levels; compare Fig. 3b. Thus, the number of samples at $\alpha = 0$ should be larger than in the basic sampling strategy, to obtain a suitable resolution of the results. In this case, the total number of simulation runs is given by $N^K + (J - 1) \times 2 + 1$, with N being the number of samples at $\alpha = 0$.

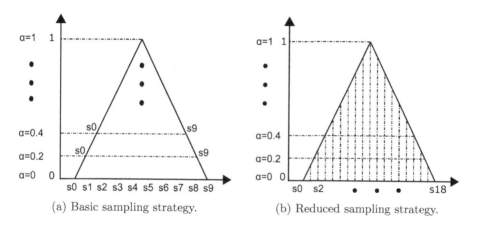

(a) Basic sampling strategy. (b) Reduced sampling strategy.

Fig. 3. Sampling strategies. (a) Equidistant samples (here 10) are independently taken on each level. (b) Reuses the samples at $\alpha = 0$ (here 19) for all levels, if the fall into the corresponding cut, complemented by two samples at each level, determined by the cut with the triangular shape. In both cases, there is only one sample for $\alpha = 1$.

Reproducibility. The simulation procedure of $\mathcal{FPN}$ described in Algorithm 1 has been previously implemented in Matlab [6]. However, Matlab requires programming skills and needs quite a lot of experience. To overcome this issue, we extended *Snoopy* [3], a unifying platform-independent and user-friendly Petri net tool, for modelling and simulating of $\mathcal{FPN}$.

Please note, we choose to represent biochemical networks as Petri nets. However, our approach can be equally applied to any related modelling approach for biochemical reaction networks involving fuzzy kinetic parameters.

3   Use Cases

This section assumes that the reader has some basic understanding of modelling and simulating of biochemical reaction networks using Petri nets in *Snoopy*. Otherwise, the textbook chapter [1] might be a good starting point to acquire this knowledge. The workflow is basically the same for all $\mathcal{FPN}$ classes. Here, we choose to discuss it by an $\mathcal{FCPN}$ example.

For reasons of space, all following references to figures or tables relate to the Appendix, which can be retrieved from *Snoopy*'s website, see Sect. 4.

$\mathcal{FCPN}$ **modelling** starts with creating a new 'Fuzzy Continuous Petri Net' file. This can first of all be done by exploiting *Snoopy*'s export feature, which permits to conveniently convert existing models into related net classes; compare the export relation given in Fig. 1. These exports involve some obvious adjustments, e.g., converting a $\mathcal{CPN}$ into an $\mathcal{FCPN}$ just creates a special case of an $\mathcal{FCPN}$ with all kinetic parameters being crisp, compare Fig. 4. An $\mathcal{FCPN}$ model can also be built from scratch by creating a new (empty) $\mathcal{FCPN}$ file; compare Fig. 5. $\mathcal{FCPN}$ modelling is similar to that of standard $\mathcal{CPN}$, meaning continuous places and transitions have to be introduced and interconnected as required to represent the given biochemical reaction network. Figure 6 shows an $\mathcal{FCPN}$ example called decay dimerisation model which is adopted from [4]. The $\mathcal{FCPN}$ model shares the structure with its crisp $\mathcal{CPN}$ counterpart.

The crucial difference consists in the constants; a user can now also introduce constants of TFN data type which define Triangular Fuzzy Numbers. These fuzzy numbers can then be used as kinetic parameters when setting up reaction rates. Figure 7 illustrates constant definitions in *Snoopy* according to the constants shown in Table 1, where k_3 and k_4 are defined as TFNs by three positive real numbers each.

Model Simulation. The $\mathcal{FCPN}$ simulation dialog, see Fig. 8, consists of the same settings as the standard $\mathcal{CPN}$ simulation dialog, but extended by the 'Fuzzy Setting' sub-section which consists of the following items:

1. **alpha levels:** specifies the number of α levels; the default value is 10.
2. **discretisation points:** specifies the number of sample points per level; the default value is 10.
3. **sampling strategy:** the user can choose between two options: "Basic sampling" and "Reduced Sampling".

Once the user clicks the start simulation button, *Snoopy* will notify the user about the number of simulations to be launched. When the simulation has finished, the user can view the simulation results by double clicking on the default view which displays the uncertain band of the selected species (upper viewer) and their composed membership function (lower viewer) at time point 0 (by default). Moreover, the user can view the composed membership function of the selected species at specific time points using the scroll bar or entering the time point directly in the field 'Time Point' located at the lower right corner of the viewer window; see Fig. 9.

$\mathcal{FSPN}$ modelling and simulation is similar to that of $\mathcal{FCPN}$. The main difference consists in choosing a suitable stochastic simulator, e.g. Gillespie's stochastic simulation algorithm. Figure 10 shows one $\mathcal{FSPN}$ example called yeast polarisation model adopted from [6], and Fig. 11 gives two uncertain bands of the species G_a and G_bg and their corresponding membership functions at simulation time point 15.

$\mathcal{FHPN}$ modelling and simulation can be easily achieved by following the same workflow as demonstrated for $\mathcal{FCPN}$ and $\mathcal{FSPN}$.

4 Installation and Future Work

Snoopy is a free and platform-independent software, which can be download from its official website https://www-dssz.informatik.tu-cottbus.de/DSSZ/Software/ Snoopy. *Snoopy* operates under Windows, Mac/OSX and for selected Linux distributions. $\mathcal{FPN}$ test cases can be retrieved from https://www-dssz.informatik. tu-cottbus.de/DSSZ/Software/Examples.

Future work will include the generation of configuration scripts to delegate the time-consuming simulation step to the command-line tool Spike [2], which will run the simulations parallelised on a server.

Acknowledgement. *Snoopy* uses software libraries that have been developed by former staff members and numerous student projects at Brandenburg Technical University (BTU), chair Data Structures and Software Dependability.

This work has been supported by National Natural Science Foundation of China (61873094), Science and Technology Program of Guangzhou, China (201804010246), and Natural Science Foundation of Guangdong Province of China (2018A030313338).

References

1. Blätke, M., Heiner, M., Marwan, W.: BioModel engineering with petri nets. In: Chapter 7, pp. 141–193. Elsevier Inc., March 2015
2. Chodak, J., Heiner, M.: Spike - reproducible simulation experiments with configuration file branching. In: Bortolussi, L., Sanguinetti, G. (eds.): CMSB 2019. LNBI, vol. 11773, pp. 315–321. Springer, Heidelberg (2019)
3. Heiner, M., Herajy, M., Liu, F., Rohr, C., Schwarick, M.: Snoopy – a unifying Petri net tool. In: Haddad, S., Pomello, L. (eds.) PETRI NETS 2012. LNCS, vol. 7347, pp. 398–407. Springer, Heidelberg (2012). https://doi.org/10.1007/978-3-642-31131-4_22
4. Liu, F., Chen, S., Heiner, M., Song, H.: Modelling biological systems with uncertain kinetic data using continuous Petri nets. BMC Syst. Biol. **12**, 64–74 (2018)
5. Liu, F., Heiner, M., Gilbert, D.: Fuzzy Petri nets for modelling of uncertain biological systems. Brief. Bioinf. (2018). https://doi.org/10.1093/bib/bby118
6. Liu, F., Heiner, M., Yang, M.: Fuzzy stochastic Petri nets for modeling biological systems with uncertain kinetic parameters. PLoS ONE **11**(2), e0149674 (2016). https://doi.org/10.1371/journal.pone.0149674
7. Zadeh, L.: Fuzzy sets. Inf. Control **8**, 338–353 (1965)

Compartmental Modeling Software: A Fast, Discrete Stochastic Framework for Biochemical and Epidemiological Simulation

Christopher W. Lorton[1]([⊠]), Joshua L. Proctor[1], Min K. Roh[1], and Philip A. Welkhoff[2]

[1] Institute for Disease Modeling, Bellevue, WA 98005, USA
CMS@idmod.org
[2] Bill and Melinda Gates Foundation, Seattle, WA 98109, USA

Abstract. The compartmental modeling software (CMS) is an open source computational framework that can simulate discrete, stochastic reaction models which are often utilized to describe complex systems from epidemiology and systems biology. In this article, we report the computational requirements, the novel input model language, the available numerical solvers, and the output file format for CMS. In addition, the CMS code repository also includes a library of example model files, unit and regression tests, and documentation. Two examples, one from systems biology and the other from computational epidemiology, are included that highlight the functionality of CMS. We believe the creation of computational frameworks such as CMS will advance our scientific understanding of complex systems as well as encourage collaborative efforts for code development and knowledge sharing.

Keywords: Stochastic simulation · Compartmental · Open source

1 Introduction

Developing fast, efficient, and scalable computational frameworks is integral to investigating a broad set of epidemiological and biological systems. Here, we present a new open-source computational framework, called the compartmental modeling software (CMS), which enables the simulation of discrete, stochastic reaction models. The CMS framework is highly flexible: the new model description language enables rapid model development; a broad set of available numerical algorithms allows users to optimize simulations based on model structure or computational speed requirements; and the standardized model output empowers a wide-variety a visualization tools. In this article, we report the functionality of the CMS framework while highlighting the key components of the software.

C. W. Lorton, J. L. Proctor and M. K. Roh—Co-first author.

L. Bortolussi and G. Sanguinetti (Eds.): CMSB 2019, LNBI 11773, pp. 308–314, 2019.
https://doi.org/10.1007/978-3-030-31304-3_18

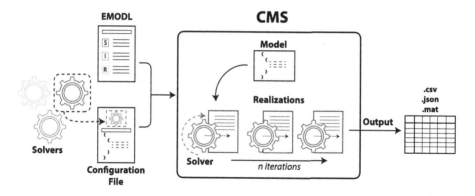

Fig. 1. An overview of the CMS software.

Open-source frameworks have been previously developed for discrete, stochastic compartmental modeling [20], including advancements that account for event handling [25], access via open-source software such as Python [6], and adaptations to cloud-based platforms [15]. Our new CMS framework broadens the scope and scale of previous open-source software. For example, CMS includes more recently developed rare-event probability estimation algorithms [13,24]. Moreover, spatial simulation algorithms [9,14,22] are also integrated to accommodate spatial diffusion processes. Reaction propensities, time delays [8,10], and state- and time- dependent events can be fully customized in the CMS framework. Coupling these features with software documentation, unit and regression tests, and object-oriented development, CMS is a novel and flexible framework to allow for modeling of complex, physical systems.

2 Compartmental Modeling Software

In this section, we briefly explain the major components of CMS. An overview of the CMS schematics is given in Fig. 1. For more detail, see the CMS documentation [1] as well as the GitHub repository for the associated code [2].

2.1 Platform and Computational Requirements

CMS was developed in C# and C++ and tested with NUnit framework [4]. It runs on 64-bit Windows 7 or later with any 64-bit Intel CPU. The CMS Visual Studio solution file has been updated to work with Visual Studio 2017 and targets the .NET Framework version 4.6.

2.2 Execution Pathways

Command line invocation is the most common usage for CMS. However, CMS can be executed in any language or scripting tool that can load Microsoft .NET technology. For example, CMS can be integrated into MATLAB or into Python through a package such as pythonnet [5].

2.3   Input Language and Configuration

A custom model description language, named the epidemiological modeling language (EMODL), was created to support the unique features of CMS and accommodate future expansion. The EMODL syntax was developed to be human interpretable allowing users to quickly learn, adapt, and develop models for simulation as compared to mark-up languages such as SBML. Moreover, the EMODL syntax allows for efficient formulation of time- and state-based events as well as delays common to epidemiological models. It also supports custom propensity formulation, enabling the inclusion of empirically measured infectious periods for epidemiological models, in addition to the traditional propensities for mass-action kinetics. Runtime and solver-specific parameters are listed in a JavaScript Object Notation (JSON) [3] formatted configuration file. Parameters such as solver name, ensemble size, length of simulation, random number generator and output format are included in this configuration file.

2.4   Discrete Stochastic Solvers in CMS

CMS offers a comprehensive suite of stochastic solvers. It contains a total of 16 solvers, whose capabilities range from exact sampling of the true underlying probability distribution of the master equation (ME), fast approximation of the ME, spatial (network) simulations, and rare event probability estimation. We briefly describe each category below. In each subsection, we refer the reader to the original publication for mathematical derivations and computational complexity.

Exact Methods. The time evolution of systems are fully described by the solution of the ME, which provides the probability of every possible system state configuration at any given time. The ME is analytically intractable for most systems, but exact numerical solvers can estimate these probabilities numerically by sampling from the ME. Gillespie's stochastic simulation algorithm (SSA) [18] is the most well-known exact method for simulating systems in stochastic chemical kinetics. CMS offers two known implementations of SSA—direct method and first reaction method—that are theoretically equivalent to each other. CMS also features Gibson and Bruck's next reaction method [17] as well as the SSA with time delays [8,10].

Approximate Methods. While the exact methods produce accurate time trajectories, explicitly simulating every reaction may be prohibitively slow for some systems. Approximate algorithms have been developed to accelerate the simulation at the expense of accuracy. CMS includes three approximate solvers—two implementations of τ-leaping [12] and one of R-leaping [7].

Tau-leaping is the most popularly used approximate method, and CMS offers two implementations for choosing a time step (τ)—adaptive and fixed. The adaptive time step selection mechanism also includes a check to avoid negative population [11] by reverting to SSA when appropriate. The fixed time step method assumes that the chosen τ is small enough to produce accurate trajectories. When this assumption is not met, negative population counts can occur.

Spatial Simulation Methods. Spatial simulation in CMS is possible via three different solvers—inhomogeneous SSA (ISSA) [22], diffusive finite state projection (DFSP) [14], and fractional diffusion (FD) [9]. ISSA divides a system into homogeneous subvolumes, and diffusive transfers are treated as a unimolecular reaction. Therefore, it can be prohibitively slow when fast diffusion is present. The other two methods are more efficient than the ISSA when diffusion occurs frequently with respect to the number of reaction events. DFSP solves the diffusion master equation by adapting the Finite State Projection (FSP) method [23], while FD is based on Lie-Trotter operator splitting of the diffusion and reaction terms. Unlike ISSA and DFSP, fractional diffusion allows for jumping to a distant locale with non-zero probability.

Rare Event Probability Methods. In addition to generation of time trajectories, CMS allows for efficient estimation of a rare event probability via the doubly weighted SSA (dwSSA) [13] and the state-dependent dwSSA (sdwSSA) [24]. Both algorithms utilize importance sampling, whose optimal parameters are determined by the information-theoretic concept of cross entropy. While the dwSSA assigns a single importance sampling parameter per reaction, the sdwSSA creates a list of state-dependent importance sampling parameters in order to further reduce the variance in the rare event probability estimate.

Exploratory Methods. The CMS framework is designed to enable efficient prototyping of new methods. A new solver can be easily implemented by extending the base solver. Four prototype methods are included in the CMS, and we refer to the documentation page [1] for further details. We note that solvers in this category are included as an example of ongoing method development and are not currently supported by the developers.

2.5 Output Files

There are three output formats available in the CMS—comma-separated values (CSV), JSON [3], and MATLAB (MAT). By default, CMS creates trajectories.csv in the output directory, with the realization index appended to each observable name specified in the EMODL file. Output-related options, such as compression and heading, can be specified in the configuration file.

3 Examples

3.1 Schlögl Process

The Schlögl process is a canonical example of a chemical system exhibiting bistability. This model consists of the following four reactions:

$$B_1 + 2X \underset{k_2}{\overset{k_1}{\rightleftharpoons}} 3X \quad \text{and} \quad B_2 \underset{k_4}{\overset{k_3}{\rightleftharpoons}} X.$$

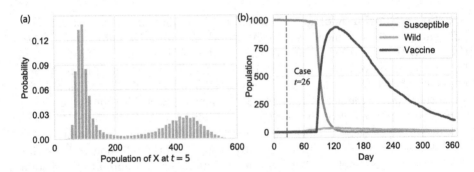

Fig. 2. (a) illustrates the bistable distribution of population of species X at the final time $t = 5$. (b) illustrates one realization of a stochastic Polio outbreak, a paralyzed child (case) detection, and the delayed start of a vaccination campaign.

We take the system description including parameters and initial condition from [19] that produces bistable behavior in X: $k_1 = 3 \times 10^{-7}$, $k_2 = 10^{-4}$, $k_3 = 10^{-3}$, $k_4 = 3.5$, and $\mathbf{x}_0 = [10^5, 2 \times 10^5, 250]$, where $\mathbf{x}_0$ denotes the initial population of B_1, B_2, and X. Using the SSA solver, an ensemble of $N = 1\mathrm{e}5$ simulations were generated to produce the distribution of X shown in Fig. 2(a).

3.2 Vaccination Campaigns for Eradicating Poliomyelitis

Globally, the number of poliomyelitis cases has dramatically decreased over the past two decades and may be the second human infectious disease to be eradicated [21]. The broad functionality of CMS can be utilized to model current vaccination questions such as programmatic mobilization time to arrest a small polio outbreak. We implement an established discrete, stochastic compartmental model of the spread of Polio [16]. We add two essential components to this model: a probabilistic case detection matching the 1 to 200 case to infection rate for polio and a vaccination campaign that is initiated after a fixed time duration following case detection.

We simulate the model starting with a population of 1000 individuals and one imported infection. When a case is detected, the model enforces a sixty day operational delay before a vaccination campaign begins. Figure 2(b) illustrates one realization of this model; note that a single case is detected on day 26 and vaccination begins on day 86. Due to the vaccination in this scenario, we do not detect another paralyzed child. To generate this realization, we utilize the SSA numerical algorithm, allow for a simulation duration of 365 days, and sample the state of the system each day. The model file, config file, and output file for this example can be found in the examples folder at [2].

4 Conclusion

Simulating biological and epidemiological processes with low population counts, such as nearing elimination of an infectious disease, requires the ability to simulate discrete, stochastic reaction models. The compartmental modeling software (CMS) is a novel, extensible framework used to generate an ensemble of trajectories that approximate the true underlying probability distribution described by the model and initial condition. CMS was designed to enable rapid model development with a custom model language, allow for user flexibility in choosing model-specific numerical solvers, and output the model trajectories into easy to visualize formats. Moreover, CMS is an open-source project allowing for community development; the object-oriented programming and class structure of the code allows for intuitive modifications of the code-base such as the inclusion of new numerical solvers or random number generators.

A number of challenges face the widespread adoption of our framework as a modeling tool. The source code has been written in C# which is not widely utilized in university settings. Also, CMS does not currently make use of multi-threading, multi-core CPUs, or GPU resources. Despite these limitations, CMS can be deployed across multiple virtual machines since the memory and disk requirements are minimal. In the near term, we plan on adding a Docker build file to the GitHub repository for portability of the tool to Linux and MacOS as well as for development and a reproducible environment for execution; we are also developing an API from Python to enable users to call into the CMS executable. More broadly, we believe computational tools such as CMS will help provide insights into realistic biological and epidemiological systems.

Acknowledgements. JLP, MKR, CWL, and PW would like to thank Bill and Melinda Gates for their active support of the Institute for Disease Modeling and their sponsorship through the Global Good Fund. The authors would also like to thank Mandy Izzo for her assistance illustrating Fig. 1.

References

1. CMS Documentation. http://idmod.org/docs/cms/
2. CMS Repository. https://github.com/InstituteforDiseaseModeling/IDM-CMS
3. JSON organization. http://www.json.org
4. NUnit 3.6.1. https://github.com/nunit/nunit/releases/3.6.1
5. Python for .NET. http://pythonnet.github.io/
6. Abel, J.H., Drawert, B., Hellander, A., Petzold, L.R.: GillesPy: a python package for stochastic model building and simulation. IEEE Life Sci. Lett. **2**(3), 35–38 (2016)
7. Auger, A., Chatelain, P., Koumoutsakos, P.: R-leaping: accelerating the stochastic simulation algorithm by reaction leaps. J. Chem. Phys. **125**(8), 084103 (2006). https://doi.org/10.1063/1.2218339
8. Barrio, M., Burrage, K., Leier, A., Tian, T.: Oscillatory regulation of hes1: discrete stochastic delay modelling and simulation. PLoS Comput. Biol. **2**(9), e117 (2006). https://doi.org/10.1371/journal.pcbi.0020117
9. Bayati, B.S.: Fractional diffusion-reaction stochastic simulations. J. Chem. Phys. **138**(10), 104117 (2013). https://doi.org/10.1063/1.4794696
10. Cai, X.: Exact stochastic simulation of coupled chemical reactions with delays. J. Chem. Phys. **126**(12), 124108 (2007). https://doi.org/10.1063/1.2710253

11. Cao, Y., Gillespie, D.T., Petzold, L.R.: Avoiding negative populations in explicit poisson tau-leaping. J. Chem. Phys. **123**(5), 054104 (2005). https://doi.org/10.1063/1.1992473

12. Cao, Y., Gillespie, D.T., Petzold, L.R.: Efficient step size selection for the tau-leaping simulation method. J. Chem. Phys. **124**(4), 044109 (2006). https://doi.org/10.1063/1.2159468

13. Daigle, B.J., Roh, M.K., Gillespie, D.T., Petzold, L.R.: Automated estimation of rare event probabilities in biochemical systems. J. Chem. Phys. **134**(4), 044110 (2011). https://doi.org/10.1063/1.3522769

14. Drawert, B., Lawson, M.J., Petzold, L., Khammash, M.: The diffusive finite state projection algorithm for efficient simulation of the stochastic reaction-diffusion master equation. J. Chem. Phys. **132**(7), 074101 (2010). https://doi.org/10.1063/1.3310809

15. Drawert, B., Trogdon, M., Toor, S., Petzold, L., Hellander, A.: Molns: a cloud platform for interactive, reproducible, and scalable spatial stochastic computational experiments in systems biology using pyurdme. SIAM J. Sci. Comput. **38**(3), C179–C202 (2016)

16. Eichner, M., Dietz, K.: Eradication of poliomyelitis: when can one be sure that polio virus transmission has been terminated? Am. J Epidemiol. **143**(8), 816–822 (1996)

17. Gibson, M.A., Bruck, J.: Efficient exact stochastic simulation of chemical systems with many species and many channels. J. Phys. Chem. A **104**(9), 1876–1889 (2000). https://doi.org/10.1021/jp993732q

18. Gillespie, D.T.: Exact stochastic simulation of coupled chemical reactions. J. Phys. Chem. **81**(25), 2340–2361 (1977). https://doi.org/10.1021/j100540a008

19. Gillespie, D.T.: Markov Processes: An Introduction for Physical Scientists. ACADEMIC PR INC, Cambridge (1991). https://www.ebook.de/de/product/3655742/danieltgillespiemarkovprocessesanintroductionforphysicalscientists.html

20. Hucka, M., et al.: The systems biology markup language (sbml): a medium for representation and exchange of biochemical network models. Bioinformatics **19**(4), 524–531 (2003)

21. Kew, O., Pallansch, M.: Breaking the last chains of poliovirus transmission: progress and challenges in global polio eradication. Annu. Rev. Virol. **5**, 427–451 (2018)

22. Lampoudi, S., Gillespie, D.T., Petzold, L.R.: The multinomial simulation algorithm for discrete stochastic simulation of reaction-diffusion systems. J. Chem. Phys. **130**(9), 094104 (2009). https://doi.org/10.1063/1.3074302

23. Munsky, B., Khammash, M.: The finite state projection algorithm for the solution of the chemical master equation. J. Chem. Phys. **124**(4), 044104 (2006). https://doi.org/10.1063/1.2145882

24. Roh, M.K., Daigle, B.J., Gillespie, D.T., Petzold, L.R.: State-dependent doubly weighted stochastic simulation algorithm for automatic characterization of stochastic biochemical rare events. J. Chem. Phys. **135**(23), 234108 (2011). https://doi.org/10.1063/1.3668100

25. Sanft, K.R., Wu, S., Roh, M., Fu, J., Lim, R.K., Petzold, L.R.: Stochkit2: software for discrete stochastic simulation of biochemical systems with events. Bioinformatics **27**(17), 2457–2458 (2011)

Spike – Reproducible Simulation Experiments with Configuration File Branching

Jacek Chodak$^{(\boxtimes)}$ and Monika Heiner

Computer Science Institute, Brandenburg Technical University,
Postbox 10 13 44, 03013 Cottbus, Germany
{jacek.chodak,monika.heiner}@b-tu.de
https://www-dssz.informatik.tu-cottbus.de

Abstract. This paper presents Spike - a command line tool for continuous, stochastic & hybrid simulation of biochemical reaction networks represented as (coloured) Petri nets. It supports import from and export to various Petri net data formats, and also imports SBML models. Spike's abilities include the configuration of models by changing stoichiometries (arc weights), initial conditions (markings) and kinetic parameters. It also unfolds coloured stochastic/continuous/hybrid Petri nets. To comply with the demand for reproducible simulation experiments, Spike builds on a script language in a human-readable format. Its core features permit the design of a set of simulation experiments by a single configuration file. These simulation experiments can be executed in parallel on a multi-core machine; distributed execution is in preparation.

Keywords: Continuous · Stochastic · Hybrid ·
Coloured (hierarchical) Petri nets · Parallel simulation ·
Configuration · Reproducibility · Parameter scanning

1 Objectives

Spike is a command line tool for the efficient execution of multiple simulation experiments of biochemical reaction networks, which we happen to represent as Petri nets interpreted in the stochastic, continuous or hybrid paradigm. Simulation of biochemical models can be time and memory consuming. Thus, simulations should be delegated for performance reasons to be executed on a server. Additionally, when experiments require to run multiple simulations, the time spent can be particularly long, when the individual simulations are merely executed one after another. Frequently, it is required to prepare a set of simulation experiments in order to find appropriate model parameters (e.g., initial conditions, kinetic parameters) or simulator options (e.g., simulator type, length of simulation traces, resolution of the traces recorded). Doing this manually, by preparing a new simulation run for each new model and/or simulator

© Springer Nature Switzerland AG 2019
L. Bortolussi and G. Sanguinetti (Eds.): CMSB 2019, LNBI 11773, pp. 315–321, 2019.
https://doi.org/10.1007/978-3-030-31304-3_19

configuration is time consuming and potentially error-prone. The reproducibility of the entire experiment suffers, if one of the runs is not well documented.

Spike has been designed to address all these issues. To achieve this, it builds on a human-readable configuration script, supporting the efficient specification of multiple model configurations as well as multiple simulator configurations in a single file. Each specific model and simulator configuration determines a specific simulation experiment, for which Spike creates a separate branch, ready to be executed on a server, with all branches treated as parallel processes. According to our experience, storing configurations in self-contained scripts allows for a simplified workflow and reproducible simulations in a user-friendly way.

2   Functionality

Spike is a slim, but powerful brother of Snoopy [2]; it is the latest addition to the PetriNuts family of tools for modelling, analysing and simulating a variety of related models, for which we use Petri nets as umbrella modelling paradigm. Spike deals with quantitative Petri nets, comprising stochastic, continuous and hybrid Petri nets, which are specifically tailored to the investigation of biochemical reaction networks. Correspondingly, Spike is capable to run three basic types of simulations: stochastic, continuous and hybrid, each comes with several algorithms, among them are: Gillespie's SSA, tau leaping, and delta leaping for stochastic simulations; a number of basic and stiff ODEs solvers for continuous simulation; HRSSA and accelerated HRSSA for hybrid simulation (for a complete list and related references see the Appendix of this paper on Spike's website). Simulation of coloured stochastic, continuous and hybrid Petri nets is supported by automatically unfolding them to their uncoloured counterparts.

A given model is simulated according to the specified simulation type, despite of place and transition types in the model. That means, all places and transitions are converted to the appropriate type. For example, if a user wants to run stochastic simulation on a continuous model, all continuous places (variables, species) are converted to discrete places, and all continuous transitions (reactions) to stochastic transitions; and vice versa for stochastic models to be simulated continuously.

Spike also offers transformation between various exchange data formats and some basic model reductions.

Simulation. The main focus of Spike lays on efficient and reproducible simulation. Spike's core features allow, among others, to configure the model (via parameters specifying arc weights, initial marking, kinetic parameters) and the simulator (via the usual, simulator-dependent options) over sets of arguments (parameter/option scanning). An argument is a value passed to a parameter or option. The set of argument sets triggers a so-called branching process. A new configuration branch is created for each argument set (if there is more than one). The set of configuration branches can be executed sequentially or in parallel.

The simulation results can be saved in CSV files which can be used later for analysis or visualisation. They may comprise any user-defined combinations of traces over place markings, transition rates, and observers (auxiliary variables).

Conversion. Spike supports the following data formats and conversion among them according to Table 1:

- ANDL and CANDL - human-readable formats for Petri nets and coloured Petri nets, respectively, used internally by the PetriNuts framework,
- SBML (Systems Biology Markup Language) - an XML-based representation format designed to exchange computational models within the systems biology community [3],
- PNML - an XML-based interchange format for qualitative Petri nets [4] used within the Petri net community,
- ERODE - a tool for the evaluation and reduction of chemical reaction networks [1].

Table 1. Data format conversions currently supported by Spike.

From	To
ANDL	PNML, ERODE
CANDL	ANDL, PNML, ERODE
SBML	ANDL, PNML, ERODE
ERODE	ANDL, PNML

Reduction. Spike is also able to structurally reduce a model by pruning clean siphons (a set of empty places, the marking of which will never be changed because any reaction which would cause a change depends on this set) and constant places (places, occurring only as side conditions). In both cases, clean siphons and constant places can be calculated by Spike or loaded from a file. It is also possible to save results of the calculation to a file, which can be used later by Spike or for other analysis purposes.

Further reductions may be applied by converting a model to the ERODE format, if the model is to be read as ordinary differential equations (ODEs). Reductions of a model can have a significant impact on simulation runtime.

Reproducibility. To comply with the demand for reproducible simulations, Spike reads a script which allows for model and simulator configuration. The script is human-readable and does not require any special tools for editing – a simple text editor is enough. The configuration script includes:

- definition of (named) constants, which can get a specific value or a set of values, either specified by a list or an interval; supported data types: boolean, integer, real, string;

- specification of model parameters and simulator options, using either constants or (direct) values as arguments; values can be given as a single specific value or a set of values, specified by a list or an interval;
- definition of observers (auxiliary variables) allow for extra measures by defining numerical functions; depending on the type of observer, it can involve, beside constants, places, transitions or simultaneously places and transitions;
- specification of multiple exports of simulation results by use of regular expressions over the nodes of which the simulation traces are to be recorded; it is possible to combine the results of places, transitions and observers, coloured and uncoloured, in one CSV file.

These features permit, among others, to simulate a given model configuration with different simulator configurations, or to simulate a set of model configurations by a given simulator configuration (parameter scanning); see Sect. 4.

In agreement with its aim of reproducibility, Spike does not assume any default values; all simulator options need to be given.

3   Architecture

Spike has a modular structure, where the modules are basically decoupled from each other. This allows for easily adding new features.

Modules communicate with each other using command patterns and a queue of commands which is globally accessible. Each module has its own list of commands with specific parameters, which must be registered to the queue during initialisations of a module. Table 2 shows a summary of all commands currently available in Spike.

Table 2. List of Spike's modules with their commands.

Module	Command	Description
Main	version	display version of Spike
CLI	help	display help for a given command
Configuration	exe	execute configuration script
Converter	load	load a model from a given file
	save	save a model to a given file
	prune	prune a model
	eval	evaluate constants
	unfold	unfold a coloured model
Simulation	sim	run a simulation of the model

Commands are processed in a sequential way. Each command is executed by the module which is responsible for it. Let's consider the following use case illustrated in Fig. 1 – the execution of a simple configuration script. When the command "exe" is at the head of the command queue, the module Configure will execute it. During execution, the configuration module communicates with other modules by appending new commands to the queue.

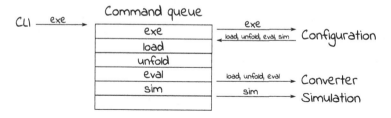

Fig. 1. Flow of commands through Spike's modules when a user types the command "exe".

When Spike is executing the configuration script, a set of configuration branches is possibly created and then executed sequentially or in parallel, depending on the available hardware. Each branch is treated as a separate process (Spike instance). Spike creates two types of processes. One so-called master process and one or more slave processes. The processes communicate with each other via network sockets. Communication is asynchronous and follows the message queue pattern.

The master process acts as broker and owner of the simulation experiment. It takes care of creating the slave processes. Each slave process is responsible for executing exactly one branch of the Spike configuration script. The number of slave processes running in parallel depends on an option passed to Spike via its command line interface (CLI). If only one slave process is allowed, then each simulation branch will be execute sequentially; that means, the master process will wait for the end of the execution of one branch, before starting the execution of the next one. Otherwise, the master process will start slave processes at most in the number specified by the Spike option. If the number of branches exceeds the number of allowed slave processes, the master process will postpone the start of new ones, until one of currently running processes will have finished its task.

Starting a slave process by Spike does not mean that the number of running threads (parallel tasks executed internally by a process) is equal to the number of processes. Each process can create a number of threads. The number of threads may depend on the applied simulation algorithm. For example, stochastic simulation may involve multi-threading to execute in parallel the independent individual runs, which are later averaged.

4   Use Cases

Spike permits to run simulations on a server as well as on the user side. It can be done in batch mode or by integration of Spike as a service. Algorithm 1

Algorithm 1. Use case: multiple simulator configurations.

---

1 Load model;
2 Determine model configuration;
3 Determine set of simulator configurations;
4 **for each** *simulator configuration* **do**
5 | Create new configuration branch;
6 | Run simulation;
7 | Save results of the simulation;
8 **end**

Algorithm 2. Use case: multiple parameter scanning.

---

1 Load model;
2 Determine simulator configuration;
3 **for each** *unique combination of parameter values* **do**
4 | Determine model configuration;
5 | Create new configuration branch;
6 | Run simulation;
7 | Save results of the simulation;
8 **end**

---

illustrates a typical scenario, which allows, e.g., to compare how a model behaves under different simulation algorithms, e.g. for performance comparison, or under different configurations of a given simulation algorithm, e.g. number of stochastic runs to be averaged. In turn, Algorithm 2 presents the use case of parameter scanning in order to compare the effect of different parameter values for a given model, while keeping the simulator options constant.

Of course, both scenarios can be combined. The detailed discussion of more scenarios exceeds the given space limit.

5   Installation and Future Work

Spike is written in C++ and available for Linux, Mac/OSX and Windows. Binaries are statically linked and can be downloaded from Spike's website https://www-dssz.informatik.tu-cottbus.de/DSSZ/Software/Spike, where one also finds the Appendix of this paper, installation instruction and a set of examples.

Spike is still under development. Future work will incorporate sophisticated model reduction, model decomposition, and distributed simulation, either for a set of simulations or a decomposed model; we are open for further suggestions.

Acknowledgement. Spike uses software libraries (data format conversions, simulation algorithms) which have been previously developed by former staff members and numerous student projects at Brandenburg Technical University (BTU), chair Data Structures and Software Dependability.

References

1. Cardelli, L., Tribastone, M., Tschaikowski, M., Vandin, A.: ERODE: a tool for the evaluation and reduction of ordinary differential equations. In: Legay, A., Margaria, T. (eds.) TACAS 2017. LNCS, vol. 10206, pp. 310–328. Springer, Heidelberg (2017). https://doi.org/10.1007/978-3-662-54580-5_19
2. Heiner, M., Herajy, M., Liu, F., Rohr, C., Schwarick, M.: Snoopy – a unifying petri net tool. In: Haddad, S., Pomello, L. (eds.) PETRI NETS 2012. LNCS, vol. 7347, pp. 398–407. Springer, Heidelberg (2012). https://doi.org/10.1007/978-3-642-31131-4_22
3. Hucka, M.: Systems biology markup language (SBML). Encycl. Comput. Neurosci., pp. 2943–2944 (2015)
4. Petri Net Markup Language (PNML): Systems and software engineering - High-level Petri nets - Part 2: Transfer format (2009). ISO/IEC 15909-2:2011

KAMIStudio: An Environment for Biocuration of Cellular Signalling Knowledge

Russ Harmer[ID] and Eugenia Oshurko[✉][ID]

Univ Lyon, EnsL, UCBL, CNRS, LIP, 69342 Lyon Cedex 07, France
{russell.harmer,ievgeniia.oshurko}@ens-lyon.fr

Abstract. In this paper we present KAMIStudio, an environment for biocuration of cellular signalling knowledge based on the KAMI framework. The environment provides an interface for the aggregation of decontextualized knowledge about individual protein-protein interactions, its interactive visualization, instantiation into signalling models and the subsequent generation of Kappa scripts that can be further used to study the dynamics of the modelled systems.

1 Introduction

Cellular signalling underlies many fundamental processes of living cells from responses to a changing environment to cell proliferation and apoptosis. Signalling abnormalities are responsible for common and serious diseases such as cancer and diabetes. However, the immense complexity of cellular signalling systems makes them extremely hard to model and analyse. These systems are comprised of a large number of complex interacting agents which makes traditional modelling approaches (ODE-based, reaction-based) simply unfeasible. The rule-based modelling approach (proposed by Kappa [2] and BioNetGen [3]) overcomes the problem of explosion in the number of agent species and even allows for automated discovery of signalling pathways leading to some events of interest. However, manually building and maintaining large rule-based models remains cumbersome, prone to errors and, in practice, impossible for large models.

While different rules express conditions for individual interactions between agents, these interactions are not necessarily independent and may share interaction mechanisms (e.g. generic mechanisms of binding through conserved domains such as SH2). At the same time, the representation level proposed by rule-based modelling languages does not capture interaction mechanisms, therefore an update of knowledge about an interaction mechanism may require manual identification and update of all the related rules (e.g. that describe bindings through the same domain). Moreover, this representation does not capture conditions on the presence or absence of conserved protein domains and specific key residues that may alter the interaction capabilities of different agent variants (e.g. splice

L. Bortolussi and G. Sanguinetti (Eds.): CMSB 2019, LNBI 11773, pp. 322–328, 2019.
https://doi.org/10.1007/978-3-030-31304-3_20

variants that do not have a particular domain cannot perform interactions that require this domain). This compromises reuse of the knowledge expressed with rule-based models and its adaptation to different contexts, for example, to model systems that contain modified agents such as mutants or splice variants.

To tackle these problems, KAMI proposes a bio-curation framework that aims to de-contextualize protein-protein interaction (PPI) knowledge [6]. It enables the collation of knowledge about *potential* individual PPIs and their necessary conditions and the semi-automatic *aggregation* of this knowledge into a coherent corpus identifying interaction agents and mechanisms according to some body of grounding knowledge. It then allows the reuse of this knowledge for the automatic generation of models in different systems, by specifying which agents are present in these systems, that determines which interaction mechanisms are realizable. We refer to the latter process as *instantiaton* of knowledge in different contexts from which we automatically generate executable models such as Kappa scripts.

KAMI distinguishes two types of 'knowledge bodies': *corpora* and *models*. Corpora contain de-contextualized knowledge: agents of interactions are called *protoforms* and represent a neighbourhood in the sequence space of a gene. By associating regions, residues and states to a specific protoform we represent a feasible neighbourhood of its variants. Interactions in a corpus represent potential interactions and the necessary conditions for them to occur. Models contain knowledge instantiated in given contexts: agents are concrete proteins and rules describe the interactions between these proteins. The knowledge representation system is based on graphs [6]. PPIs are encoded with small graphs called *nuggets* whose nodes represent agents, their components and actions. The body of grounding knowledge is represented with the *action graph* which defines the kinds of entities and interactions that can exist in a model. Every node in a nugget maps homomorphically to some node in the action graph and by this it is identified with a kind of entity or an action mechanism.

The bio-curation pipeline implemented in KAMI, illustrated in Fig. 1, is discussed in detail in Sect. 2. In Sect. 3, we present KAMIStudio, a standalone web-based application that provides a curation environment based on the KAMI framework. It allows to create and store corpora, instantiate models, input and visualize knowledge as well as automatically generate executable Kappa models.

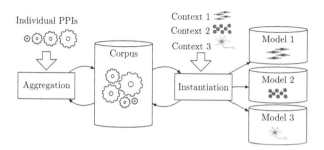

Fig. 1. Biocuration pipeline of KAMI

Attempts to facilitate model building and its decoupling from knowledge curation have been made in MetaKappa [5], rxncon [7] and INDRA [4,8]. However, to our knowledge, none of them manage to solve the above problems.

Notably, INDRA aims at similar problems but does so with a significantly different approach: while KAMI provides a semantically rigorous framework for curation of de-contextualized knowledge about *generic* mechanisms of PPIs (at the last step of which resides the generation of concrete models), INDRA allows the extraction of (contextualized) knowledge about *concrete* PPIs into a 'pool' of independent statements and employs various techniques (both systematic and ad hoc) to assemble these statements into models.

2   Main Features

KAMIStudio provides features for semi-automatic curation of large corpora of cellular signalling knowledge including: interactive visualization of knowledge stored in corpora and models; input of individual PPIs to a corpus through intuitive forms as well as batch import from JSON-formatted interactions resulting in the automatic aggregation of the new knowledge to the corpus; an interface for specifying protein variants; automatic instantiation of corpora into models using protein variants; and automatic generation of Kappa scripts from models.

Visualization. KAMIStudio provides capabilities for interactive visualization of KAMI corpora and models. The user can interact with graphs in various ways: click on graph elements to view (and modify) the attached meta-data, zoom, pan, drag the nodes. Moreover, using the meta-data attached to the graph elements KAMIStudio provides cross-referencing to the common databases such as UniProt and InterPro. Such interactive capabilities may provide some additional insights on the knowledge, e.g. on the structure of the underlying PPI network, its connected components or its hub nodes and may also suggest manual edits necessary to make the data consistent with the modeller's viewpoint.

Aggregation. The approach to automatic knowledge aggregation used in KAMI is based on the identification of agents of interactions, their components *and* the mechanisms of interactions. Unlike the identification of interaction agents and their components, the problem of identification of interaction mechanisms is highly non-trivial and is performed automatically only for the interactions of specific conserved protein domains whose semantics is hard-wired in KAMI's background knowledge (currently protein kinase and SH2 domains). In practice, a mechanism in KAMI is encoded with a single element of a knowledge corpora (for example, a single binding or modification node of the action graph), and by default such an element is created for every interaction provided in a nugget. The strategy of automatic aggregation is conservative and for any two individual interactions KAMI assumes that *"the interaction mechanisms are not known to be the same"* (unless it is known according to the background knowledge, discussed in more details in [6]). KAMIStudio provides an interface for manual intervention in the aggregation process and allows the user to select different action nodes and merge them, by which stating *"I know that these interactions are instances of the same mechanism"*.

The identification of two mechanisms as "the same" has a direct influence on the dynamics of the underlying system (usually expressed as two interactions being conflicting). Figure 2 illustrates two nuggets aggregated in two different ways, i.e. producing the action graph with two different interaction mechanisms (2a) or with the same mechanism (2b).

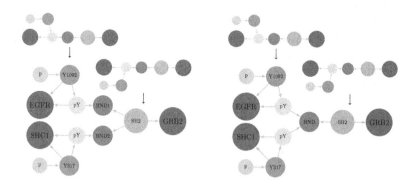

(a) Different interaction mechanisms   (b) Single interaction mechanism

Fig. 2. Schematic example of two nugget graphs aggregated into two different action graphs. Small unlabeled graphs on the top represent nuggets, large labeled graphs on the bottom—two different action graphs, arrows from the nuggets to the action graphs—graph homomorphism. The action graphs are interpreted as (a) the interactions described by nuggets are not known to share the same mechanism and (b) interactions share the same mechanism.

Instantiation. Different protein variants (or isoforms) derived from the same gene arise as the result of alternative splicing and mutations whose principal targets are regions and residues respectively. KAMIStudio provides an intuitive interface for specifying and storing protein variants. The knowledge instantiation process consists in an update of a given knowledge corpora, given variant definitions in concrete *contexts* (e. g. different cell types, wild type vs. mutants). This update is performed automatically in KAMI and may lead to the invalidation of some interactions, i. e. these interactions no longer take place in a given context. KAMIStudio allows to select subsets of variants for automatic instantiation of concrete signalling models. By default (i.e. if no variants are specified), KAMI uses "wild-type variants" which are obtained from the canonical gene sequences (retrieved from corresponding UniProt entries). In addition, KAMIStudio allows the user to specify custom wild-type variants that will be used as the default variants during the instantiation. The models produced as the result are stored and can be further modified manually and used to generate Kappa.

Kappa Generation. The Kappa generation capabilities provided in KAMIStudio are compatible with version 4 of the Kappa language. It consists of two main steps: the generation of agent signatures and the generation of rules.

To generate Kappa agents, KAMIStudio inspects the action graph and generates a distinct agent per protoform. It encodes protein variants derived from the same protoform with a dedicated Kappa-site called `variant` in order to optimize the simulation performance of KaSim4 [1]. Then, for each agent, it explores all the derived variants and creates a site per (not necessarily directly) adjacent state node. As the state nodes in KAMI represent binary on/off states, every such site is of the form `site_name{0 1}`. After this, KAMI adds a site per adjacent KAMI-site node and binding node (in both cases the nodes are not required to be directly adjacent, but can be adjacent to some components of the current variant). For example, the gene EGFR with two variants `WT` and `p60`, phosphorylable residue Y1092 and a binding site `pY` would correspond to the Kappa agent signature `EGFR(variant{WT p60}, pY1092{0 1}, pY_site)`. To generate Kappa rules KAMI examines nuggets together with their mapping to the action graph (example in Fig. 3). As it was previously mentioned, for a given agent every adjacent binding action (therefore every binding mechanism) gives a rise to a separate Kappa site. This represents the main subtlety of the Kappa generation process as for every binding nugget in KAMI we need to identify the site corresponding to the interaction mechanism of the binding. In KAMI's knowledge representation framework interaction rates are encoded in the interaction nodes of nuggets. However, KAMI does not enforce them to be specified as these rates for some interactions may be unknown or depend on the context. Therefore, to generate valid Kappa, KAMI allows the user to specify default rates for binding, unbinding and modification interactions in a model; these rates are used to generate Kappa rules for nuggets whose rates are not available.

```
EGFR(pY10921 pY[.]), GRB2(SH2_site1[.])->          EGFR(pY10921 pY[.]), GRB2(SH2_site[.]) ->
    EGFR(pY10921 pY[1]), GRB2(SH2_site1[1])            EGFR(pY10921 pY[1]), GRB2(SH2_site[1])
SHC1(pY3171 pY[.]), GRB2(SH2_site2[.]) ->          SHC1(pY3171 pY[.]), GRB2(SH2_site[.]) ->
    SHC1(pY3171 pY[1]), GRB2(SH2_site2[1])             SHC1(pY3171 pY[1]), GRB2(SH2_site[1])
```

    (a) Kappa rules for Fig. 2a   (b) Kappa rules for Fig. 2b

Fig. 3. The automatically generated Kappa rules for the example of Fig. 2.

3   Technical Description

KAMIStudio is a web-based application: its server can be started locally and its functionality can be used in a browser via the provided client. The knowledge representation and update-related backend is based on the Python libraries `ReGraph` and `KAMI`[1]. To store data, KAMIStudio uses two noSQL database technologies: Neo4j and MongoDB. The full version can be installed from the source

---

[1] https://github.com/Kappa-Dev/ReGraph, https://github.com/Kappa-Dev/KAMI.

https://github.com/Kappa-Dev/KAMIStudio and run locally (detailed installation instructions can be found in the github repository). In addition, a read-only demo is available online at http://kamistudio.ens-lyon.fr/.

The online demo contains three example corpora: *EGFR signalling* built from a subset of individual PPIs involved in the EGFR signalling pathway, *pYNET 20* and *pYNET 200* built from respectively 20 and 200 random PPIs involving tyrosine phosphorylations and bindings of SH2 domains to phosphotyrosine-containing sites. The demo also contains three models that can be used to generate Kappa scripts. The first model is an instantiation of the *EGFR signalling* corpus using splice variants and mutants of genes EGFR and GRB2. The two other models represent instantiations of *pYNET 20* and *pYNET 200* using the wild-type variants. These models are built by aggregation of independent PPIs without pre-conceived pathways in mind. Superficial look at the action graph of the *pYNET 20* model reveals a number of disconnected components most of which correspond to individual PPIs, which suggests to the modeller some gaps in the collected knowledge. On the other hand, the action graph of the *pYNET 200* model starts exhibiting a large connected component, which suggests the potential emergence of pathways.

4   Future Work

Together with the bio-curation framework of KAMI and its corresponding Python library, KAMIStudio is currently in an active development phase. Therefore, a great of amount of work remains to be done to make it a full-blown curation environment. Among the main features envisaged are a query language for browsing corpora and models, a richer annotation system for represented PPIs, a system of version control and static analysis of corpora (e.g. nugget relations, reachability of a molecular species). In addition, we plan to provide richer means of representation and generation of Kappa that would allow, for example, to accommodate knowledge on kinetic refinements, negative conditions of PPIs, etc.

References

1. Boutillier, P., Ehrhard, T., Krivine, J.: Incremental update for graph rewriting. In: Yang, H. (ed.) ESOP 2017. LNCS, vol. 10201, pp. 201–228. Springer, Heidelberg (2017). https://doi.org/10.1007/978-3-662-54434-1_8
2. Danos, V., Feret, J., Fontana, W., Harmer, R., Krivine, J.: Rule-based modelling of cellular signalling. In: Caires, L., Vasconcelos, V.T. (eds.) CONCUR 2007. LNCS, vol. 4703, pp. 17–41. Springer, Heidelberg (2007). https://doi.org/10.1007/978-3-540-74407-8_3
3. Faeder, J.R., Blinov, M.L., Hlavacek, W.S.: Rule-based modeling of biochemical systems with bionetgen. In: Systems Biology, pp. 113–167. Springer, Cham (2009). https://doi.org/10.1007/978-1-59745-525-1_5

4. Gyori, B.M., Bachman, J.A., Subramanian, K., Muhlich, J.L., Galescu, L., Sorger, P.K.: From word models to executable models of signaling networks using automated assembly. Mol. Syst. Biol. **13**(11), 954 (2017)
5. Harmer, R.: Rule-based modelling and tunable resolution. EPTCS **9**, 65–72 (2009)
6. Harmer, R., Le Cornec, Y.S., Légaré, S., Oshurko, I.: Bio-curation for cellular signalling: the KAMI project. In: CMSB 2017, pp. 3–19 (2017)
7. Romers, J.C., Krantz, M.: rxncon 2.0: a language for executable molecular systems biology. BioRxiv, p. 107136 (2017)
8. Todorov, P.V., Gyori, B.M., Bachman, J.A., Sorger, P.K.: INDRA-IPM: interactive pathway modeling using natural language with automated assembly. Bioinformatics, btz289 (2019)

A New Version of DAISY to Test Structural Identifiability of Biological Models

M. P. Saccomani[1]($\boxtimes$)(iD), G. Bellu[2], S. Audoly[2], and L. d'Angió[2]

[1] Department of Information Engineering, University of Padova, Padova, Italy
mariapia.saccomani@unipd.it
[2] Department of Mathematics, University of Cagliari, Cagliari, Italy
https://www.dei.unipd.it/persona/0A7F06976D2D39F73AEDD095D97CF91A

Abstract. Often ODE models in systems biology, medical research, epidemiology, ecology and many other areas, contain unknown parameters which need to be estimated from experimental data. Identifiability deals with the uniqueness of the relation between model parameters and ODE solution thus being a prerequisite for the well-posedness of parameter estimation. In this paper a novel extension of the software tool DAISY (Differential Algebra for Identifiability of SYstems) is presented. DAISY performs structural identifiability analysis for linear and nonlinear dynamic models described by polynomial or rational ODE's. The major upgrades of this new version regard the ability to include in the identifiability analysis either known and unknown model initial conditions, the possibility of entering a parameter estimate to calculate all the equivalent parameter solutions, the portability to MacOS platforms and an user-friendly interface. These upgrades make DAISY surely more general and easy to use. Practical examples are presented. DAISY is available at the web site daisy.dei.unipd.it.

Keywords: Identifiability software · Global identifiability · Biological models · Nonlinear ODE systems

1 Introduction

Biological and biomedical systems dynamics are generally modeled by nonlinear ordinary differential equations (e.g. Michaelis-Menten equation), whose parameter identification is often difficult due to experimental limitations and/or ill-posedness. Given the increased model complexity, many models may be parameterized redundantly, so that multiple or even an infinite number of equivalent parameter values may lead to the same input-output behaviour. Equivalent parameterizations can describe the same trajectory of the measurable data, but may generally be associated with different dynamic evolution of the internal (not directly measurable) variables. Such a situation is undesirable since mathematical models are principally used to predict unmeasurable quantities.

© Springer Nature Switzerland AG 2019
L. Bortolussi and G. Sanguinetti (Eds.): CMSB 2019, LNBI 11773, pp. 329–334, 2019.
https://doi.org/10.1007/978-3-030-31304-3_21

A fundamental prerequisite for parameter estimation in dynamical models, is the *structural identifiability* (see e.g. [1,2]). Structural identifiability analysis of nonlinear systems is in general very difficult since it requires to check the solvability of an unusually large system of nonlinear algebraic equations. Nevertheless, checking the uniqueness of the parameter solution (*structural global identifiability*) is crucial especially before investing resources in performing delicate experiments which may otherwise provide unreliable numerical estimate of the unknown parameters [3]. Note that identifiability analysis does not require experimental data as it should just check a mathematical property of a model.

The primary goal of DAISY is to bring to systems biology and biomedical research a piece of software for checking identifiability which, although being based on a rather sophisticated set of mathematical tools, does not require knowledge of higher mathematics and computer algebra by the user and yet allows to tackle problems which are hard and computationally intensive in a transparent way, without requiring any knowledge of high-level programming languages.

In the literature software checking identifiability of biological models generally deals with *practical identifiability*. These statistical model fitting software [4,5] check local identifiability but cannot however provide a mathematically rigorous answer to the uniqueness problem. Recently software to check structural global identifiability, based on analytic calculations, have been proposed together with some comparisons among their performances [8,9].

DAISY is based on differential algebra and provides an analytic method to check structural identifiability [1]. It is difficult to define up to which model size DAISY survives because it does not depend only on the number of ODE equations, but also on that of the unknown parameters and on the number and types of nonlinearities present in the ODE. The new version of DAISY successfully deals with models for which the previous version failed. We applied DAISY to many biological models of the recent literature.

In this paper we present the recent upgrades which have been implemented in the new version of DAISY. The principal goals of this new version are:

1. to include in the identifiability analysis the known as well as the unknown initial conditions;
2. the possibility to enter the parameter estimate obtained with a whatever optimization algorithm to obtain all the equivalent parameter solutions (which equivalently describe the output function);
3. the portability to the MacOS platform (not only Windows);
4. the new friendly interface which makes for a much easier use of DAISY.

2   Checking a Priori Identifiability

Consider a nonlinear dynamic system described in state space form

$$\dot{\mathbf{x}}(t) = \mathbf{f}(\mathbf{x}(t), \boldsymbol{\theta}) + \sum_{i=1}^{m} \mathbf{g}_i(\mathbf{x}(t), \boldsymbol{\theta})u_i(t), \tag{1}$$

$$\mathbf{y}(t) = h(\mathbf{x}(t), \mathbf{u}(t), \boldsymbol{\theta}) \tag{2}$$

with initial condition $x(0) = x_0$, with state $x(t) \in \mathbb{R}^n$, input $u(t) \in \mathbb{R}^q$, output $y(t) \in \mathbb{R}^m$ and constant unknown parameter vector θ belonging to some open subset $\Theta \subseteq \mathbb{R}^p$. Here functions f, $g_1, \ldots, g_m$ and h are assumed to be vectors of *rational functions* in x. Also we assume that u is a free variable not depending on y. The affine structure in u is not essential and could be relaxed. Equality constraints (linear or nonlinear) on θ may be also present.

Let $y = \psi_{x_0}(\theta, u)$ be the input-output map of the system (1, 2) started at the initial state x_0 (we assume that this map exists).

One says that the system (1, 2) is *structurally globally (or uniquely) identifiable from input-output data* if, for at least a generic set of points $\theta^* \in \Theta$, there exists (at least) one input function u such that the equation

$$\psi_{x_0}(\theta, u) = \psi_{x_0}(\theta^*, u) \tag{3}$$

has only one solution $\theta = \theta^*$ for all x_0 in a generic subset of $\mathbb{R}^n$.

If Eq. (3) has a finite (more than one) or an infinite number of solution θ the system (1, 2) is *locally* or *non identifiable* respectively.

To check structural identifiability DAISY uses a method based on differential algebra. We refer the reader to [1,6] where the original algorithm is described in detail. By a suitable elimination technique, this algorithm permits to calculate the *characteristic set*, a minimal set of differential polynomials which provides the *input-output relation* of the model: a set of r polynomial equations involving only the known variables (u, y) and their time derivatives, thereby describing all input-output pairs satisfying the original dynamic model. The coefficients of the input-output relation are known. In particular they are (nonlinear) algebraic functions of the unknown parameters of the original model. The resulting system of nonlinear equations in the unknown model parameters may be solved by the Buchberger's algorithm. This is a computer algebra algorithm which calculates the Groebner basis, a set of polynomials with specific properties which determines if the system admits only one solution, or a finite or an infinite number of solutions for each parameter, thus allowing one to distinguish between global or local or non identifiability.

3   The New Version of DAISY

In this section the upgrades of the new version of the algorithm are illustrated, while for the description of the algorithm itself and of its implementation the reader is referred to [6,7].

3.1   Identifiability with Known and Unknown Initial Conditions

The calculation of the characteristic set is independent from the initial conditions. In the new version of DAISY, the initial conditions are included in the algorithm in both cases of known and unknown initial conditions. In particular, the initial condition values (numerical if known, symbolic if unknown) are substituted where they appear in the polynomials of the *characteristic set* of the model. These polynomials are evaluated at time $t = 0$ and included in the *exhaustive summary*. In this case the assumptions of algebraic observability [2] (automatically checked by DAISY itself), and of accessibility [1] are required.

Note that a model can change its identifiability properties depending on its initial conditions. For example, consider the following three compartments model

$$\begin{cases} \dot{x}_1 = -(k_{21} + k_{31} + k_{01})x_1 + k_{12}x_2 + k_{13}x_3 + u1 & x_1(0) = x_{10} \\ \dot{x}_2 = k_{21}x_1 - k_{12}x_2 & x_2(0) = x_{20} \\ \dot{x}_3 = k_{31}x_1 - k_{13}x_3 & x_3(0) = x_{30} \\ y = x_1 \end{cases} \quad (4)$$

with known initial conditions the model is globally identifiable, while with only the known constraint $x_2(0) = x_3(0)$ the model becomes locally identifiable. In this case, the parameter estimation from the experimental data possibly provides only one of the two parameter solutions. By ignoring the second one, one can misinterpret the biological results. This shows the relevance of including initial conditions in the identifiability test, as the new version of DAISY does.

3.2   Calculation of All the Equivalent Parameter Solutions

A new important feature of DAISY is the introduction of a *flag* which gives to the user the possibility of assigning a specific numerical value to the parameter vector in order to do the calculations required by the algorithm. Instead, the previous version of DAISY a randomly chosen value was automatically assigned to the parameter vector. This is because, in the structural identifiability, the goal is to know the number of parameter solutions, not their values. The novelty allows to use DAISY in conjunction with *practical* identifiability algorithms and this provides very interesting results [3]:

1. in case of local identifiability (finite number of solutions), given a parameter estimate obtained with a whatever optimization procedure, the numerical values of all the remaining solutions equivalently describing the output function [10];

2. in case of nonidentifiable models, the analytic relations between the correlated parameters; thus by knowing one solution, one can analytically know all the others equivalently describing the output function.

Because of this improvement, a joint use of the two different identifiability methodologies, namely *structural* and *practical* identifiability, which are traditionally regarded as disjoint, has been recently proposed [10]. Practically, these findings can constitute a rational and powerful tool for the biological investigator to disentangle the various causes of non identifiability assessed with sensitivity-based approaches, and to provide reliable results.

3.3   The New Interface

The new version of the identifiability software tool DAISY is freely available at the web site daisy.dei.unipd.it. Implementations of DAISY are now available on most variants not only of Microsoft Windows (as the old one was) but also of Apple Macintosh. This will surely enlarge the DAISY users community (so far, about 600 people have downloaded DAISY).

The novel extension of DAISY is available, after registration, together with a README file where instructions for installation are reported, a User Manual where there are detailed instructions about its usage and a folder MOD with the input and the output files of some examples. Almost all these instructions are also available to the user in the interface without the need of registration. DAISY is written in the symbolic language REDUCE [11] thus the user needs to install REDUCE on its computer. Instructions to do this are given in the DAISY interface.

3.4   A Case Study

To give an idea of what kind of models DAISY can deal with, a recently proposed HIV model [12] is used here as an example:

$$
\begin{cases}
\dot{C} &= k_1 C + k_3 CI - k_2 CXI/(k_8 + XI) - k_5 C\,H \\
\dot{CI} &= (k_1 - k_3)CI + k_2 CXI/(k_8 + XI) \\
\dot{CH} &= (k_1 - k_4)CH + k_5 C\,H - k_2 CHXI/(k_8 + XI) + k_3 CHI \\
\dot{CHI} &= (k_1 - k_3 - k_4)CHI + k_2 CHXI/(k_8 + XI) \\
\dot{H} &= k_6 CH - k_7 H \\
\dot{XI} &= -k_9 XI \\
y_1 &= CH + CHI \\
y_2 &= C + CI \\
y_3 &= H
\end{cases}
\tag{5}
$$

where C, CI, CH, CHI, H, XI are the state variables, y_1, y_2 and y_3 the measured outputs, $\theta = [k_1, k_2, k_3, k_4, k_5, k_6, k_7, k_8, k_9]$ is the unknown parameter vector. To check structural identifiability with DAISY, the dynamical model should be provided in a separate file in a specific format. This file should contain the ODE defining the dynamic system $(1, 2)$ together with their known or unknown initial conditions, an ordered list of input, output and state variables, a list of unknown parameters and, if present, equality constraints among the parameters. In less then 10 s. DAISY shows that the model is globally identifiable in a PC with at least an i5 CPU and a RAM of 4.0 GB (for lack of space the output of DAISY is not reported). This result implies that the parameter estimation problem is well-posed and thus the cost function defined by an optimization algorithm has only one global minimum. Obviously, it is worth noting that if a model is structurally identifiable, it may nevertheless turn out to be practically non-identifiable. In this case the inability to unequivocally estimate model parameters may be caused by a number of distinct reasons, among which: (1) excessive noise in the measurements, (2) poor or very sparse sampling schedules, (3) poorly designed experiments, where measurement locations or inputs are missing or insufficiently informative. However, if the model turns out to be practically non identifiable, only by first checking structural identifiability it is possible to know if the problem lays on a too complex model-experiment structure or on the above reasons related to experimental data.

4 Conclusions

We have described a novel extension of DAISY (Differential Algebra for Identifiability of SYstems), a general software tool allowing biomedical researchers to perform global identifiability analysis for linear and nonlinear dynamic models. In particular, DAISY effectively facilitates the solution to the underappreciated problem of determining if unique parameter estimation from the experimental data is theoretically possible. Although DAISY is a computer-algebra code implementing a differential algebra algorithm, high-level programming languages, mathematical and computer algebra skills are not a prerequisite for using the software. These upgrades make DAISY surely more general and easy to use.

References

1. Saccomani, M.P., Audoly, S., D'Angiò, L.: Parameter identifiability of nonlinear systems: the role of initial conditions. Automatica **39**, 619–632 (2003)
2. Ljung, L., Glad, S.T.: On global identifiability for arbitrary model parameterizations. Automatica **30**(2), 265–276 (1994)
3. Saccomani, M.P., Thomaseth, K.: The union between structural and practical identifiability makes strength in reducing oncological model complexity: a case study. Complexity (2018). Article ID 2380650, 10 p
4. Balsa-Canto, E., Banga, J.R.: AMIGO, a toolbox for advanced model identification in systems biology using global optimization. Bioinformatics **27**(16), 2311–2313 (2011). https://doi.org/10.1093/bioinformatics/btr370
5. Hoops, S., et al.: COPASI: a COmplex PAthway SImulator. Bioinformatics **22**, 3067–74 (2006)
6. Bellu, G., Saccomani, M.P., Audoly, S., D'Angiò, L.: DAISY: a new software tool to test global identifiability of biological and physiological systems. Comput. Method Progr. Biomed. **88**, 52–61 (2007)
7. Saccomani, M.P., Audoly, S., Bellu, G., D'Angiò, L.: Examples of testing global identifiability of biological and biomedical models with the DAISY software. Comput. Biol. Med. **40**(4), 402–407 (2010)
8. Hong, H., Ovchinnikov, A., Pogudin, G., Yao, C.: SIAN: software for structural identifiability of ODE models. Bioinformatics arXiv:1812.10180v1 (in press)
9. Ligon, T.S., et al.: GenSSI 2.0: multi-experiment structural identifiability analysis of SBML models. Bioinformatics **34**(8), 1421–1423 (2017). 10.1093/bioinformatics/btx735
10. Thomaseth, K., Saccomani, M.P.: Local identifiability analysis of nonlinear ODE models: how to determine all candidate solutions. IFAC PapersOnLine **51–2**, 529–534 (2018)
11. REDUCE Computer Algebra System at SourceForge. http://reduce-algebra.sourceforge.net. Accessed 28 Sep 2015
12. Browne, E.P., Letham, B., Rudin, C.A.: Computational model of inhibition of HIV-1 by interferon-alpha. PLoS ONE **11**(3), e0152316 (2016). https://doi.org/10.1371/journal.pone.0152316

Extended Abstracts (Posters and Highlight Talks)

Semi-quantitative Abstraction and Analysis of Chemical Reaction Networks (Extended Abstract)

Milan Češka[1]($\boxtimes$) and Jan Křetínský[2]

[1] Brno University of Technology, FIT, IT4I Centre of Excellence,
Brno, Czech Republic
ceskam@fit.vutbr.cz
[2] Technical University of Munich, Munich, Germany

Introduction. Chemical Reaction Networks (CRNs) are a versatile language widely used for *modelling and analysis* of biochemical systems [4] as well as for high-level *programming* of molecular devices [1,14]. Motivated by numerous potential applications ranging from system biology to synthetic biology, various techniques allowing simulation and formal analysis of CRNs have been proposed [2,7,10,13], and embodied in the design process of biochemical systems [6,11,12]. The time-evolution of CRNs is governed by the Chemical Master Equation (CME), which describes the probability of the molecular counts of each chemical species. Many important biochemical systems lead to complex dynamics that includes *state space explosion, stochasticity, stiffness, and multimodality* of the population distributions [9,15], and that fundamentally limits the class of systems the existing techniques can effectively handle. More importantly, biologist and engineers often seek for plausible explanations why the system under study has or has not the required behaviour. In many cases, a set of system simulations/trajectories or population distributions are not sufficient and the ability to provide an accurate explanation for the temporal or steady-state behaviour is another major challenge for the existing techniques.

In order to cope with the computational complexity of the analysis and in order to obtain explanations of the behaviour, we shift the focus from quantitatively precise results to a more qualitative analysis, closer to how a human would behold the system. Yet we insist on providing at least rough timing information on the behaviour as well as rough classification of probability of different behaviours at the extent of "very likely", "few percent", "barely possible", so that we can conclude on issues such as time to extinction or bimodality of behaviour. This gives rise to our *semi-quantitative* approach. We stipulate that analyses in this framework reflect quantities in orders of magnitude, both for time duration and probabilities, but not more than that. This paradigm shift is

---

This work has been accepted to the 31st International Conference on Computer-Aided Verification (CAV'19). The full version of the paper is available at [3]. The work has been supported by the Czech Science Foundation grant No. GA19-24397S, the IT4Innovations excellence in science project No. LQ1602, and the German Research Foundation (DFG) project KR 4890/2-1 "Statistical Unbounded Verification".

© Springer Nature Switzerland AG 2019
L. Bortolussi and G. Sanguinetti (Eds.): CMSB 2019, LNBI 11773, pp. 337–341, 2019.
https://doi.org/10.1007/978-3-030-31304-3_22

reflected on two levels: (1) We abstract systems into semi-quantitative models. (2) We analyse systems in a semi-quantitative way. While each of the two can be combined with a traditional abstraction/analysis, when combined together they provide powerful means to understand systems' behaviour with virtually no computational cost.

Semi-quantitative Models. The states of the models contain information on the current amount of objects of each species as an interval spanning often several orders of magnitude, unless instructed otherwise. For instance, if an amount of a certain species is to be closely monitored (as a part of the input specification/property of the system) then this abstraction can be finer. Similarly, whenever the analysis of a previous version of the abstraction points to the lack of precision in certain states, preventing us to conclude which of the possible behaviours is prevalent, the corresponding refinement can take place. Further, the rates of the transitions are also captured only with such imprecision. The crucial point allowing for existence of such models that are small, yet faithful, is our concept of *acceleration*. It captures certain *sequences* of transitions. It eliminates most of the non-determinism that paralyses other types of abstractions, which are too over-approximative, unable to conclude anything, but safety properties.

Semi-quantitative Analysis. Instead of performing exact transient or steady-state analysis, we can consider most probable transitions and then carefully lift this to most probable temporal behaviours. Technically, this is done by *alternating between transient and steady-state analysis* where only some rates and transitions are taken into account at different iterations. In order to further facilitate the resulting insight of the human on the result of the analysis, we provide an algorithm to perform this analysis with virtually no computation effort and thus possibly manually. The trivial computations immediately pinpoint why certain behaviours occur. Moreover, less likely behaviours can also be identified easily, to any desired degree of probability (dozens of percent, promilles etc.).

Summary. The first step of our approach yields tiny models, allowing for a synoptic observation of the model; due to their size these models can be either analysed easily using standard means, or can be subject to the second step. The second step provides an efficient approximative analysis, which is also very illustrative due to the limited use of quantities. It can be applied to any system; however, it is particularly interesting in connection with the models coming from the first step since (i) no extra effort (size, computation) is wasted on overly precise treatment that is ignored by the other step, and (ii) together they yield an understandable explanation of the behaviour. An entertaining feature of this paradigm is that the stiffer (with rates at hugely different time scales) the system is the easier it is to analyse.

To demonstrate the capabilities of our approach, we consider three challenging and biologically relevant case studies that have been used in literature to evaluate state-of-the-art methods for the CRN analysis. It has been shown that many approaches fail, either due to time-outs or incapability to capture

Table 1. Gene expression. The rates are in h^{-1}.

$D_{off} \xrightarrow{0.05} D_{on}$	$D_{on} \xrightarrow{0.05} D_{off}$	$D_{on} \xrightarrow{10} D_{on} + RNA$	$RNA \xrightarrow{1} \emptyset$
$RNA \xrightarrow{4} RNA + P$	$P \xrightarrow{1} \emptyset$	$P + D_{off} \xrightarrow{0.0015} P + D_{on}$	

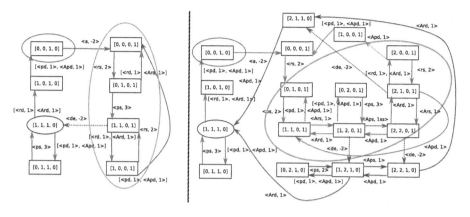

Fig. 1. Pruned abstraction for the gene expression model using the coarse discretisation (left) and after the refinement (right). The state vector is $[P, RNA, D_{off}, D_{on}]$. (Color figure online)

differences in behaviours, and some tailored ones require considerable computational effort, e.g. an hour of computation. Our experiments clearly show that the proposed approach can deliver results that yield qualitatively same information, more understanding and can be computed in minutes by hand (or within a fraction of a second by computer).

Demonstration: Analysis of Stochastic Gene Expression Model [8]. The CRN underlying the stochastic gene expression model is described in Table 1. As discussed in [5,10], the system oscillates between two phases characterised by the D_{on} state and the D_{off} state, respectively. Biologists are interested in how the distribution of the D_{on} and D_{off} states is aligned with the distribution of RNA and protein P.

In order to demonstrate the refinement step and its effect on the accuracy of the model, we start with a very coarse abstraction. It distinguishes only the zero population and the non-zero populations. The pruned abstract model obtained using our approach is depicted in Fig. 1 (left). The full one before pruning is shown in Fig. 6 [3, Appendix].

The proposed analysis of the model identifies the key trends in the system dynamic. The red transitions, representing iterations 1–3 of the semi-quantitative analysis, capture the most probable paths in the system. The green component includes states with DNA on where the system oscillates. The component is reached via the blue state with D_{off} and no RNAs/P. The blue state is promptly reached from the initial state and then the system waits (roughly 100 h according our rate abstraction) for the next DNA activation. The component is left via a deactivation in the iteration 4 (the blue dotted transition). The estimation of the exit time is 100 h. The deactivation is then followed by fast red transitions leading to the blue state, where the system waits for the next activation. We obtain an oscillation between the blue state and the green component, representing the expected oscillation between the D_{on} and D_{off} states.

As expected, this abstraction does not clearly predict the bimodal distribution on the RNA/P populations – the green component includes states with both the zero and the non-zero population of the mRNA and the protein. In order to obtain a more accurate analysis of the system, we refine the population discretisation using a single level threshold for P and DNA, that is equal to 100 and 10, respectively (the rates in the CRN indicate that the population of P reaches higher values).

Figure 1 (right) depicts the pruned abstract model with the new discretisation (the full model is depicted in Fig. 7 [3, Appendix]. We again obtain the oscillation between the green component representing DNA_{on} states and the blue DNA_{off} state. The states in the green component more accurately predicts that in the DNA_{on} states the populations of RNA and P are high and drop to zero only for short time periods. The figure also shows orange transitions within the iteration 2 that extend the green component by two states. Note that the system promptly returns from these states back to the green component. After the deactivation in the iteration 4, the system takes (within the same iteration) the fast transitions (solid blue) leading to the blue component where system waits for another activation and where the mRNA/protein populations decrease. The expected time spent in states on blue solid transitions is small and thus we can reliably predict the bimodal distribution of the mRNA/P populations and its correlation with the DNA state. The refined abstraction also reveals that the switching time from the DNA_{on} mode to the DNA_{off} mode is lower. These predications are in accordance with the results obtained in [10]. See in Fig. 8 [3, Appendix] that is adopted from [10] and illustrates these results.

To conclude this case study, we observe a very aligned agreement between the results obtained using our approach with virtually no computational cost and results in [10] obtained via advanced and time consuming numerical methods.

References

1. Cardelli, L.: Two-domain DNA strand displacement. Math. Struct. Comput. Sci. **23**(02), 247–271 (2013)
2. Cardelli, L., Kwiatkowska, M., Laurenti, L.: A Stochastic hybrid approximation for chemical kinetics based on the linear noise approximation. In: Bartocci, E., Lio, P., Paoletti, N. (eds.) CMSB 2016. LNCS, vol. 9859, pp. 147–167. Springer, Cham (2016). https://doi.org/10.1007/978-3-319-45177-0_10
3. Češka, M., Křetínský, J.: Semi-quantitative abstraction and analysis of chemical reaction networks. Tech. Rep. abs/1905.09914 (2019)
4. Chellaboina, V., Bhat, S.P., Haddad, W.M., Bernstein, D.S.: Modeling and analysis of mass-action kinetics. IEEE Control Syst. Mag. **29**(4), 60–78 (2009)

5. Gandhi, S.J., Zenklusen, D., Lionnet, T., Singer, R.H.: Transcription of functionally related constitutive genes is not coordinated. Nat. Struct. Mol. Biol. **18**(1), 27 (2011)

6. Giacobbe, M., Guet, C.C., Gupta, A., Henzinger, T.A., Paixão, T., Petrov, T.: Model checking gene regulatory networks. In: Baier, C., Tinelli, C. (eds.) TACAS 2015. LNCS, vol. 9035, pp. 469–483. Springer, Heidelberg (2015). https://doi.org/10.1007/978-3-662-46681-0_47

7. Gillespie, D.T.: Exact stochastic simulation of coupled chemical reactions. J. Phys. Chem. **81**(25), 2340–2361 (1977)

8. Golding, I., Paulsson, J., Zawilski, S.M., Cox, E.C.: Real-time kinetics of gene activity in individual bacteria. Cell **123**(6), 1025–1036 (2005)

9. Goutsias, J.: Quasiequilibrium approximation of fast reaction kinetics in stochastic biochemical systems. J. Chem. Phys. **122**(18), 184102 (2005)

10. Hasenauer, J., Wolf, V., Kazeroonian, A., Theis, F.: Method of conditional moments (MCM) for the chemical master equation. J. Math. Biol. **69**(3), 1–49 (2013)

11. Heath, J., Kwiatkowska, M., Norman, G., Parker, D., Tymchyshyn, O.: Probabilistic model checking of complex biological pathways. Theor. Comput. Sci. **391**(3), 239–257 (2008)

12. Lakin, M.R., Parker, D., Cardelli, L., Kwiatkowska, M., Phillips, A.: Design and analysis of DNA strand displacement devices using probabilistic model checking. J. R. Soc. Interface **9**(72), 1470–1485 (2012)

13. Salis, H., Kaznessis, Y.: Accurate hybrid stochastic simulation of a system of coupled chemical or biochemical reactions. J. Chem. Phys. **122**(5), 054103 (2005)

14. Soloveichik, D., Seelig, G., Winfree, E.: DNA as a universal substrate for chemical kinetics. Proc. Nat. Acad. Sci. U.S.A. **107**(12), 5393–5398 (2010)

15. Van Kampen, N.G.: Stochastic Processes in Physics and Chemistry, vol. 1. Elsevier, Amsterdam (1992)

Bayesian Parameter Estimation for Stochastic Reaction Networks from Steady-State Observations

Ankit Gupta[1], Mustafa Khammash[1], and Guido Sanguinetti[2]([envelope])

[1] Department of Biosystems Science and Engineering, ETH Zürich,
Zürich, Switzerland
[2] School of Informatics, University of Edinburgh, Edinburgh, Scotland
gsanguin@inf.ed.ac.uk

Stochasticity is a fundamental feature of biology at the single cell level. Quantitative experimental data ranging from microscopy to single-cell transcriptomic is continually expanding our understanding of the role of stochasticity in gene expression and other cellular processes. Computational modelling has played a fundamental role in elucidating the potential function of stochasticity in biological dynamics, creating a fertile field of interaction between the computational and life sciences (see e.g. [7]).

While methods for forward simulation of stochastic processes are well developed, the *inverse problem* of parametrizing a stochastic reaction network from data is still a very active field of research (see e.g. [6] for a recent review). In this abstract, we are focussed on Bayesian methods, due to their ability to perform full uncertainty quantification and treat noise in a principled way. These methods have seen intensive research particularly for the case of time-series data, where powerful techniques based on particle filtering/particle Markov chain Monte Carlo (MCMC) can be naturally deployed. Unfortunately, time series data from single cells is still relatively rare compared to steady state data, which are relatively under-studied in the statistical literature.

Here we build on recent developments both in the statistics and stochastic literature to provide an effective and accurate Bayesian inference algorithm for steady state data. The main ingredients of our approach are the stationary Finite State Projection (sFSP) of [4], and the random truncation approach of [3]. In the following, we introduce the basic concepts, describe the novel algorithm and briefly illustrate its performance on a case study of a toggle-switch network.

We consider a reaction network consisting of species $\mathcal{S}_1, \ldots, \mathcal{S}_M$ involved in K reactions where each reaction $k = 1, \ldots, K$ has the form

$$\sum_{i=1}^{M} \nu_{ik} \mathcal{S}_i \longrightarrow \sum_{i=1}^{M} \nu'_{ik} \mathcal{S}_i. \tag{1}$$

For each reaction k, the vector $\zeta_k = (\nu'_{1k} - \nu_{1k}, \ldots, \nu'_{Mk} - \nu_{Mk}) \in \mathbb{Z}^M$ denotes the *stoichiometry vector* and a pre-selected *propensity* function $\lambda_k : \mathbb{N}_0^M \to [0, \infty)$ specifies the rate of firing of the reaction. We assume that the propensities depend

© Springer Nature Switzerland AG 2019
L. Bortolussi and G. Sanguinetti (Eds.): CMSB 2019, LNBI 11773, pp. 342–346, 2019.
https://doi.org/10.1007/978-3-030-31304-3_23

on a parameter vector θ; in the standard mass-action formulation, these are the reaction rates, but more complex scenarios (e.g. Michaelis-Menten or Hill dynamics) are possible too. In the commonly used *continuous-time Markov chain* (CTMC) model of a reaction network [1], the state at time t is simply the vector of species copy-numbers $X(t) = (X_1(t), \ldots, X_M(t)) \in \mathbb{N}_0^M$ at time t and at state $X(t) = x$, reaction k fires at rate $\lambda_k(x)$ and moves the state to $(x + \zeta_k)$. The probability vector time-evolution is given by the famous *Chemical Master Equation* (CME) which is expressible as

$$\frac{dp}{dt} = Q^T p(t). \tag{2}$$

where $Q = [Q_{ij}]$ is the bi-infinite transition rate matrix.

Even though the CME is a first-order linear system of ODEs, solving it analytically is infeasible as the number of ODEs is infinite. Many different approximation schemes have been constructed to solve the CME [6]; here we focus on the *Finite State Projection* (FSP) [5] algorithm that approximately solves the CME by considering the reaction dynamics on a finite truncated state-space $\mathcal{E}_n \subset \mathcal{E}$ where all the outgoing transitions lead to an absorbing state. It is known that over finite time-periods the absorption probability can be made arbitrarily small by choosing a large enough truncated state-space and this yields an accurate estimation of the CME solution.

For fixed parameters θ, the CME gives the probability of observing a particular state $X(t) = x$ at time t. In this paper, we are interested in solving the *inverse problem*: Given observations of the state of the CTMC $\mathcal{D}$, what is the implied probability distribution over the parameter vector θ?

In the classical Bayesian setting, the posterior distribution for the unknown parameter θ given the data $\mathcal{D}$ is proportional to the product of the likelihood and the prior distribution. If the likelihood $\ell(\theta)$ can be computed then one can use *Markov Chain Monte Carlo* (MCMC) approaches (like the Metropolis-Hastings algorithm) to obtain a sample from the posterior distribution and consequently infer θ. Unfortunately, computing the likelihood $\ell(\theta)$ is equivalent to solving the CME, which is infeasible. Approximation schemes such as the FSP introduce a bias in the estimation of the likelihood. It is unclear how this bias is reflected in the quality of parameter estimates; empirical evidence (not shown) suggest that in certain cases the bias can lead to very large errors in the estimation. Georgoulas et al. [3] devised a scheme that enables this unbiased estimation by randomizing the FSP truncation. To see how this works, suppose $\{\mathcal{E}_n : n = n_0, n_0 + 1, \ldots\}$ is an increasing family of truncated state-spaces that converge to the full state-space $\mathcal{E}$ as $n \to \infty$ (i.e. $\cup_{n=1}^{\infty} \mathcal{E}_n = \mathcal{E}$) and let $p_\theta^{(n)}(t)$ denote the FSP-based approximation of $p_\theta(t)$ with truncated state-space $\mathcal{E}_n$. Moreover for some $a \in (0, 1)$ let η be a geometric random variable with p.m.f.

$$\mathbb{P}(\eta = m) = (1 - a)a^m \qquad \text{for} \qquad m = 0, 1, \ldots. \tag{3}$$

Then it can be shown that

$$p_\theta(t) = \mathbb{E}\left[p_\theta^{(n_0)}(t) + \sum_{j=0}^{\eta} a^{-j} \left(p_\theta^{(j+n_0+1)}(t) - p_\theta^{(j+n_0)}(t) \right) \right] \tag{4}$$

which implies that the random vector inside the r.h.s. expectation is an unbiased estimator for $p_\theta(t)$. Therefore, in the MCMC approach, using this random variable to estimate each likelihood yields unbiased samples from the true posterior distribution.

Neither the FSP nor the Bayesian algorithm of [3] can however be used when the experimental data $\mathcal{D}$ is at steady-state, which is often the case in biological studies. Intuitively, this is because all trajectories of the system will hit the absorbing state in the run up to steady state, so that any finite state projection will collapse at steady state.

Recently the *stationary Finite State Projection (sFSP)* [4] was developed to solve this issue. It modifies the FSP by directing all the outgoing transitions to a designated state-space within the truncated state-space $\mathcal{E}_n$. Under certain stability conditions for the original CTMC, sFSP provides an accurate estimate of the true stationary distribution and since sFSP relies on solving a linear-algebraic system of equations (rather than ODEs) it feasibly applies to a larger class of networks than FSP. In spite of the accuracy of sFSP, a small bias in the estimate is unavoidable due to the finiteness of the truncated state-space.

Unfortunately, the sFSP does not share the monotonic behaviour of the FSP, meaning that increasing the size of the truncated space may decrease the probability mass associated with some states. This means that the random truncation approach of Georgoulas et al. [3] can fail in this steady-state setting because it can lead to negative estimates of the stationary probabilities which are meaningless. In other words, if $\pi_\theta^{(n)}$ denotes the sFSP-based approximation of π_θ with truncated state-space $\mathcal{E}_n$, then the vector $\pi_\theta^{(j+n_0+1)} - \pi_\theta^{(j+n_0)}$ can have negative components.

In order to avoid negative components in the estimated stationary distribution we devise another random truncation approach. This approach is based on the simple observation that for any $a \in (0,1)$

$$\pi_\theta = \lim_{n \to \infty} \frac{\sum_{j=0}^{n} a^{-j} \pi_\theta^{(j+n_0)}}{\sum_{j=0}^{n} a^{-j}}$$

Hence if η is a geometric random variable (3) with parameter a close to 1 then

$$\pi_\theta^{(\eta)} = \frac{\sum_{j=0}^{\eta} a^{-j} \pi_\theta^{(j+n_0)}}{\sum_{j=0}^{\eta} a^{-j}}$$

serves as an accurate estimator for the true stationary distribution π_θ. Using $\pi_\theta^{(\eta)}$ we can estimate the likelihood and obtain samples from the posterior via a MCMC procedure to infer the unknown parameter θ. A theoretical analysis of

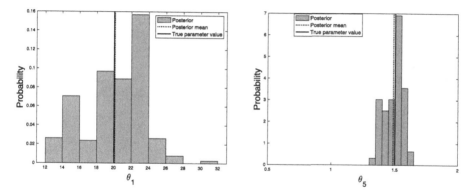

Fig. 1. Inference results for the toggle-switch model [2] using the sFSP-based random truncation approach.

the bias and variance of this estimator is deferred to another venue. Our empirical results indicate that this sFSP-based random truncation approach works better than an analogous approach where the truncation size is fixed a priori.

We now consider the non-linear toggle switch network proposed by Gardner [2]. It has two species S_1 and S_2 with four simple reactions

$$ \emptyset \xrightarrow{\lambda_1} S_1, \quad S_1 \xrightarrow{\lambda_2} \emptyset, \quad \emptyset \xrightarrow{\lambda_3} S_2 \text{ and } S_2 \xrightarrow{\lambda_4} \emptyset, $$

where the propensity functions λ_i-s follow mass action kinetics. We set the parameters as $\theta_1 = 20, \theta_2 = 3.5, \theta_3 = 50, \theta_4 = 2.7, \theta_5 = 1.5$ and $\theta_6 = 1$. We obtain stationary S_2 data $\mathcal{D}$ of size 1000 from SSA trajectories. The measurement noise variance is $\sigma_m^2 = 0.25$. We infer both θ_1 and θ_5 from this noisy data $\mathcal{D}$ using our approach and the results are shown in Fig. 1.

The results on this simple example network indicate a promising level of accuracy and a reasonable computational complexity. A more thorough exploration of the practical usability of the algorithm is a priority for future work.

References

1. Anderson, D., Kurtz, T.: Continuous time Markov chain models for chemical reaction networks. In: Koeppl, H., Setti, G., di Bernardo, M., Densmore, D. (eds.) Design and Analysis of Biomolecular Circuits. Springer, New York (2011). https://doi.org/10.1007/978-1-4419-6766-4_1
2. Gardner, T.S., Cantor, C.R., Collins, J.J.: Construction of a genetic toggle switch in Escherichia coli. Nature **403**(6767), 339–342 (2000)
3. Georgoulas, A., Hillston, J., Sanguinetti, G.: Unbiased Bayesian inference for population Markov jump processes via random truncations. Stat. Comput. **27**(4), 991–1002 (2017)
4. Gupta, A., Mikelson, J., Khammash, M.: A finite state projection algorithm for the stationary solution of the chemical master equation. J. Chem. Phys **147**(15), 154101 (2017)

5. Munsky, B., Khammash, M.: The finite state projection algorithm for the solution of the chemical master equation. J. Chem. Phys. **124**(4), 044104 (2006)
6. Schnoerr, D., Sanguinetti, G., Grima, R.: Approximation and inference methods for stochastic biochemical kinetics–a tutorial review. J. Phys. A Math Theor. **50**(9), 093001 (2017)
7. Székely Jr., T., Burrage, K.: Stochastic simulation in systems biology. Comput. Struct. Biotechnol. J. **12**(20–21), 14–25 (2014)

Wasserstein Distances for Estimating Parameters in Stochastic Reaction Networks

Kaan Öcal[1,2]([⊠]), Ramon Grima[2], and Guido Sanguinetti[1]

[1] School of Informatics, University of Edinburgh, Edinburgh, UK
{kaan.ocal,g.sanguinetti}@ed.ac.uk
[2] School of Biological Sciences, University of Edinburgh, Edinburgh, UK
ramon.grima@ed.ac.uk

Keywords: Wasserstein distance · Bayesian optimization · Chemical master equation · Parameter estimation

Modern experimental methods such as flow cytometry and fluorescence in-situ hybridization (FISH) allow the measurement of cell-by-cell molecule numbers for RNA, proteins and other substances for large numbers of cells at a time, opening up new possibilities for the quantitative analysis of biological systems. Of particular interest is the study of biological reaction systems describing processes such as gene expression, cellular signalling and metabolism on a molecular level. It is well established that many of these processes are inherently stochastic [1–3] and that deterministic approaches to their study can fail to capture properties essential for our understanding of these systems [4,5]. Despite recent technological and conceptual advances, modelling and inference for stochastic models of reaction networks remains challenging due to additional complexities not present in the deterministic case. The Chemical Master Equation (CME) [6] in particular, while frequently used to model many types of reaction networks, is difficult to solve exactly, and parameter inference in practice often relies on a variety of approximation schemes whose accuracy can vary widely and unpredictably depending on the context [6–8].

Methods for inferring parameters from population snapshot data often rely on continuum approximations to the CME or on computing moments of the resulting distributions [9–12], two approaches with limited applicability in practice. Continuum approximations can break down in the presence of low copy number species such as mRNA (often present at copy numbers of less than 20 in cells [13]), while computing moments usually requires the use of moment closure approximations, validity conditions for which are not well-understood and still an active topic of research [8,14,15]. Direct likelihood-based parameter estimation [16,17] typically requires approximating the CME numerically in order to compute the likelihoods, which is computationally expensive for larger systems.

We propose a new method to estimate parameters with the CME by simulating it at various parameter settings and computing the discrepancy between simulated observations and experimental data in order to construct a probabilistic

© Springer Nature Switzerland AG 2019
L. Bortolussi and G. Sanguinetti (Eds.): CMSB 2019, LNBI 11773, pp. 347–351, 2019.
https://doi.org/10.1007/978-3-030-31304-3_24

model of where the correct parameters are most likely to be found. We introduce the use of Wasserstein distances for quantifying the discrepancy between simulated and experimentally measured distributions and train a Gaussian Process to learn these distances for all parameter values. We subsequently use Bayesian optimization to minimize the Wasserstein distance by iteratively computing the next best parameter set to simulate, keeping the number of simulations that need to be run at a minimum. Our approach is suitable for any simulator-based model of reaction networks for which alternative inference methods may not be readily available, e.g. Brownian Dynamics.

Wasserstein distances [18,19], also called Earth Mover's distances in the literature, are a family of distance metrics between probability distributions based on the amount of probability mass that has to be moved in order to convert one probability distribution into another (Fig. 1b). They provide an intuitive and interpretable quantification of similarity between probability distributions, are well-suited for computations in the context of reaction networks and overcome several limitations of standard measures such as KL divergences and the total variation distance. We perform parameter estimation by minimizing the distance between experimental data and simulator output, called the Wasserstein loss in the sequel, in order to obtain distributional fits which capture both qualitative and quantitative aspects of the input data.

Since the functional dependence of the Wasserstein loss with respect to the parameters is not available in closed form and can only be evaluated pointwise at considerable computational cost by running the Stochastic Simulation Algorithm [20] we train a Gaussian Process to learn the Wasserstein loss for all parameter settings, incorporating uncertainty about regions of parameter space that have not been explored. We then use Bayesian optimization to repeatedly identify parameter settings which are likely to be optimal, compute the Wasserstein loss at the chosen parameters and update the Gaussian Process with the obtained data until the correct parameters have been identified. This approach is able to find parameter settings minimizing the Wasserstein loss efficiently in terms of the number of simulations needed and, unlike classical optimization methods, is robust to the observation noise typically present in sampling-based computations.

We tested our method on the classical three-stage gene expression model studied in [21], given by the following reactions:

$$
\begin{array}{lll}
G \xrightarrow{\rho_m} G + M & M \xrightarrow{\rho_p} M + P & M \xrightarrow{\delta_m} \emptyset \\
G \underset{\sigma_a}{\overset{\sigma_d}{\rightleftharpoons}} G^* & & P \xrightarrow{1} \emptyset
\end{array}
\tag{1}
$$

Here G and G^* denote a gene in its activated and inactivated state, M the corresponding mRNA and P the translated protein. Based on simulated observations of joint mRNA and protein counts in the steady state we infer the five parameters of this model using our approach. We are able to obtain parameter estimates closely reproducing the input observations as displayed in Fig. 1a. We found that our method performs well for a range of models, including a genetic feedback loop

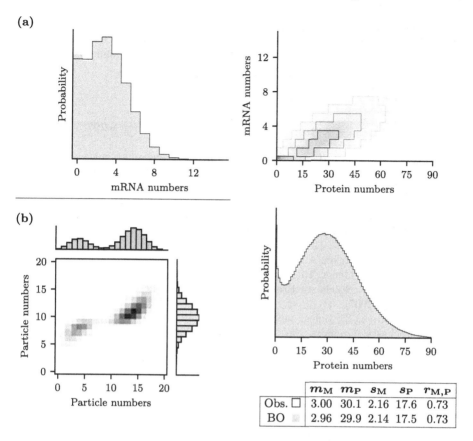

Fig. 1. (a) Inference for the three-stage gene expression model, showing the joint mRNA/protein distributions for observed data (contours) and the parameters estimated using Bayesian optimization (shaded). The table shows the means (m) and standard deviations (s) of mRNA and protein numbers as well as their Pearson correlation coefficient (r). (b) Visualization of a transport plan between two one-dimensional distributions. Wasserstein distances measure how much probability mass needs to be moved to convert one distribution into the other by finding an optimal transport plan between the two.

inspired from [22] to which classical moment-based inference methods cannot be applied.

Bayesian optimization has been successfully applied for identifying parameters in cosmology [23], genomic prediction [24] and in the context of maximum likelihood estimation for general Markov processes [25]. Given the large number of computational models in biology we hope that our adaptation of Bayesian optimization to the context of biochemical reaction networks will provide a stepping stone for other inference problems in the field and allow scientists to use expensive simulation-based models more effectively in the future.

Acknowledgments. This work was supported in part by the EPSRC Centre for Doctoral Training in Data Science, funded by the UK Engineering and Physical Sciences Research Council (grant EP/L016427/1) and the University of Edinburgh.

References

1. Elowitz, M.B.: Stochastic gene expression in a single cell. Science **297**(5584), 1183–1186 (2002)
2. Choi, P.J., Cai, L., Frieda, K., Xie, X.S.: A stochastic single-molecule event triggers phenotype switching of a bacterial cell. Science **322**(5900), 442–446 (2008)
3. Kiviet, D.J., Nghe, P., Walker, N., Boulineau, S., Sunderlikova, V., Tans, S.J.: Stochasticity of metabolism and growth at the single-cell level. Nature **514**(7522), 376–379 (2014)
4. Morton-Firth, C.J., Bray, D.: Predicting temporal fluctuations in an intracellular signalling pathway. J. Theor. Biol. **192**(1), 117–128 (1998)
5. McAdams, H.H., Arkin, A.: It's a noisy business! Genetic regulation at the nanomolar scale. Trends Genet. **15**(2), 65–69 (1999)
6. van Kampen, N.: Stochastic Processes in Physics and Chemistry, 3rd edn. Elsevier, Amsterdam (2007)
7. Cao, Z., Grima, R.: Linear mapping approximation of gene regulatory networks with stochastic dynamics. Nat. Commun. **9**(1), 3305 (2018)
8. Schnoerr, D., Sanguinetti, G., Grima, R.: Comparison of different moment-closure approximations for stochastic chemical kinetics. J. Chem. Phys. **143**(18), 185101 (2015)
9. Zechner, C., et al.: Moment-based inference predicts bimodality in transient gene expression. Proc. Nat. Acad. Sci. **109**(21), 8340–8345 (2012)
10. Ruess, J., Lygeros, J.: Moment-based methods for parameter inference and experiment design for stochastic biochemical reaction networks. ACM Trans. Model. Comput. Simul. **25**(2), 8:1–8:25 (2015)
11. Fröhlich, F., Thomas, P., Kazeroonian, A., Theis, F.J., Grima, R., Hasenauer, J.: Inference for stochastic chemical kinetics using moment equations and system size expansion. PLOS Comput. Biol. **12**(7), e1005030 (2016)
12. Cinquemani, E.: Identifiability and reconstruction of biochemical reaction networks from population snapshot data. Processes **6**(9), 136 (2018)
13. Marguerat, S., Schmidt, A., Codlin, S., Chen, W., Aebersold, R., Bähler, J.: Quantitative analysis of fission yeast transcriptomes and proteomes in proliferating and quiescent cells. Cell **151**(3), 671–683 (2012)
14. Schnoerr, D., Sanguinetti, G., Grima, R.: Validity conditions for moment closure approximations in stochastic chemical kinetics. J. Chem. Phys. **141**(8), 084103 (2014)
15. Schilling, C., Bogomolov, S., Henzinger, T.A., Podelski, A., Ruess, J.: Adaptive moment closure for parameter inference of biochemical reaction networks. Biosystems **149**, 15–25 (2016)
16. Neuert, G., Munsky, B., Tan, R.Z., Teytelman, L., Khammash, M., Oudenaarden, A.V.: Systematic identification of signal-activated stochastic gene regulation. Science **339**(6119), 584–587 (2013)
17. Munsky, B., Li, G., Fox, Z.R., Shepherd, D.P., Neuert, G.: Distribution shapes govern the discovery of predictive models for gene regulation. Proc. Nat. Acad. Sci. **115**(29), 7533–7538 (2018)

18. Villani, C.: Optimal Transport: Old and New. Grundlehren der mathematischen Wissenschaften. Springer, Berlin (2009). https://doi.org/10.1007/978-3-540-71050-9

19. Peyré, G., Cuturi, M.: Computational optimal transport. Found. Trends Mach. Learn. **11**(5–6), 355–607 (2019)

20. Gillespie, D.T.: A general method for numerically simulating the stochastic time evolution of coupled chemical reactions. J. Comput. Phys. **22**(4), 403–434 (1976)

21. Shahrezaei, V., Swain, P.S.: Analytical distributions for stochastic gene expression. Proc. Nat. Acad. Sci. **105**(45), 17256–17261 (2008)

22. Cao, Z., Grima, R.: Accuracy of parameter estimation for auto-regulatory transcriptional feedback loops from noisy data. J. Roy. Soc. Interface **16**(153), 20180967 (2019)

23. Leclercq, F.: Bayesian optimisation for likelihood-free cosmological inference. Phys. Rev. D **98**(6), 063511 (2018)

24. Tanaka, R., Iwata, H.: Bayesian optimization for genomic selection: a method for discovering the best genotype among a large number of candidates. Theor. Appl. Genet. **131**(1), 93–105 (2018)

25. Bortolussi, L., Sanguinetti, G.: Learning and designing stochastic processes from logical constraints. In: Joshi, K., Siegle, M., Stoelinga, M., D'Argenio, P.R. (eds.) QEST 2013. LNCS, vol. 8054, pp. 89–105. Springer, Heidelberg (2013). https://doi.org/10.1007/978-3-642-40196-1_7

On Inferring Reactions from Data Time Series by a Statistical Learning Greedy Heuristics

Julien Martinelli[1,2], Jeremy Grignard[1,3], Sylvain Soliman[1], and François Fages[1(✉)]

[1] Inria Saclay-Île de France, Palaiseau, France
francois.fages@inria.fr
[2] INSERM U935, Villejuif, France
[3] Institut de Recherches Servier, Croissy sur Seine, France

Abstract. With the automation of biological experiments and the increase of quality of single cell data that can now be obtained by phosphoproteomic and time lapse videomicroscopy, automating the building of mechanistic models from these data time series becomes conceivable and a necessity for many new applications. While learning numerical parameters to fit a given model structure to observed data is now a quite well understood subject, learning the structure of the model is a more challenging problem that previous attempts failed to solve without relying quite heavily on prior knowledge about that structure. In this paper, we consider mechanistic models based on chemical reaction networks (CRN) with their continuous dynamics based on ordinary differential equations, and finite time series about the time evolution of concentration of molecular species for a given time horizon and a finite set of perturbed initial conditions. We present a greedy heuristics unsupervised statistical learning algorithm to infer reactions with a time complexity for inferring one reaction in $\mathcal{O}(t.n^2)$ where n is the number of species and t the number of observed transitions in the traces. We evaluate this algorithm both on simulated data from hidden CRNs, and on real videomicroscopy single cell data about the circadian clock and cell cycle progression of NIH3T3 embryonic fibroblasts. In all cases, our algorithm is able to infer meaningful reactions, though generally not a complete set for instance in presence of multiple time scales or highly variable traces.

1 Introduction

Recent breakthroughs in Machine Learning are paving the way for new kinds of algorithms for analysing data and making diagnosis and predictions in biology and medicine. While capable of making accurate predictions, the direct application of machine learning methods do not provide however a biological understanding of the underlying processes nor explanation for the predictions, and may be not accepted in the biomedical domain. For these reasons, a lot of work aims at providing explanations for the predictions made as output of neural networks or other machine learning algorithms trained on data.

© Springer Nature Switzerland AG 2019
L. Bortolussi and G. Sanguinetti (Eds.): CMSB 2019, LNBI 11773, pp. 352–355, 2019.
https://doi.org/10.1007/978-3-030-31304-3_25

Another approach is to try to learn mechanistic models that will make predictions instead of learning directly the predictions from the data. Building mechanistic models of cell processes is however a hard work which necessitates to determine the biochemical mechanisms that are responsible for the high level functions of the cell and its behaviors in normal and perturbed conditions. Automating this process would enable new applications such as automated experiment design or patient-tailored therapeutics.

The main difficulty is to be able to discriminate between causality and correlations in data time series [4]. Most of the work on network inference concerns either undirected interaction graph models, or influence models such as gene regulatory network but then requiring prior knowledge on the structure of the network. The DREAM challenges are used to measure progress in this field.

Less work is devoted to the learning of reaction models and chemical reaction networks (CRNs). In [1], this problem is defined as the minimization of a fitness criterion based on the compatibility of the learned mechanistic model with the observed traces. An evolutionary algorithm is proposed via a two-step iterative procedure: first a set of reactions is inferred, then mass action law kinetic parameters are estimated.

In this paper, we present a greedy heuristics with low computational complexity for inferring reactions from data time series. This unsupervised statistical learning algorithm does not require prior knowledge nor training. We consider at most binary reactions with mass-action, Michaelian or order 4 Hill kinetics. Based on a pairing of the variations of molecular species in each observed transition, the algorithm repeatedly infers the reaction that minimizes the standard deviation of the inferred rate function among all the observed transitions where the reaction can occur. Once inferred, the contributions of that reaction to state change in the set of observed transitions are subtracted before inferring the next reaction. Figure 1 shows the flowchart and low complexity of this algorithm [3].

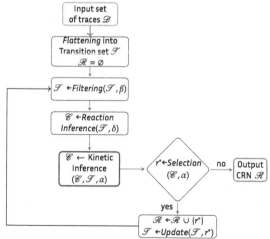

Proposition 1. *The time complexity for inferring one reaction is $\mathcal{O}(t.n^2)$ where t is the number of observed transitions in the traces and n is the number of variables.*

Fig. 1. Flowchart of our CRN learning algorithm and complexity.

2   Evaluation on Simulation Traces

In the context of evaluating the learning algorithm on simulation traces, the hidden CRN used to generate the traces can be used to compare the learned CRN in terms of correct reactions (true positives), wrong reactions (false positives) and missing reactions (false negatives). On a simple chain of 4 reactions with mass action law kinetics over 5 molecular species, our algorithm is able to reconstruct the CRN from a single simulation trace (Fig. 2) with a low sensitivity to statistical learning thresholds.

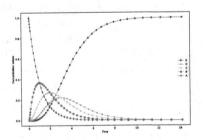

Hidden CRN	Learned CRN
$A \overset{1}{\Rightarrow} B$	$A \overset{1.07}{\Rightarrow} B$
$B \overset{1}{\Rightarrow} C$	$B \overset{1.09}{\Rightarrow} C$
$C \overset{1}{\Rightarrow} D$	$C \overset{1.04}{\Rightarrow} D$
$D \overset{1}{\Rightarrow} E$	$D \overset{0.99}{\Rightarrow} E$

Fig. 2. Chain example: simulation trace and learned CRN.

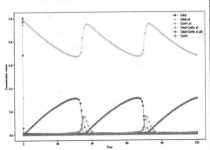

Hidden CRN	Learned CRN
$\emptyset \overset{0.015}{\Rightarrow} cy$	$\emptyset \overset{0.66}{\Rightarrow} cy1 + cdcy2$
$cy + cd1 \overset{200}{\Rightarrow} cdcy2$	$\emptyset \overset{0.01}{\Rightarrow} cdcy2$
$cdcy2 \overset{0.018}{\Rightarrow} cdcy1$	$cdcy2 \overset{0.1152}{\Rightarrow} cdcy1$
$cdcy2 + 2 * cdcy1$	$cdcy2 \overset{0.05}{\Rightarrow} cy1$
$\overset{180}{\Rightarrow} 3 * cdcy1$	
$cdcy1 \overset{1}{\Rightarrow} cy1 + cd$	$cdcy1 \overset{1.62}{\Rightarrow} \emptyset$
$cy1 \overset{0.6}{\Rightarrow} \emptyset$	$cy1 \overset{0.4}{\Rightarrow} cdcy1$
$cd1 \overset{100}{\Rightarrow} cd$	$cd1 \overset{11259}{\Rightarrow} cd$
$cd \overset{10000}{\Rightarrow} cd1$	$cd \overset{5912}{\Rightarrow} cd1$

Fig. 3. Cell cycle model of Tyson [5] and learned reactions from the canonical trace where cd is present.

On the simulation trace depicted in Fig. 3 of the yeast cell cycle model of Tyson [5], our algorithm infers reactions corresponding to the observable slow dynamics of that system. In particular, the discrepancies concerning the synthesis reaction of the cyclin can be very well explained by the existence of multiple time scales in this model. When it is produced, the Cyclin is indeed immediately complexed with Cdc and phosphorylated by very fast reactions. Therefore the free state of the Cyclin cannot be observed and what is inferred is the synthesis of the fast equilibrium state where the Cyclin is in complex form. On the other hand, the autocatalysis reaction cannot be recovered since our algorithm does not consider stoichiometric coefficients other than $\{-1, 0, 1\}$.

3   Evaluation on Videomicroscopy Data

Figure 4 shows the results on videomicroscopy data obtained over 3 days of NIH3T3 embryonic mice fibroblasts, using FUCCI markers of the cell cycle, and Rev-erb-α marker of the circadian clock, with a total of 91 tracked cell traces and 26000 observed state transitions [2]. It is remarkable that, despite the very high variability of cell behaviors, the stochasticity of the cell cycle, and the noise of measurements, meaningful reactions coupling the cell cycle progression with the circadian clock marker could be inferred from this dataset, in just 5 mn CPU time on a laptop.

On-going work concerns strategies to automatically adapt the threshold parameters of this statistical algorithm to the quality of the trace dataset.

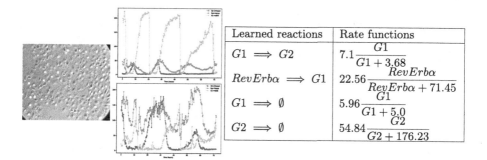

Learned reactions	Rate functions
$G1 \implies G2$	$7.1 \dfrac{G1}{G1 + 3.68}$
$RevErb\alpha \implies G1$	$22.56 \dfrac{RevErb\alpha}{RevErb\alpha + 71.45}$
$G1 \implies \emptyset$	$5.96 \dfrac{G1}{G1 + 5.0}$
$G2 \implies \emptyset$	$54.84 \dfrac{G2}{G2 + 176.23}$

Fig. 4. Inferred reactions on videomicroscopy data of embryonic NIH3T3 fibroblasts [2].

References

1. Choi, K., Hellerstein, J., Wiley, H.S., Sauro, H.M.: Inferring reaction networks using perturbation data. Bio arXiv (2018)
2. Feillet, C., et al.: Phase locking and multiple oscillating attractors for the coupled mammalian clock and cell cycle. Proc. Nat. Acad. Sci. U.S.A. **111**(27), 9828–9833 (2014)
3. Martinelli, J., Grignard, J., Soliman, S., Fages, F.: A statistical unsupervised learning algorithm for inferring reaction networks from time series data. In: ICML Workshop on Computational Biology. Long Beach (June 2019)
4. Pearl, J.: Causality: Models Reasoning and Inference, 2nd edn. Cambridge University Press, New York (2009)
5. Tyson, J.J.: Modeling the cell division cycle: cdc2 and cyclin interactions. Proc. Nat. Acad. Sci. **88**(16), 7328–7332 (1991)

Barbaric Robustness Monitoring Revisited for STL* in Parasim

David Šafránek[✉], Matej Troják, Vojtěch Brůža, Tomáš Vejpustek,
Jan Papoušek, Martin Demko, Samuel Pastva, Aleš Pejznoch, and Luboš Brim

Faculty of Informatics, Masaryk University, Brno, Czech Republic
safranek@fi.muni.cz

Abstract. In our previous work, we have introduced an extension of signal temporal logic called STL* that allows expressing freezing of values referred within temporal operators. The extension is important especially to express several aspects of signals that cannot be expressed in plain STL (e.g., presence of local extremes and their mutual relationships, nontrivial oscillatory behaviour such as damped oscillations, etc.). In this short paper, we address the tool Parasim that includes an implementation of the algorithm for computing robustness with respect to an STL* specification. The tool is in its current version considered as a prototype implementation of the algorithms for STL* robust monitoring of ODE models.

1 Introduction

The problem of evaluating robustness is important to obtain deeper understanding of the role of parameters on the presence of a specified behaviour [12]. A robustness measure provides a tool that significantly helps to compare several models presenting the specified behaviour [2]. There are many examples in systems biology literature that have used robustness analysis to get mechanistic insights into a certain phenomenon [1,11].

There exist two major approaches of defining and analysing robustness in the context of ODE models and temporal logic specifications. The behaviour-oriented approach has been explored in [8] for Metric Temporal Logic (MTL), further extended in [6] for Signal Temporal Logic (STL) and implemented in the tool Breach [5] (standing for "barbaric reachability", technically being based on clever parameter space sampling by utilising sensitivity of systems dynamics wrt parameters). Another way to look at perturbations is from the perspective of utilising LTL (or CTL) with first-order constraints over reals [10] and using the validity domain of a formula as the choice for robustness measure.

In our previous work, we have introduced an extension of signal temporal logic called STL* that allows to express freezing of values referred within temporal [3]. The extension is important especially to express several

This work has been partially supported by the Czech Science Foundation grant No. 18-00178S.

© Springer Nature Switzerland AG 2019
L. Bortolussi and G. Sanguinetti (Eds.): CMSB 2019, LNBI 11773, pp. 356–359, 2019.
https://doi.org/10.1007/978-3-030-31304-3_26

aspects of signals that cannot be expressed in plain STL (e.g., presence of local extremes and their mutual relationships, non-trivial oscillatory behaviour such as damped oscillations, etc.). Based on [8], we have defined quantitative semantics of STL* [4] allowing to compute robustness of STL* formulae for a given ODE model and a bounded parameter perturbation.

In this short paper, we describe the tool Parasim and the related technology. The most interesting feature of the tool that distinghuishes it from the tools mentioned above is definitely the algorithm for computing robustness with respect to an STL* specification.

2 Background

We briefly summarise the notions related with STL* introduced in [3].

Definition 1 (Signal). *Let $n \in \mathbb{N}$ and $T = [0, \tau]$ where $\tau \in \mathbb{R}_{\geq 0}$. Then $\mathfrak{x} : T \to \mathbb{R}^n_{\geq 0}$ is a bounded-time continuous signal and T its time domain.*

Definition 2. *Let $\mathcal{I}$ be a finite set of indices of freezing operators. Syntax of STL* is defined by the following grammar:*

$$\varphi ::= \mathfrak{p} \mid true \mid \neg\varphi \mid \varphi_1 \vee \varphi_2 \mid \varphi_1 \mathbf{U}_I \varphi_2 \mid *_i \varphi$$

where $i \in \mathcal{I}$, true denotes the true constant, $\mathfrak{p}$ is a linear predicate over continuous model variables and $I \subseteq \mathbb{R}_{\geq 0}$ is a closed non-singular interval.

Quantitative semantics of a formula φ interpreted on a signal $\mathfrak{x}$, denoted by $\rho(\varphi, \mathfrak{x}, t, t^*)$, is given for each time point $t \in T$ and a frozen time vector t^*. Further we denote $\rho(\varphi, \mathfrak{x}) = \rho(\varphi, \mathfrak{x}, 0, 0)$ the robustness of the entire signal $\mathfrak{x}$ starting at the point $\mathfrak{x}(0)$ measured with respect to formula φ. The value $\rho(\varphi, \mathfrak{x})$ under-approximates the distance of $\mathfrak{x}$ from the set of signals where φ has different truth value. Note that predicates in STL* are restricted to be linear. This allows solving the robustness of predicates by convex programming (see [4] for details). The algorithm for computing the robustness for a given bounded time signal $\mathfrak{x}(t)$ expects the signal to be sampled into a discrete-time series of values. Numerical simulations have exactly such shape (they can be considered as piece-wise constant signals where the value changes only finitely many times). The algorithm traverses the formula starting from predicates and following the inductive definition of robustness. For every subformula, the signal is processed by analysing all the sampled points.

3 Parasim Description

Parasim is a Java-based open-source tool with graphical user interface for computing robustness of an STL* formula in an ODE model with respect to parameter perturbations. Given an SBML model, a formula and a perturbation set in the form of a hyperrectangle in $\mathbb{R}^n$, Parasim samples the perturbation set into points and for each point simulates the model and computes robustness of the resulting signal with respect to STL* robustness measure defined in [4]. In the neighbourhood of signals with low robustness, additional points are sampled. To reduce analysis execution time, formula optimising algorithms are implemented and incorporated into the robustness monitoring procedure.

Parasim accepts the user input in the form of Extensible Markup Language documents (XML). The input is organised in terms of *projects* stored in separated data folders of the local file system. The project is determined by an SBML model, a *property* file containing STL* formulae, a *simulation configuration* file including a list of settings of the ODE solver, and a *perturbation configuration* file including a list of settings of perturbations – each perturbation is set by a particular region of initial conditions and values of perturbed parameters. The experiment session is then determined by selecting a model, a particular property, a simulation configuration, and a perturbation set. The Parasim GUI is depicted in Fig. 1.

After invoking the analysis procedure, a sample of simulations is computed depending on the particular iteration limit specifying the number of refinements of the perturbation space sampling. Every simulation is associated with a value of the computed robustness for the particular property. Visualisation of the results is provided in terms of interactive 2D projections showing the points in the perturbation space (green/red points show areas where the property is satisfied/violated, the corresponding positive/negative robustness degree is shown by the colour shade).

One of the advantages of Parasim is its modular architecture which enables its efficient extension. The recent version employs Octave [7] and the Simulation Core library [9].

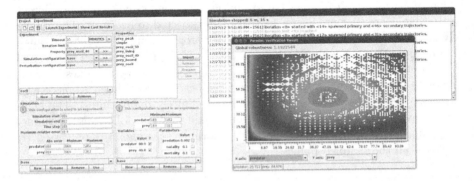

Fig. 1. (left) Project specification in Parasim GUI. (right) Results of robustness monitoring visualised on a two-dimensional projection of the sampled parameter space. (Color figure online)

4 Conclusions and Future Work

In the proposed poster presentation, we would like to address the main aspects of the tool Parasim as overviewed in this abstract. Detailed information on Parasim and its user documentation are freely available on the Parasim website[1]. The tool is in its current version considered as a prototype implementation of the algorithms for STL* robust monitoring of ODE models. The tool has not been published before in terms of a dedicated paper or a poster presentation. For future work, we plan to publish online version of the tool that will include parallel implementation of the algorithms running in multicore environments.

References

1. Barkai, N., Leibler, S.: Robustness in simple biochemical networks. Nature **6636**(387), 913–917 (1997)
2. Bates, D.G., Cosentino, C.: Validation and invalidation of systems biology models using robustness analysis. IET Syst. Biol. **5**(4), 229–244 (2011)
3. Brim, L., Dluhoš, P., Šafránek, D., Vejpustek, T.: STL*: extending signal temporal logic with signal-value freezing operator. Inf. Comput. **236**, 52–67 (2014)
4. Brim, L., Vejpustek, T., Šafránek, D., Fabriková, J.: Robustness analysis for value-freezing signal temporal logic. arXiv:1309.0867 (2013)
5. Donzé, A.: Breach, a toolbox for verification and parameter synthesis of hybrid systems. In: Touili, T., Cook, B., Jackson, P. (eds.) CAV 2010. LNCS, vol. 6174, pp. 167–170. Springer, Heidelberg (2010). https://doi.org/10.1007/978-3-642-14295-6_17
6. Donzé, A., Maler, O.: Robust satisfaction of temporal logic over real-valued signals. In: Chatterjee, K., Henzinger, T.A. (eds.) FORMATS 2010. LNCS, vol. 6246, pp. 92–106. Springer, Heidelberg (2010). https://doi.org/10.1007/978-3-642-15297-9_9
7. Eaton, J.W., Bateman, D., Hauberg, S., Wehbring, R.: GNU Octave Version 3.8.1 Manual: A High-level Interactive Language for Numerical Computations. CreateSpace Independent Publishing Platform, Scotts Valley (2014)
8. Fainekos, G., Pappas, G.J.: Robustness of temporal logic specifications for continuous-time signals. Theoret. Comput. Sci. **410**(42), 4262–4291 (2009)
9. Keller, R., et al.: The systems biology simulation core algorithm. BMC Syst. Biol. **7**(1), 55 (2013)
10. Rizk, A., Batt, G., Fages, F., Soliman, S.: A general computational method for robustness analysis with applications to synthetic gene networks. Bioinformatics **25**, 169–178 (2009)
11. Steuer, R., Waldherr, S., Sourjik, V., Kollmann, M.: Robust signal processing in living cells. PLoS Comput. Biol. **7**(11), e1002218 (2011)
12. Streif, S., et al.: Robustness analysis, prediction and estimation for uncertain biochemical networks. IFAC Proc. Volumes **46**(32), 1–20 (2013)

---

[1] https://github.com/sybila/parasim/.

Symmetry Breaking for GATA-1/PU.1 Model

Lenka Přibylová[(✉)] and Barbora Losová

Department of Mathematics and Statistics, Faculty of Science, Masaryk University,
Kotlářská 2, 611 37 Brno, Czech Republic
pribylova@math.muni.cz

Abstract. This paper explains a substantial feature of symmetry breaking of dynamical systems that include bistability from the mathematical point of view to highlight important consequences of this phenomenon to biochemical and system biology studies since symmetry breaking as a bifurcation itself can serve as a source of branching. We take hematopoietic stem cells modeling as a particular case.

Keywords: Bistability · Symmetry breaking · Pitchfork bifurcation · GATA-1/PU.1 model · Biochemical switch

1 Symmetry Breaking Phenomenon

Symmetry breaking is an important, but neglected, phenomenon that occurs in various types of models. As a particular case we take GATA-1/PU.1 biochemical switch model[1] in a symmetric nondimensionalized form proposed by Roeder and Glauche in [6].

Let us consider the following model

$$\frac{\mathrm{d}g}{\mathrm{d}\tau} = -g + \frac{sg^2 + uk_up^2}{1 + g^2 + k_up^2 + (k_r + \epsilon)gp},$$
$$\frac{\mathrm{d}p}{\mathrm{d}\tau} = -p + \frac{sp^2 + uk_ug^2}{1 + p^2 + k_ug^2 + k_rgp}, \tag{1}$$

where s, u, k_u and k_r are nonnegative parameters related to reaction rate constants and g and p are nondimensional state variables describing concentrations of transcription factors GATA-1 and PU.1 and τ describes the nondimensional time. Parameter ϵ in the neighbourhood of zero breaks the symmetry of the

---

[1] There is an increasing number of studies that develop qualitative as well as quantitative mechanistic models of hematopoietic stem cells at different levels of resolution, see e.g. [1,4–6,8] and others.

This work was supported by grant Mathematical and statistical modeling number MUNI/A/1503/2018.

L. Bortolussi and G. Sanguinetti (Eds.): CMSB 2019, LNBI 11773, pp. 360–363, 2019.
https://doi.org/10.1007/978-3-030-31304-3_27

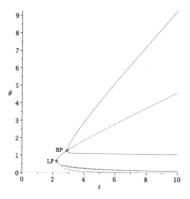

Fig. 1. The manifold of steady states of the model (1), s is a bifurcation parameter, $\epsilon = 0$, $k_r = 0.5$, $k_u = 0.8$, $u = 1$.

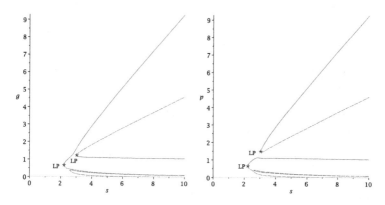

Fig. 2. The manifolds of steady states of the model (1), s is a bifurcation parameter, $\epsilon = -0.001$, $k_r = 0.5$, $k_u = 0.8$, $u = 1$.

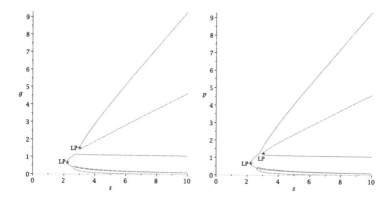

Fig. 3. The manifolds of steady states of the model (1), s is a bifurcation parameter, $\epsilon = 0.001$, $k_r = 0.5$, $k_u = 0.8$, $u = 1$.

model. We computed equilibrium manifold and bifurcation points of (1) analytically using the Gröbner basis method in both symmetric ($\epsilon = 0$) and asymmetric cases. The equilibrium manifolds are implicitly given as

$$
\begin{aligned}
g\,\big(&10000\,\epsilon^3 g^7 + 10000\,\epsilon^2 g^8 - 10000\,\epsilon^2 g^7 s + 11000\,\epsilon^2 g^7 + 8000\,\epsilon^2 g^6 s - 250\,\epsilon\,g^8 \\
&+ 27000\,\epsilon\,g^7 s - 39250\,\epsilon\,g^6 s^2 + 12500\,\epsilon\,g^5 s^3 + 26500\,\epsilon^2 g^6 - 20500\,\epsilon^2 g^5 s \\
&- 11380\,\epsilon\,g^7 + 20000\,\epsilon\,g^6 s + 8000\,\epsilon\,g^5 s^2 + 1495\,g^8 - 9620\,g^7 s + 34375\,g^6 s^2 \\
&- 36250\,g^5 s^3 + 10000\,g^4 s^4 + 18760\,\epsilon\,g^6 + 16000\,\epsilon\,g^5 s - 35000\,\epsilon\,g^4 s^2 + 2470\,g^7 \\
&- 17960\,g^6 s + 300\,g^5 s^2 + 30000\,g^4 s^3 - 10000\,g^3 s^4 + 12500\,\epsilon^2 g^4 - 3480\,\epsilon\,g^5 \\
&- 23000\,\epsilon\,g^4 s + 10000\,\epsilon\,g^3 s^2 + 7591\,g^6 + 5605\,g^5 s + 14400\,g^4 s^2 - 20000\,g^3 s^3 \\
&+ 12450\,\epsilon\,g^4 + 22500\,\epsilon\,g^3 s + 9932\,g^5 - 64900\,g^4 s - 2200\,g^3 s^2 + 20000\,g^2 s^3 \\
&+ 11400\,\epsilon\,g^3 - 10000\,\epsilon\,g^2 s + 25308\,g^4 + 8550\,g^3 s + 10000\,g^2 s^2 + 19904\,g^3 \\
&- 31800\,sg^2 - 10000\,gs^2 + 21300\,g^2 - 8000\,gs + 4000\,g + 8000\big) = 0,
\end{aligned}
$$

$$(2)$$

and

$$
\begin{aligned}
p\,\big(&10000\,\epsilon^2 p^8 - 20000\,\epsilon^2 p^7 s + 10000\,\epsilon^2 p^6 s^2 - 10000\,\epsilon^2 p^7 + 20000\,\epsilon^2 p^6 s \\
&- 10000\,\epsilon^2 p^5 s^2 - 250\,\epsilon\,p^8 - 23500\,\epsilon\,p^7 s + 43750\,\epsilon\,p^6 s^2 - 20000\,\epsilon\,p^5 s^3 \\
&+ 20000\,\epsilon^2 p^6 - 20000\,\epsilon^2 p^5 s + 4800\,\epsilon\,p^7 + 23200\,\epsilon\,p^6 s - 50000\,\epsilon\,p^5 s^2 \\
&+ 20000\,\epsilon\,p^4 s^3 + 1495\,p^8 - 9620\,p^7 s + 34375\,p^6 s^2 - 36250\,p^5 s^3 + 10000\,p^4 s^4 \\
&- 20000\,\epsilon^2 p^5 + 20000\,\epsilon^2 p^4 s - 25100\,\epsilon\,p^6 - 16900\,\epsilon\,p^5 s + 40000\,\epsilon\,p^4 s^2 \\
&+ 2470\,p^7 - 17960\,p^6 s + 300\,p^5 s^2 + 30000\,p^4 s^3 - 10000\,p^3 s^4 + 10000\,\epsilon^2 p^4 \\
&+ 14600\,\epsilon\,p^5 + 42200\,\epsilon\,p^4 s - 40000\,\epsilon\,p^3 s^2 + 7591\,p^6 + 5605\,p^5 s + 14400\,p^4 s^2 \\
&- 20000\,p^3 s^3 - 10000\,\epsilon^2 p^3 - 26850\,\epsilon\,p^4 - 20000\,\epsilon\,p^3 s + 9932\,p^5 - 64900\,p^4 s \\
&- 2200\,p^3 s^2 + 20000\,p^2 s^3 + 7800\,\epsilon\,p^3 + 20000\,\epsilon\,p^2 s + 25308\,p^4 + 8550\,p^3 s \\
&+ 10000\,p^2 s^2 + 19904\,p^3 - 31800\,sp^2 - 10000\,ps^2 + 21300\,p^2 - 8000\,ps \\
&+ 4000\,p + 8000\big) = 0.
\end{aligned}
$$

$$(3)$$

We used software Maple for Gröbner basis computations. For more details about Gröbner basis see [2], usage in applied bifurcation theory can be found in [3].

It can be proved that in the symmetric case the system undergoes a pitchfork bifurcation (as can be seen from Fig. 1, stable branches are depicted with solid lines and unstable branches with dashed lines.). Figures 2 and 3 illustrate the symmetry breaking near the pitchfork bifurcation point. Roeder and Glauche in [6] simulated the asymmetric case numerically, they showed a case with a similar bifurcation diagram with two basins of attraction separated by the unstable branch. We have fully analytical results. The similarity of bifurcation diagrams is a very important result implied by the fact that solutions of (1) depend on parameters continuously, so it is not a coincidence. Near the pitchfork bifurcation point, the steady state branches can merge from fold or transcritical bifurcation points. Symmetry breaking itself serves as a bifurcation phenomenon (see e.g. section 20.3 Stability of bifurcations under perturbations in [9] or in [7]).

2 Conclusion

From the mathematical point of view, the symmetric case is structurally unstable and there is zero probability that the parameters are symmetric in an experimentally measured system. But that doesn't mean that the symmetric model is irrelevant. The mean parameter values of the population of stem cells should be close to this symmetric case and their variance guarantees the differentiation and maturation of both types of the cells on the two branches that merge from the fold or transcritical bifurcation points. On the other hand, a disorder in hematopoietic differentiation could be caused by a shift out of the pitchfork parameters neighborhood. Branching would not be possible too far from the pitchfork bifurcation, so this shift can model a hematopoietic differentiation failure in principal. Moreover, it is in full agreement with typical results of adaptive dynamics that explain evolutionary branching of evolutionary stable equilibria as pitchfork bifurcation phenomena. Symmetry breaking seems to be generic since the parameter of symmetricity is necessarily a bifurcation parameter of higher codimension. For biochemical dynamical systems (or evolving systems generally) it implies that such branching mechanism happens for any generic symmetry breaking of the pitchfork bifurcation and it could be expected in evolutionary and maturing systems, including stem cells dynamics.

References

1. Bokes, P., King, J.R., Loose, M.: A bistable genetic switch which does not require high co-operativity at the promoter: a two-timescale model for the PU.1-GATA-1 interaction. Math. Med. Biol. J. IMA **26**(2), 117–132 (2009)
2. Cox, D., Little, J., OShea, D.: Ideals, Varieties, and Algorithms: An Introduction to Computational Algebraic Geometry and Commutative Algebra. Springer, Cham (2013). https://doi.org/10.1007/978-3-319-16721-3
3. Hajnová, V., Přibylová, L.: Bifurcation manifolds in predator-prey models computed by Gröbner basis method. Math. Biosci. **312**, 1–7 (2019)
4. May, G., et al.: Dynamic analysis of gene expression and genome-wide transcription factor binding during lineage specification of multipotent progenitors. Cell Stem Cell **13**(6), 754–768 (2013)
5. Olariu, V., Peterson, C.: Kinetic models of hematopoietic differentiation. Wiley Interdisc. Rev. Syst. Biol. Med. **11**(1), e1424 (2019)
6. Roeder, I., Glauche, I.: Towards an understanding of lineage specification in hematopoietic stem cells: a mathematical model for the interaction of transcription factors GATA-1 and PU.1. J. Theor. Biol. **241**(4), 852–865 (2006)
7. Seydel, R.: Practical Bifurcation and Stability Analysis, vol. 5. Springer, New York (2009). https://doi.org/10.1007/978-1-4419-1740-9
8. Tian, T., Smith-Miles, K.: Mathematical modeling of GATA-switching for regulating the differentiation of hematopoietic stem cell. BMC Syst. Biol. **8**, S8 (2014). BioMed Central
9. Wiggins, S.: Introduction to Applied Nonlinear Dynamical Systems and Chaos, vol. 2. Springer, New York (2003). https://doi.org/10.1007/b97481

Scalable Control of Asynchronous Boolean Networks

Cui Su[1], Soumya Paul[2], and Jun Pang[1,2(✉)]

[1] Interdisciplinary Centre for Security, Reliability and Trust,
University of Luxembourg, Esch-sur-Alzette, Luxembourg
jun.pang@uni.lu
[2] Faculty of Science, Technology and Communication, University of Luxembourg,
Esch-sur-Alzette, Luxembourg

Abstract. We summarise our recent research results on developing efficient and scalable control methods for gene regulatory networks modelled as asynchronous Boolean networks. Our methods compute a minimal subset of nodes of a given Boolean network, such that (different types of) perturbations of these nodes, in one step or a sequence of steps, can drive the network (from an initial state) to a target steady state.

1   Introduction

The ground-breaking discovery of cell reprogramming has overturned the conventional thinking that cell differentiation was irreversible. With cell reprogramming techniques, it is possible to reprogram cell fates in many different ways, such as trans-differentiation, de-differentiation, retro-differentiation, etc. This has opened up a great opportunity for regenerative medicine to treat the most devastating diseases, such as Parkinson's disease, Alzheimer's disease, etc.

A big obstacle for the application of *in vivo* cell reprogramming is the effective identification of target genes. Numerical-experimental methods are infeasible [13], due to the combinatorial complexity of target genes and high experimental costs. This gives rise to the need of computational modelling of gene regulatory networks (GRNs), which makes it possible to analyse GRNs with formal reasoning and tools. Among various modelling frameworks, Boolean networks (BNs) have distinct advantages: being simple and yet able to capture important properties of nonlinear dynamical GRNs [2]. In BNs, genes are modelled as binary nodes, being either 'expressed' or 'not expressed'. Activation/inhibition regulations between genes are described by Boolean functions. The updating of nodes can be either *synchronous* or *asynchronous*. The steady states of GRNs are described as *attractors*, to one of which the system eventually settles down.

Asynchronous BNs are considered more realistic than synchronous BNs, because only the asynchronous updating allows for the biological processes occurring on different time-scales [7]. Owing to the non-determinism of asynchronous BNs, the control methods for synchronous BNs [3,15] are not applicable to asynchronous BNs. Recently, a stable motif based method was developed to control

L. Bortolussi and G. Sanguinetti (Eds.): CMSB 2019, LNBI 11773, pp. 364–367, 2019.
https://doi.org/10.1007/978-3-030-31304-3_28

asynchronous BNs [14]. However, this method does not guarantee the minimality of the control sets, which may result in unnecessary experimental costs.

The limitations of the existing methods motivate us to develop scalable, efficient and practical methods to control asynchronous BNs.

2 Control Problems in Boolean Networks

Attractors of a BN characterise cell phenotypes, which are biologically observable states [2]. Only the control of attractors is meaningful. Thus, the control objective for GRNs, in the context of asynchronous BNs, can be described as: finding a subset of nodes, called driver nodes, such that the control of these nodes can drive the network (from a source attractor) to a target attractor. If the source state is known, we call it source-target control; otherwise, we call it target control.

The control (perturbation) of a node means to change the expression of the node to either '1' or '0'. Based on the amount of time that the control is applied, we distinguish the following three types of controls:

(a) instantaneous control – the control is applied instantaneously;
(b) temporary control – the control is applied for finite steps and then released;
(c) permanent control – the control is applied for all the following time steps.

Thanks to the rapid advances in cell reprogramming, these three controls can be realised in biological experiments with different bimolecular tools. In particular, for the source-target control, based on the number of control steps, we also have:

(a) one-step control – perturbations are applied simultaneously for one step;
(b) sequential control – perturbations are applied in a sequence of steps.

3 Results

The major challenge in the control of asynchronous BNs lies in the infamous state space explosion problem: the state space grows exponentially with respect to the number of nodes of a BN. It prohibits the *efficiency, scalability* and *minimality* of the control methods. To cope with this problem, we employ the 'divide and conquer' strategy to explore both the structural and dynamical properties of a BN. As shown in Table 1, we have developed efficient methods to solve the minimal one-step source-target control with instantaneous, temporary and permanent perturbations [8,9,11], the minimal sequential source-target control with instantaneous perturbations [4,5], as well as the target control with instantaneous perturbations [1]. Among these methods, sequential source-target control identifies a sequence of intermediate states and the associated perturbations. At each control step, we apply a set of perturbations, wait until the network reaches the intermediate state and then apply another set of perturbations. Based on the status of intermediate states, we have developed a general sequential control [5], where any state (transient states or steady states) can act as intermediate states,

Table 1. Different control strategies for asynchronous BNs.

		Instantaneous control	Temporary control	Permanent control
Source-target control	One-step control	[8,9]	[11]	
	Sequential control	[4,5]		
Target control	One-step control	[1]		

and an attractor-based sequential control [4], where only steady states can play the role of intermediate states.

Our methods are implemented as part of the software ASSA-PBN [6]. We have evaluated our methods on a variety of real-life biological networks.[1] The results show that our methods are quite efficient in terms of the computation time and scale well for large networks. Moreover, our methods identify the minimal control sets for different strategies, which can reduce the experimental costs to a great extent. For the source-target control, both sequential control methods [4,5] require less perturbations than the one-step control [8,9]. The attractor-based sequential control [4] is considered more practical than the general sequential control [5], since it uses biologically observable states as intermediate states and thus only requires partial observability of the system. Considering different types of perturbations, the temporary perturbation is preferable than the instantaneous and permanent perturbations due to: (1) temporary control requires the least number of perturbations, which not only leads to less experimental costs, but also makes the experiments easier to conduct; and (2) temporary perturbations are eventually released and thus less invasive. In this way, we can avoid some unknown side effects compared to permanent perturbations.

Furthermore, in order to make the results more practical, experimental constraints need to be incorporated. We have made the following improvements to integrate practical and experimental concerns.

- Our methods can avoid certain perturbations (perturbing a gene from 'expressed' to 'not expressed' and/or the reverse direction) during the computation. In this case, we can avoid perturbing genes that are essential for cell survival and genes that are hard or expensive to perturb.
- For attractor-based sequential source-target control, our method can avoid undesired attractors as intermediate states, such as apoptosis.
- Considering the minimal solution may not be the best solution, an upper bound of the number of perturbations can be set as a prerequisite. Then our methods can compute all the solutions within that upper bound.

4   Discussion and Future Work

Due to the diversity of GRNs, it is less likely to find one strategy that suits different networks. Furthermore, the cost and success rate of different cell

---

[1] We refer details on their evaluation to our previous works [1,4,5,8,9,11].

reprogramming techniques vary a lot. Taking these into consideration, we can compute a bunch of reprogramming paths with different control methods. Afterwards, biologists can make a choice from the provided solutions based on specific biological systems and experimental settings.

Given the strengths of temporary perturbations, currently we are working on sequential source-target control and target control with temporary perturbations. In future, we plan to extend our work to the control of PBNs [10,12].

References

1. Baudin, A., Paul, S., Su, C., Pang, J.: Controlling large Boolean networks with single-step perturbations. Bioinformatics **35**, i558–i567 (2019)
2. Kauffman, S.A.: Homeostasis and differentiation in random genetic control networks. Nature **224**, 177–178 (1969)
3. Kim, J., Park, S., Cho, K.: Discovery of a kernel for controlling biomolecular regulatory networks. Sci. Rep. **3**, 2223 (2013)
4. Mandon, H., Su, C., Haar, S., Pang, J., Paulevé, L.: Sequential reprogramming of Boolean networks made practical. In: Bortolussi, L., Sanguinetti, G. (eds.) CMSB 2019. LNBI, pp. 3–19. Springer, Cham (2019)
5. Mandon, H., Su, C., Pang, J., Paul, S., Haar, S., Paulevé, L.: Algorithms for the sequential reprogramming of Boolean networks. IEEE/ACM Trans. Comput. Biol. Bioinf. (2019, accepted)
6. Mizera, A., Pang, J., Su, C., Yuan, Q.: ASSA-PBN: a toolbox for probabilistic Boolean networks. IEEE/ACM Trans. Comput. Biol. Bioinf. **15**(4), 1203–1216 (2018)
7. Papin, J.A., Hunter, T., Palsson, B.O., Subramaniam, S.: Reconstruction of cellular signalling networks and analysis of their properties. Nat. Rev. Mol. Cell Biol. **6**(2), 99 (2005)
8. Paul, S., Su, C., Pang, J., Mizera, A.: A decomposition-based approach towards the control of Boolean networks. In: Proceedings of the 9th ACM Conference on Bioinformatics, Computational Biology, and Health Informatics, pp. 11–20. ACM Press (2018)
9. Paul, S., Su, C., Pang, J., Mizera, A.: An efficient approach towards the source-target control of Boolean networks. IEEE/ACM Trans. Comput. Biol. Bioinf. (2019, accepted)
10. Shmulevich, I., Dougherty, E.R.: Probabilistic Boolean Networks: The Modeling and Control of Gene Regulatory Networks. SIAM Press, New York (2010)
11. Su, C., Paul, S., Pang, J.: Controlling large Boolean networks with temporary and permanent perturbations. In: Proceedings of the 23rd International Symposium on Formal Methods. LNCS. Springer (2019, accepted)
12. Trairatphisan, P., Mizera, A., Pang, J., Tantar, A.A., Sauter, T.: optPBN: an optimisation toolbox for probabilistic Boolean networks. PLoS ONE **9**(7), e98001 (2014)
13. Wang, L.Z., et al.: A geometrical approach to control and controllability of nonlinear dynamical networks. Nat. Commun. **7**, 11323 (2016)
14. Zañudo, J.G.T., Albert, R.: Cell fate reprogramming by control of intracellular network dynamics. PLoS Comput. Biol. **11**(4), e1004193 (2015)
15. Zhao, Y., Kim, J., Filippone, M.: Aggregation algorithm towards large-scale Boolean network analysis. IEEE Trans. Autom. Control **58**(8), 1976–1985 (2013)

Transcriptional Response of SK-N-AS Cells to Methamidophos (Extended Abstract)

Akos Vertes[1], Albert-Baskar Arul[1], Peter Avar[1], Andrew R. Korte[1], Lida Parvin[1], Ziad J. Sahab[1], Deborah I. Bunin[2], Merrill Knapp[2], Denise Nishita[2], Andrew Poggio[2], Mark-Oliver Stehr[2], Carolyn L. Talcott[2(✉)], Brian M. Davis[3], Christine A. Morton[3], Christopher J. Sevinsky[3], and Maria I. Zavodszky[3]

[1] Department of Chemistry, George Washington University,
Washington, DC 20052, USA
[2] SRI International, Menlo Park, CA 94025, USA
carolyn.talcott@gmail.com
[3] GE Global Research, Niskayuna, NY 12309, USA

Abstract. Transcriptomics response of SK-N-AS cells to methamidophos (an acetylcholine esterase inhibitor) exposure was measured at 10 time points between 0.5 and 48 h. The data was analyzed using a combination of traditional statistical methods, machine learning techniques, and methods to infer causal relations between time profiles. We identified several processes that appeared to be upregulated in cells treated with methamidophos including: unfolded protein response, response to cAMP, calcium ion response, and cell-cell signaling. The data confirmed the expected consequence of acetylcholine buildup. Transcripts with potentially key roles were identified by anomaly detection using convolutional autoencoders and Generative Adversarial Networks, and causal networks relating these transcripts were inferred using Siamese convolutional networks and time warp causal inference.

1 Introduction

Rapid determination of the mechanism of action (MoA) of an unknown or novel xenobiotic (toxin, drug, pathogen) and its consequences is important both scientifically and for biodefense. Time series data generated by omics experimental techniques provides a wealth of data about change in relative concentrations of transcripts, proteins and metabolites. For example, chemically perturbing cells can result in thousands of mRNAs with at least a 2 fold expression change. The challenge is to get the most information purely from the data, before augmenting the conclusions with knowledge from databases and literature. Thus, it is important to consider not only what changes, but how it changes over time, to identify key responders and how they organize into cellular processes.

Sponsored by the US Army Research Office and the Defense Advanced Research Projects Agency; accomplished under Cooperative Agreement W911NF-14-2-0020.

© Springer Nature Switzerland AG 2019
L. Bortolussi and G. Sanguinetti (Eds.): CMSB 2019, LNBI 11773, pp. 368–372, 2019.
https://doi.org/10.1007/978-3-030-31304-3_29

As part of the DARPA Rapid Threat Assessment project we developed a suite of data analysis methods to identify candidate biological players and processes that make up the cellular response to a challenge. These included traditional statistical analysis, shape/feature analysis, Gaussian process representation, machine learning methods for identifying anomalies and methods to infer causal relations. The methods were developed using data from HepG2/C3A cells exposed to a series of different drugs each affecting different known cellular processes. To test the robustness and generality of the analysis methods we selected a different cell type (SK-N-AS human neuroblastoma cells) and toxin (the organophosphate methamidophos). We expected the biological noise to be different in a different cell type. We also expected the timing and organization of response to an organophosphate to be different from the previously tested drugs. Thus we expected the new experiment would be a step to validating that our algorithms and data analysis scheme work in a more general setting.

2 Data Analysis

Transcriptomic response of SK-N-AS cells to methamidophos (an acetylcholine esterase inhibitor) exposure was measured at 10 time points between 0.5 and 48 h. The data was analyzed using a combination of traditional statistical methods, machine learning techniques, and methods to infer causal relations between time profiles. Figure 1 shows a schematic of our data analysis process. The left branch uses $\log_2$ fold change (basic) time profiles derived from the means of the control and treated signals in the usual way. Up/down charts map transcripts to the first time point the $\log_2$ fold change magnitude passes 1. Regulation intervals delimit times that $\log_2$ fold change stays above 0.75 or below -0.75. The right branch uses time profiles obtained by Gaussian process (GP) modeling [8]. Using these time profiles, transcripts are clustered (k-means using PCA to reduce dimensionality), and ranked by contribution to PCA components and by two machine learning algorithms, one using autoencoder techniques (see [8]) and one using Generative Adversarial Nets (GANs) [2]. Transcripts are ranked highly as anomalies according to how poorly they are reconstructed from the autoencoding, or how unsure the trained GAN discriminator is that the time profile represents a transcript. Transcripts are also given a 'real/typical' ranking according to how confident the GAN discriminator is that the time profile represents a transcript. Two algorithms were used to infer potential 'causal' edges between time profiles. The Timewarp algorithm inputs basic time profiles and uses a variant of the Needleman-Wunsch alignment algorithm [5] to align time profiles. the Siamese twin causality detection algorithm [6] is based on two Siamese neural networks [1]. One Siamese network is trained to detect undirected causality; the other is trained to detect lag. Lag detection is used to direct the undirected causality edges. The results of the analyses, along with an indication of satisfaction of several significance filters, are collected in a 'feature' table that can be sorted to highlight features of interest. We used the significance filters to select a set of top ranking transcripts, Top20X, as the starting point for further identifying MoA candidates in two categories: biological processes, transcripts. To

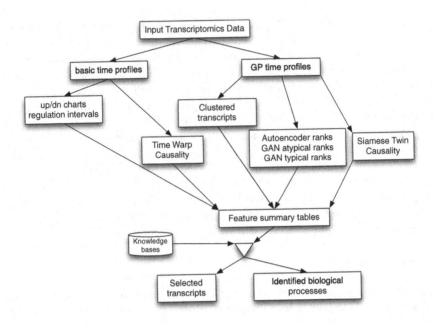

Fig. 1. Data analysis schematic

identify candidate processes we used PatherDB over representation analysis [4] combined with our GO term annotations of k-means clusters [8]. The Top20X transcripts were also annotated with specific GO terms using Uniprot and our database of curated experimental results.

3   Discussion

Our data analysis found four processes as candidate elements of the broader MoA of methamidophos: unfolded protein response (UPR), cAMP response, calcium ion related processes, and cell-cell signaling. As a form of confirmation, we also used data from other experiments and pathway databases to identify transcripts expected to be regulated in these processes. We also built a model of acetylcholine buildup as a consequence of the inhibition of acetylcholinesterase (ACHE). The resulting model suggested three downstream effects: increase in DAG (modeled by PMA response), increase in IP3, and active G-protein-coupled receptor (GPCR). Increase in IP3 induces release of calcium ion consistent with the identified calcium ion related response. The cAMP response could be connected to the acetylcholine signal through the G protein binding partners of ChRm3 or the calcium ion release [9]. G2 arrest, hypothesized based on cyclin time profiles, is a pause in the cell life cycle due to detected problems, such as DNA damage. Nine of the Top20X transcripts are annotated with DNA damage related terms. The strong UPR response suggests another reason for a G2 arrest.

Timewarp causality inference found relations among transcripts related to ER-stress/UPR, with some connections supported by ENCODE. Siamese twin causality inference found relations among transcripts related ER-stress/UPR, DNA damage, and calcium ion response. The causal relation found between DDIT3 (aka CHOP) and GDF15 in the context of ER stress is supported by work reported in Li et al. [3]. We have not yet found evidence supporting or disagreeing with the other hypothesized relations. In both cases, many of the related transcripts were highly ranked as GAN typical/real or GAN anomalies.

4 Conclusion

We applied a novel combination of Gaussian process modeling, anomaly ranking, and causal inference algorithms to analyze transcriptomics data gathered from SK-N-AS cells treated with methamidophos. Our analysis revealed potentially interesting downstream effects of methamidophos (see [7] for details). The anomaly ranking and causal inference approaches seem promising. Much more work remains to fully understand what features they are finding and the relevant biology.[1]

References

1. Bromley, J., Guyon, I., LeCun, Y., Säckinger, E., Shah, R.: Signature verification using a "siamese" time delay neural network. In: Cowan, J.D., Tesauro, G., Alspector, J. (eds.) Advances in Neural Information Processing Systems, vol. 6, pp. 737–744. Morgan-Kaufmann (1994)
2. Goodfellow, I.J., et al.: Generative adversarial nets. In: Proceedings of the 27th International Conference on Neural Information Processing Systems, NIPS 2014, vol. 2, pp. 2672–2680. MIT Press (2014)
3. Li, D., Zhang, H., Zhong, Y.: Hepatic GDF15 is regulated by CHOP of the unfolded protein response and alleviates NAFLD progression in obese mice. Biochem. Biophys. Res. Commun. **498**, 388–394 (2018)
4. Mi, H., et al.: PANTHER version 11: expanded annotation data from gene ontology and reactome pathways, and data analysis tool enhancements. Nucleic Acids Res. **45**(D1), D183–D189 (2017)
5. Needleman, S.B., Wunsch, C.D.: A general method applicable to the search for similarities in the amino acid sequence of two proteins. J. Mol. Biol. **48**(3), 443–453 (1970)

---

[1] Disclaimer. Research was sponsored by the U.S. Army Research Office and the Defense Advanced Research Projects Agency and was accomplished under Cooperative Agreement Number W911NF-14-2-0020. The views and conclusions contained in this document are those of the authors and should not be interpreted as representing the official policies, either expressed or implied, of the Army Research Office, DARPA, or the U.S. Government. The U.S. Government is authorized to reproduce and distribute reprints for Government purposes notwithstanding any copyright notation hereon.

6. Stehr, M.O., et al.: Learning causality: synthesis of large-scale causal networks from high-dimensional time series data. CoRR abs/1905.02291 (2019). http://arxiv.org/abs/1905.02291
7. Talcott, C., et al.: Transcriptional response of SK-N-AS cells to methamidophos: extended version (2019). http://www.csl.sri.com/users/clt/XYZ/methamidophosX.pdf
8. Vertes, A., et al.: Inferring mechanism of action of an unknown compound from time series omics data. In: Češka, M., Šafránek, D. (eds.) CMSB 2018. LNCS, vol. 11095, pp. 238–255. Springer, Cham (2018). https://doi.org/10.1007/978-3-319-99429-1_14
9. Willoughby, D., Cooper, D.M.F.: Organization and Ca++ regulation of adenylyl cyclases in camp microdomains. Physiol. Rev. **87**(3), 965–1010 (2007)

Separators for Polynomial Dynamic Systems with Linear Complexity

Ines Abdeljaoued-Tej[1,4](✉) , Alia Benkahla[1] , Ghassen Haddad[1,2],
and Annick Valibouze[3]

[1] Laboratory BIMS in Institute Pasteur of Tunis, LR16IPT09,
University of Tunis El Manar, Tunis, Tunisia
inestej@gmail.com
[2] Laboratory Jacques Louis Lions (LJLL), Sorbonne University, Paris, France
[3] Sorbonne University, CNRS, LIP6, LPSM, 75005 Paris, France
[4] University of Carthage, ESSAI, Tunis, Tunisia

Abstract. Computation biology helps to understand processes in organisms from interaction of molecules to complex functions of whole organs. Therefore, there is a need for mathematical methods and models that deliver logical explanations in a reasonable time. We propose herein a method based on algebraic separators, which are special polynomials abundantly studied in effective Galois theory. These polynomials are used in modelling discrete data related to cellular pathways affected in cancer and targeting therapies.

Keywords: Polynomial dynamical system · Algebraic separators ·
Finite field · Mathematical modelling · Discrete data

1 Introduction

A polynomial dynamical system is a tool used for understanding the behaviour of complex systems over time. It deals with biological systems that study and describes the interactions between micro-biological outputs. It finds its roots in symbolic computation and mathematical modelling. One of the precursors of the polynomial dynamical system is Thomas with his Boolean dynamical system [12]. In the last decade, several studies have been made, including contributions in hybrid systems biology [1,2]. The algebraic method presented in this paper adopts the techniques of Galois theory to issues of bio-systems. This approach can effectively model the important size of the biological data, with relatively simple tools: the calculation of elementary symmetric polynomials in the case of Boolean or the fundamental modulus in the polynomial case.

We present a method for creating polynomial dynamical systems (in discrete time) to capture biological data. The first use of a polynomial dynamical system (PDS) on system biology was published in [6]: the model is a deterministic graphical model which depends on the degree p of data discretization ($p = 2$ in

© Springer Nature Switzerland AG 2019
L. Bortolussi and G. Sanguinetti (Eds.): CMSB 2019, LNBI 11773, pp. 373–378, 2019.
https://doi.org/10.1007/978-3-030-31304-3_30

Boolean modelling). Thus, when n entries evolve in time, the dynamical function can be represented by n polynomials which describe a table of p^n possible states. Our researches were inspired by a classical method applied on Lagrange interpolation [5]. The proposed method is based on determining polynomial separators (which can be done using Galois theory). Finding polynomial separators has been used in classification, and this idea applied to biological modelling was presented in [14]. We were guided by the use of the Tchebotarev's fundamental modulus to develop our algorithm. Algebraic separators are directly determined by using symmetric functions, or by linear combinations of fundamental modulus. A vast literature on the subject of polynomial interpolation and ideals of points is available [3,8,9]. Our approach avoids heavy calculations of Gröbner bases and provides a polynomial model in linear time. The computation of a basic polynomial dynamical system, as detailed in this work, can perfectly be computed in parallel.

2   Methods

In the following, we fix n a positive integer and p a prime integer called the *degree of discretization*. Let $\mathbb{F}_p$ be the Galois field $\mathbb{Z}/p\mathbb{Z}$ of p elements. We consider the polynomial dynamical systems (PDS) $\mathbf{f} = (f_1, \ldots, f_n)$ of dimension n whose components f_j, $j \in [1, n]$, are polynomials of the quotient ring $\mathbb{F}_p[\mathbf{x}]/\langle x_1^p - x_1, \ldots, x_n^p - x_n \rangle$, i.e., polynomials on the n variables in the n-tuple $\mathbf{x} = (x_1, \ldots, x_n)$ with coefficients in $\mathbb{F}_p$ and degree smaller than $p - 1$ in each variable. The set $\mathbb{F}_p^n$ contains the p^n possible states of the experiment. At each time $t \in [0, m]$, the vector $\mathbf{s}_t = (s_{t,1}, \ldots, s_{t,n}) \in \mathbb{F}_p^n$ is called *the state of the system at time t*. A such trajectory of length $m + 1$ $\mathbf{s}_0 \mapsto \mathbf{s}_1 \mapsto \ldots \mapsto \mathbf{s}_m$ is called a *discrete trajectory*. The goal of this work is to compute a PDS $\mathbf{f}$ satisfying at each time $t \in [0, m-1]$: $\mathbf{f}(\mathbf{s}_t) = \mathbf{s}_{t+1}$ or, more precisely, $f_j(\mathbf{s}_t) = \mathbf{s}_{t+1,j}$ for each $j \in [1, n]$. Note that if $\mathbf{f}$ is a such PDS then $\mathbf{f}' = (f_1', \ldots, f_n')$ is also a such PDS iff for each $j \in [1, n]$, $f_j - f_j'$ belongs to $\mathrm{Id}(V) = \{h \in \mathbb{F}_p[\mathbf{x}] \mid h(\mathbf{s}) = 0, \forall \mathbf{s} \in V\}$, the ideal of the affine variety $V = \{\mathbf{s}_0, \ldots, \mathbf{s}_{m-1}\} \subset \mathbb{F}_p^n$. It is possible to separate an element of V with respect to others thanks to a polynomial of $\mathbb{F}_p[\mathbf{x}]$:

Definition 1. Let $V \subset \mathbb{F}_p^n$ and $\mathbf{s}$ be a state in V. A *separator of $\mathbf{s}$ in V* is a polynomial r in $\mathbb{F}_p[\mathbf{x}]$ such that $r(\mathbf{s}) = 1$ and $r(\mathbf{x}) = 0$ for each $\mathbf{x} \in V \backslash \{\mathbf{s}\}$.

The m respective separators of the m states in V are to interpolation in several variables what Lagrange polynomials are to interpolation in only one variable: they are used to calculate a PDS $\mathbf{f}$ that we search.

Let $\mathbf{s} = (s_1, \ldots, s_n) \in \mathbb{F}_p^n$. In $\mathbb{F}_p[\mathbf{x}]$, the maximal ideal $\mathrm{Id}(\{\mathbf{s}\})$ of $\mathbf{s}$-relations is generated by $q_1 = x_1 - s_1, q_2 = x_2 - s_2, \ldots, q_n = x_n - s_n$. The set $\mathbf{q} = \{q_1, \ldots, q_n\}$ is called by N. Tchebotarev the set of *fundamental modulus of* $\mathbf{s}$ [11]. We have for all $\mathbf{u} \in \mathbb{F}_p^n$: $\forall j \in [1, n]$ $q_j(\mathbf{u}) = 0$ iff $\mathbf{u} = \mathbf{s}$; in other words $V(\mathrm{Id}(\{\mathbf{s}\})) = \{\mathbf{s}\}$. It is the particular case of Galois theory in which the Galois group over the field $\mathbb{F}_p$ of the polynomial $(x - s_1) \cdots (x - s_n)$ is the identity group. Let J be the set of indices j for which all elements of $V = \{\mathbf{s}_0, \mathbf{s}_1, \ldots, \mathbf{s}_{m-1}\}$ have the same

j-th coordinate: $J = \{j \in [1, n] \mid \forall l \in [0, m]\ s_{l,j} = s_{0,j}\}$; on the coordinates indexed by j no separation is possible. Fix $S = \{1, \ldots, n\} \backslash J$, the minimum subset of $\{1, \ldots, n\}$ which separates V's elements, S is called *separator set of* V. We keep our calculations for a data set of genes products which vary at least once: these are S's elements. In particular, for $V = \mathbb{F}_p^n$, the separator set is $S = \{1, \ldots, n\}$. Let us consider the theorem that lights us on the algebraic computation of separators:

Theorem 1. *Let* $V = \{s_0, s_1, \ldots, s_{m-1}\} \subset \mathbb{F}_p^n$, S *be the separator set of* V, $\mathbf{s} \in V$ *and* $q_1, \ldots, q_n$ *be its fundamental modulus. Let be the following polynomials in* $\mathbb{F}_p[\mathbf{x}]$:

$$g(\mathbf{x}) = \prod_{j \in S} \prod_{l \in E} (q_j(\mathbf{x}) - l) \ \text{satisfying} \ g(\mathbf{s}) \neq 0 \quad \text{and} \quad r(\mathbf{x}) = \frac{g(\mathbf{x})}{g(\mathbf{s})} \quad (1)$$

where $E = \{q_j(\mathbf{u}) \mid j \in S,\ \mathbf{u} \in V\ :\ q_j(\mathbf{u}) \neq 0\} \subset \mathbb{F}_p \backslash \{0\}$. *Then* r *is a separator of* $\mathbf{s}$ *in* V.

Proof. Let $\mathbf{s} \in V$. As $\mathrm{Id}(\{\mathbf{s}\}) = \{q_1, \ldots, q_n\}$, we have $g(\mathbf{s}) = (\prod_{l \in E} -l)^{\mathrm{Card}(S)}$; then $g(\mathbf{s}) \neq 0$ because p is prime and $E \subset \mathbb{F}_p \backslash \{0\}$. So $r(\mathbf{s}) = 1$. Now let $\mathbf{u} \in V \backslash \{\mathbf{s}\}$. As $\mathbf{u} \neq \mathbf{s}$ there exists $j \in [1, n]$ such that $q_j(\mathbf{u}) \neq 0$; then, by definition of E, $l = q_j(\mathbf{u}) \in E$; as the $g(\mathbf{u})$'s factor $q_j(\mathbf{u}) - l$ equals 0, $r(\mathbf{u}) = 0$ also $\qquad \square$

As the separators of V's elements are independent of each other, their computation by applying Theorem 1 may be made simultaneously. So, the computation of a PDS can be done in a distributive manner. We can easily verify that computing separators in parallel have stated complexity of $O(nm)$ where n is the number of variables and m is the number of data points.

3   Boolean Case

We suppose, for simplicity, that the separator's set S of V is $\{1, \ldots, n\}$. In case $p = 2$, the set E of Theorem 1 is reduced to $\{1\}$ and a separator r of $\mathbf{s} \in V$ can be expressed in a compact form. Indeed, we can prove that $r(\mathbf{x}) = 1 + \sum_{i=1}^{n} (-1)^i e_i(\mathbf{q}) \in \mathbb{F}_2[\mathbf{x}]$, where $e_1(\mathbf{q}), \ldots, e_n(\mathbf{q})$ are the classical first n elementary symmetric functions into the elements of $\mathbf{q} = \{q_1, \ldots, q_n\}$, the set of fundamental modulus of $\mathbf{s}$. To quickly calculate the $e_i(\mathbf{q})$, we use tools developed in [13].

To illustrate our approach, we consider a data set of reduced number of genes involved in bladder cancer therapy [15]. These genes are either inhibited or activated over time: we take an example of $n = 4$ genes ($gene_1, \ldots, gene_4$) that evolve under different conditions; the genes are common to two cell lines, noted C_1 and C_2 in Table 1. For each cell line, we have a trajectory $\mathbf{s}_0 \mapsto \mathbf{s}_1 \mapsto \mathbf{s}_2$ of length $m + 1 = 3$ where the coordinates of states $\mathbf{s}_t$ are given in the columns of the following array:

Table 1. Example of $n = 4$ genes, common to two cell lines C_1 and C_2 in three steps of time.

Gene cell lines	C_1 at 0h: $\mathbf{s}_0$	C_1 at 24h: $\mathbf{s}_1$	C_1 at 72h: $\mathbf{s}_2$	C_2 at 0h: $\mathbf{s}_0$	C_2 at 24h: $\mathbf{s}_1$	C_2 at 72h: $\mathbf{s}_2$
$gene_1$	1	1	0	1	0	0
$gene_2$	0	1	1	1	1	0
$gene_3$	1	0	0	1	0	1
$gene_4$	1	1	0	0	1	0

Let us compute a model for each cell line, which allows us to compare their change and progression over time. We denote by $\mathbf{f}^{C_k}$ a PDS which describes C_k, $k = 1, 2$. We assume that the value of variables appearing in each component of a PDS is a key of the gene's behaviour for a biological system. What counts is to have a model that sticks most to the realities of data. We denote by $\mathbf{q}_t^{C_k}$ the set of fundamental modulus of the state $\mathbf{s}_t$ of C_k, $k = 1, 2$ and $r_t^{C_k}$ its separator. We have $r_t^{C_k}(\mathbf{x}) = 1 - e_1(\mathbf{q}_t^{C_k}) + e_2(\mathbf{q}_t^{C_k}) - e_3(\mathbf{q}_t^{C_k}) + e_4(\mathbf{q}_t^{C_k})$ where $\mathbf{q}_0^{C_1} = \{x_1 + 1, x_2, x_3 + 1, x_4 + 1\}, \mathbf{q}_1^{C_1} = \{x_1 + 1, x_2 + 1, x_3, x_4 + 1\}$, and $\mathbf{q}_0^{C_2} = \{x_1 + 1, x_2 + 1, x_3 + 1, x_4\}, \mathbf{q}_1^{C_2} = \{x_1, x_2 + 1, x_3, x_4 + 1\}$. We compute them in parallel: $r_0^{C_1}(\mathbf{x}) = a + x_1 x_3 x_4$, $r_1^{C_1}(\mathbf{x}) = a + x_1 x_2 x_4$, $r_0^{C_2}(\mathbf{x}) = a + x_1 x_2 x_3$ and $r_1^{C_2}(\mathbf{x}) = a + x_1 x_2 x_4 + x_2 x_3 x_4 + x_2 x_4$ where $a = x_1 x_2 x_3 x_4$ and we deduce our two PDS $\mathbf{f}^{C_k}$ by applying this formula: $f_j^{C_k} = \mathbf{s}_{1,j} r_0^{C_k}(\mathbf{x}) + \mathbf{s}_{2,j} r_1^{C_k}(\mathbf{x})$. The SageMath [10] function that implements our method returns: $\mathbf{f}^{C_1} = (a + x_1 x_3 x_4, x_1 x_2 x_4 + x_1 x_3 x_4, 0, a + x_1 x_3 x_4)$ and $\mathbf{f}^{C_2} = (0, a + x_1 x_2 x_3, a + x_1 x_2 x_4 + x_2 x_3 x_4 + x_2 x_4, a + x_1 x_2 x_3)$.

The wiring diagram and the state space graph of the given input data could be computed with tools developed in [4]. Describing a gene network in terms of polynomial dynamical system has advantages. First, it describes gene interactions in an explicitly numerical form. Second, these are causal relations between genes: for a cell line C_k a coefficient x_i in a function $f_j^{C_k}$ determines the effect of $gene_i$ on $gene_j$.

4 Future Work

Many biological systems are modelled with discrete models. Here, we use a classical method based on generalised Lagrange's interpolation. This work proposes a linear algorithm to learn polynomial dynamical systems in the frame of biological networks. It enables us to propose quickly models to biologists in a simple way: it takes into account the sparsity of biological experimental data. This paper details an approach allowing separators' computation: we present a method based on Galois theory's tools as the fundamental modulus or elementary symmetric functions. In this context, we developed an analytic method of easily readable expression and easily interpreted specific data. There is only one parameter p introduced into the model, unlike the continuous model using differential equations which must be added a number of constraints and parameters for successful modelling. The inference of gene interrelations from temporal data of gene expression follows a method using algebraic separators. Clearly, further research will be required on experimental data. Continuing research on this field appears fully justified because of the simplicity of this approach [16]. The improvement in [7] is similar to our work where the complexity is linear. We discuss the reverse engineering of biological networks using algebraic methods. The main contribution of this method is to offer an effective alternative to Gröbner basis methods by introducing algebraic separators from effective Galois theory. The strong point of this method for the mentioned above objectives are the linear complexity and the ability of parallel computation of separators.

References

1. Benkahla, A., Guizani-Tabbane, L., Abdeljaoued-Tej, I., BenMiled, S., Dellagi, K.: Systems biology and infectious diseases. In: Handbook of Research on Systems Biology Applications in Medicine, vol. 1, pp. 377–402 (2008)
2. Bortolussi, L., Policriti, A.: Hybrid Systems and Biology. Springer, Heidelberg (2008). https://doi.org/10.1007/978-3-540-68894-5_12
3. Ceria, M., Mora, T., Visconti, A.: Efficient computation of squarefree separator polynomials. In: Davenport, J.H., Kauers, M., Labahn, G., Urban, J. (eds.) ICMS 2018. LNCS, vol. 10931, pp. 98–104. Springer, Cham (2018). https://doi.org/10.1007/978-3-319-96418-8_12
4. Dimitrova, E.S., Vera-Licona, P., McGee, J., Laubenbacher, R.C.: Discretization of time series data. J. Comput. Biol. **17**, 853–868 (2010)
5. Lagrange, J.: Réflexions sur la résolution algébrique des équations (1770)
6. Laubenbacher, R.: A computer algebra approach to biological systems. In: Proceedings of the 2003 International Symposium on Symbolic and Algebraic Computation. ACM, New York (2003)
7. Lundqvist, S.: Complexity of comparing monomials and two improvements of the Buchberger-Möller algorithm. In: Calmet, J., Geiselmann, W., Müller-Quade, J. (eds.) Mathematical Methods in Computer Science. LNCS, vol. 5393, pp. 105–125. Springer, Heidelberg (2008). https://doi.org/10.1007/978-3-540-89994-5_9
8. Lundqvist, S.: Vector space bases associated to vanishing ideals of points. J. Pure Appl. Algebra **214**(4), 309–321 (2010)
9. Mora, T.: The FGLM problem and Möeller's algorithm on zero-dimensional ideals. In: Sala, M., Sakata, S., Mora, T., Traverso, C., Perret, L. (eds.) Gröbner Bases, Coding, and Cryptography, pp. 27–45. Springer, Heidelberg (2009). https://doi.org/10.1007/978-3-540-93806-4_3
10. Stein, W., Joyner, D., Developers, T.S.: SageMath (System for algebra and geometry experimentation), the Sage Mathematics Software System (2019). http://www.sagemath.org
11. Tchebotarev, N.: Gründzüge des Galois'shen Theorie. P. Noordhoff (1950)
12. Thomas, R.: Kinetic Logic: A Boolean Approach to the Analysis of Complex Regulatory Systems. Lecture Notes in Biomathematics, vol. 29. Springer, Heidelberg (1979). https://doi.org/10.1007/978-3-642-49321-8
13. Valibouze, A.: Symbolic computation with symmetric polynomials, an extension to MACSYMA. In: Kaltofen, E., Watt, S.M. (eds.) Computers and Mathematics, pp. 308–320. Springer, New York (1989). https://doi.org/10.1007/978-1-4613-9647-5_35

14. Valibouze, A., Abdeljaoued, I., BenKahla, A.: Galoisian separators for biological systems. In: Mathematics Algorithms Proofs - Formalization of Mathematics, Monastir, Tunisia (2009)
15. Van Kessel, K.E., Zuiverloon, T.C., Alberts, A.R., Boormans, J.L., Zwarthoff, E.C.: Targeted therapies in bladder cancer: an overview of in vivo research. Nat. Rev. Urol. **12**(12), 681 (2015)
16. Wang, X., Zhang, S., Dong, T.: A bivariate preprocessing paradigm for the Buchberger-Möller algorithm. J. Comput. Appl. Math. **234**(12), 3344–3355 (2010)

Bounding First Passage Times
in Chemical Reaction Networks
Poster Abstract

Michael Backenköhler[1(✉)], Luca Bortolussi[1,2], and Verena Wolf[1]

[1] Saarland University, Saarbrücken, Germany
michael.backenkoehler@uni-saarland.de
[2] University of Trieste, Trieste, Italy

1 Goal

Chemical reaction networks describe the interaction of different molecular species in a well-stirred reactor. For example,

$$\varnothing \xrightarrow{10} M \quad \text{and} \quad M + M \xrightarrow{0.1} D \tag{1}$$

describes an influx of monomers M and their dimerization to form D. Under many circumstances the system's behavior over time is best described *stochastically* by a Continuous-time Markov Chain (CTMC) $X_t, t \geq 0$ over all possible molecular counts.

The analysis of such systems is challenging because often state spaces are large or even infinite and exact solutions are rarely available. We are interested in the property of *first passage times*. This, for example, would be the time τ the population of M reaches the threshold of 10 molecules with $X_0 = 0$ at time $t = 0$ or 8 times units elapsed:

$$\tau := \inf\{t \geq 0 \mid X_t \geq 10\} \wedge 8 . \tag{2}$$

This *stopping time* is a random variable itself. We tackle the problem of bounding the *mean first-passage time* $\mathbb{E}(\tau)$, i.e. identifying tight bounds l_τ and u_τ such that

$$l_\tau \leq \mathbb{E}(\tau) \leq u_\tau . \tag{3}$$

2 Methods

We use an approach for the *generalized moment problem*, popularized by Lasserre [7], which has been applied in many contexts. Recently, it has been adapted to chemical reaction networks [2–6, 10] to bound population moments in both, the steady-state and transient context. We extend the dynamic approach to compute bounds on the first hitting time distribution. The main idea of this approach is to multiply a time-weighting term with the differential equations describing the

© Springer Nature Switzerland AG 2019
L. Bortolussi and G. Sanguinetti (Eds.): CMSB 2019, LNBI 11773, pp. 379–382, 2019.
https://doi.org/10.1007/978-3-030-31304-3_31

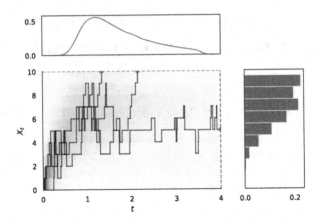

Fig. 1. The decomposition of the exit location probability measure for $\tau = \inf\{t \geq 0 \mid X_t \geq 10\} \wedge 4$. The shaded area indicates the structure of the occupation measure. 3 example trajectories are additionally plotted with their exit location highlighted. The plots are based on 50,000 SSA samples.

moment dynamics. These equations are integrated w.r.t. time [3,10]. This leads to an expected value of a martingale process of the form

$$Z_t^{(k,m)} := T^k X_t^m - 0^k x_0^m + \sum_i c_i \int_0^T t^{k_i} X_t^{m_i} \, dt \,, \tag{4}$$

where T is the time horizon, x_0 the initial counts, and c_i, m_i are determined by the differential equations describing the moment dynamics. With this process in place, we can adopt the technique presented by Lasserre in the context of option pricing models [8].

Given the stopping time τ by Doob's optional sampling theorem $\mathbb{E}(Z_t^{(k,m)}) = 0$ and thus

$$0 = \mathbb{E}\left(\tau^k X_\tau^m\right) - 0^k x_0^m + \sum_i c_i \mathbb{E}\left(\int_0^\tau t^{k_i} X_t^{m_i} \, dt\right). \tag{5}$$

This gives us constraints on the moments of the *expected occupation measure* z_{km} and the *exit location probability measure* y_{km}, where

$$z_{km} := \mathbb{E}\left(\int_0^\tau t^k X_t^m \, dt\right) \quad \text{and} \quad y_{km} := \mathbb{E}\left(\tau^k X_\tau^m\right). \tag{6}$$

We can decompose the exit location measure by conditioning on reaching the maximal time-horizon: $y_{km} = 10^m v_{1k} + 8^k v_{2m}$, where

$$v_{1k} := \mathbb{E}\left(\tau^k \mid \tau < T\right) \Pr(\tau < 8), \quad v_{2k} := \mathbb{E}\left(X_\tau^k \mid \tau = T\right) \Pr(\tau = 8).$$

This way we have three measures coupled through linear constraints. The decomposition is illustrated in Fig. 1.

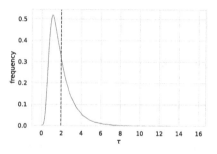

Fig. 2. First passage times for the dimerization model with $\tau = \inf\{t \geq 0 \mid X_t \geq 10\} \wedge 100$. The dashed red line denotes the sampled mean first passage time. (left) The sample distribution of τ based on 100,000 SSA samples. (right) The bounds based on constraints up to different moment orders.

It is well known, that the *moment matrix* M of a positive measure is *positive semi-definite* (PSD). For a one-dimensional distribution, for example, this means

$$M \succeq 0 \Leftrightarrow v^T M v \geq 0, \quad \forall v \in \mathbb{R}^3, \quad \text{where} \quad M = \begin{bmatrix} \mathbb{E}\left(X^0\right) & \mathbb{E}\left(X^1\right) & \mathbb{E}\left(X^2\right) \\ \mathbb{E}\left(X^1\right) & \mathbb{E}\left(X^2\right) & \mathbb{E}\left(X^3\right) \\ \mathbb{E}\left(X^2\right) & \mathbb{E}\left(X^3\right) & \mathbb{E}\left(X^4\right) \end{bmatrix}.$$

The restriction of measures to certain subsets of the state space can also be expressed in terms of PSD constraints.

We therefore have an optimization problem at hand that contains equality constraints on the moments via (5) and PSD constraints on the moments matrices. Optimization problems of this form are *semi-definite programs* and can be solved numerically.

3 Results

We derived the moment constraints (5) symbolically for all k, m up to a fixed order and solved the resulting SDP using the MOSEK solver [9] via the CVXPY modeling framework [1]. The results for the dimerization example are summarized in Fig. 2. We observe very tight bounds when the order is increased sufficiently. This implementation of this method is numerically challenging due to the extreme differences in magnitude between moments of different orders. This necessitates an appropriate scaling of moments [2,3]. The scaling becomes more difficult with increased population sizes.

Acknowledgements. This work was supported by the DFG project MULTIMODE.

References

1. Diamond, S., Boyd, S.: CVXPY: a Python-embedded modeling language for convex optimization. J. Mach. Learn. Res. **17**(83), 1–5 (2016)
2. Dowdy, G.R., Barton, P.I.: Bounds on stochastic chemical kinetic systems at steady state. J. Chem. Phys. **148**(8), 084106 (2018)
3. Dowdy, G.R., Barton, P.I.: Dynamic bounds on stochastic chemical kinetic systems using semidefinite programming. J. Chem. Phys. **149**(7), 074103 (2018)
4. Ghusinga, K.R., Lamperski, A., Singh, A.: Moment analysis of stochastic hybrid systems using semidefinite programming. arXiv preprint: arXiv:1802.00376 (2018)
5. Ghusinga, K.R., Vargas-Garcia, C.A., Lamperski, A., Singh, A.: Exact lower and upper bounds on stationary moments in stochastic biochemical systems. Phys. Biol. **14**(4), 04LT01 (2017)
6. Kuntz, J., Thomas, P., Stan, G.B., Barahona, M.: Rigorous bounds on the stationary distributions of the chemical master equation via mathematical programming. arXiv preprint: arXiv:1702.05468 (2017)
7. Lasserre, J.B.: Moments, Positive Polynomials and Their Applications, vol. 1. World Scientific, Singapore (2010)
8. Lasserre, J.B., Prieto-Rumeau, T., Zervos, M.: Pricing a class of exotic options via moments and SDP relaxations. Math. Finance **16**(3), 469–494 (2006)
9. MOSEK ApS: MOSEK Optimizer API for C 8.1.0.67 (2018). https://docs.mosek.com/8.1/capi/index.html
10. Sakurai, Y., Hori, Y.: Bounding transient moments of stochastic chemical reactions. IEEE Control Syst. Lett. **3**(2), 290–295 (2019)

Data-Informed Parameter Synthesis for Population Markov Chains

Matej Hajnal[2,4]([⊠]), Morgane Nouvian[1,3], Tatjana Petrov[2,3]([⊠]), and David Šafránek[4]([⊠])

[1] Department of Biology, University of Konstanz, Konstanz, Germany
[2] Department of Computer and Information Sciences, University of Konstanz, Konstanz, Germany
matej.hajnal@gmail.com, tatjana.petrov@gmail.com
[3] Centre for the Advanced Study of Collective Behaviour, University of Konstanz, 78464 Konstanz, Germany
[4] Systems Biology Laboratory, Faculty of Informatics, Masaryk University, Botanická 68a, 602 00 Brno, Czech Republic
safranek@fi.muni.cz

Population models are widely used to model different phenomena: animal collectives such as social insects, flocking birds, schooling fish, or humans within societies, as well as molecular species inside a cell, cells forming a tissue. Animal collectives show remarkable self-organisation towards emergent behaviours without centralised control. Quantitative models of the underlying mechanisms can directly serve important societal concerns (for example, prediction of seismic activity [5]), inspire the design of distributed algorithms (for example, ant colony algorithm [1]), or aid robust design and engineering of collective, adaptive systems under given functionality and resources, which is recently gaining attention in vision of smart cities [3,4]. Quantitative prediction of the behaviour of a population of agents over time and space, each having several behavioural modes, results in a high-dimensional, non-linear, and stochastic system [2]. Hence, computational modelling with population models is challenging, especially when the model parameters are unknown and experiments are expensive.

In this work, we investigate how to obtain the parameters for single agent behaviour, based on data collected for a population. Measurements for different population sizes are especially important when studying social feedback: an adaptation of individual's behaviour to the changing context of the population. For example, honeybees protect their colonies against vertebrates by releasing an alarm pheromone to recruit a large number of defenders into a massive sting-

This work has been presented at Hybrid Systems and Biology - HSB 2019. TP's research is supported by the Ministry of Science, Research and the Arts of the state of Baden-Württemberg, and the DFG Centre of Excellence 2117 'Centre for the Advanced Study of Collective Behaviour' (ID: 422037984), MH's research is supported by Young Scholar Fund (YSF), project no. $P83943018FP430\_/18$. MN's research is supported by the Mentorship grant from the Zukunftskolleg. DŠ's research is supported by the Czech Grant Agency grant no. GA18-00178S.

L. Bortolussi and G. Sanguinetti (Eds.): CMSB 2019, LNBI 11773, pp. 383–386, 2019.
https://doi.org/10.1007/978-3-030-31304-3_32

ing response [6]. However, these workers will then die from abdominal damage caused by the sting tearing loose [7]. In order to achieve a balanced trade-off towards efficient defence, yet no critical worker loss, each bee's response to the same amount of pheromone may vary greatly, depending on its social context, which, in the case of bees, has been experimentally validated.

To tackle this problem, we assume a simple communication scheme among identical individuals, such that n individuals together form a discrete-time Markov chain (DTMC) M with at most n parameters. Each population eventually reaches one of its terminal strongly connected components (tSCC) in the underlying MC. A graphical representation of a population model for three agents is given in Fig. 1. We employ the theoretical steady-state assumption that is commonly used in biological modelling scenarios: we assume that the experimental observations can be taken when the steady state is reached, hence that experimental measurements allow us to estimate probabilities of reaching any of the tSCCs in the form of a confidence interval (for any desired confidence level α). We assume $\mathcal{V}$ denotes a set of model parameters, each defined over domain $[0, 1]$. Our major goal is to synthesise a *viable parameter space* Θ, $\Theta \subseteq [0, 1]^{|\mathcal{V}|}$, such that the following condition is satisfied:

$$\theta \in \Theta \text{ if and only if } M(\theta) \models \bigwedge_{\text{all } tSCCs} \varphi(tSCC \mid \mathsf{data}) \tag{1}$$

where $\varphi(tSCC \mid \mathsf{data})$ expresses that reaching a tSCC is achieved within the confidence interval estimated from experimental data. In contrast to traditional

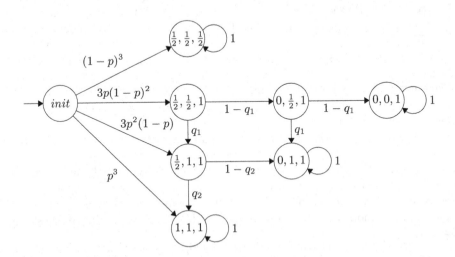

Fig. 1. Parametric Markov chain representing population of three bees with parameters $\mathcal{V} = \{p, q_1, q_2\}$. Parameter p represents the initial probability that an agent solves the task, while q_i representes the probability of success in the second attempt. A vector labelling states represents state of the individual agent (1 denotes success, $\frac{1}{2}$ denotes the second chance, 0 denotes no success)

red = unsafe region, green = safe region, white = in between
alpha:0.95, n_samples:100, max_recursion_depth:12,
min_rec_size:0.0001, achieved_coverage:0.963623046875, alg3
It took Freya 16.0 second(s)

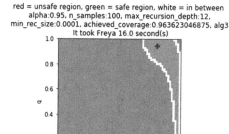

red = unsafe region, green = safe region, white = in between
alpha:0.95, n_samples:1500, max_recursion_depth:14,
min_rec_size:0.0001, achieved_coverage:0.99609375, alg3
It took Freya 3.5 second(s)

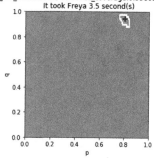

Fig. 2. An example of visualisation of parameter space refinement with two different intervals inferred from the data in the respective column. Parameter point from which the data were obtained is $p = 0.81, q = 0.92$ - shown as a blue cross. (Color figure online)

parameter inference techniques which return a single estimate, the parameter synthesis approach gives a quantitative characterisation of the entire domain of satisfying parameter values.

We propose and implement a workflow for obtaining the viable parameter space for a simple population model. Analysis of the parameter space consists of two steps: first, we obtain a symbolic characterisation of the distribution over tSCCs in form of multivariate rational functions, leveraging the existing tools for parameter synthesis. Second, we employ CEGAR-like reasoning (candidate region generation and checking) for determining the viable parameter space, until the desired proportion of the domain (called coverage) is reached. Refined space – result of this part can be visualised in the case of 2-dimensional space as a green area – see Fig. 2. We implemented several variants of the search algorithm, and tested the performance on synthetic data sets. It is worth noting that, for showcasing the framework, we here implemented a workflow for a specific class of population models which have acyclic underlying transition system and each tSCC contains one state (being strongly inspired by the propagation of alarm pheromones in honeybees). The framework is applicable to any general DTMC and any persistence or repeated reachability temporal logic property.

References

1. Dorigo, M., Birattari, M., Blum, C., Clerc, M., Stützle, T., Winfield, A.F.T. (eds.): ANTS 2008. LNCS, vol. 5217. Springer, Heidelberg (2008). https://doi.org/10.1007/978-3-540-87527-7
2. Giardina, I.: Collective behavior in animal groups: theoretical models and empirical studies. HFSP J. **2**(4), 205–219 (2008)
3. Hillston, J.: Challenges for quantitative analysis of collective adaptive systems. In: Abadi, M., Lluch Lafuente, A. (eds.) TGC 2013. LNCS, vol. 8358, pp. 14–21. Springer, Cham (2014). https://doi.org/10.1007/978-3-319-05119-2_2

4. Loreti, M., Hillston, J.: Modelling and analysis of collective adaptive systems with CARMA and its tools. In: Bernardo, M., De Nicola, R., Hillston, J. (eds.) SFM 2016. LNCS, vol. 9700, pp. 83–119. Springer, Cham (2016). https://doi.org/10.1007/978-3-319-34096-8_4
5. Mai, M., et al.: Monitoring pre-seismic activity changes in a domestic animal collective in central Italy. In: EGU General Assembly Conference Abstracts, vol. 20, p. 19348 (2018)
6. Nouvian, M., Reinhard, J., Giurfa, M.: The defensive response of the honeybee Apis mellifera. J. Exp. Biol. **219**(22), 3505–3517 (2016)
7. Shorter, J.R., Rueppell, O.: A review on self-destructive defense behaviors in social insects. Insectes Sociaux **59**(1), 1–10 (2012)

Author Index

Abdeljaoued-Tej, Ines   373
Allart, Emilie   266
Antoneli, Fernando   60
Arul, Albert-Baskar   368
Assaf, George   302
Audoly, S.   329
Avar, Peter   368

Backenköhler, Michael   42, 379
Ballarini, Paolo   207
Bartocci, Ezio   120
Bellu, G.   329
Benkahla, Alia   373
Bentriou, Mahmoud   207
Bokes, Pavol   140
Bortolussi, Luca   42, 379
Boutillier, Pierre   296
Breton, Marc   188
Brim, Luboš   356
Brůža, Vojtěch   356
Bunin, Deborah I.   368

Češka, Milan   337
Chodak, Jacek   315
Cournède, Paul-Henry   207

d'Angió, L.   329
Dalchau, Neil   224
Davis, Brian M.   368
Degrand, Elisabeth   78
Delaplace, Franck   20
Demko, Martin   356

Fages, François   78, 352
Feng, Lu   188

Goldfeder, Judah   289
Grignard, Jeremy   352
Grima, Ramon   347
Guet, Calin   155
Gupta, Ankit   342

Haar, Stefan   3
Haddad, Ghassen   373
Hajnal, Matej   383
Harmer, Russ   322
Heiner, Monika   302, 315
Hemery, Mathieu   78
Henzinger, Thomas A.   155
Hillston, Jane   120
Hodgkinson, Arran   60

Igler, Claudia   155
Innocentini, Guilherme C. P.   60
Ivanov, Sergiu   20

Khammash, Mustafa   342
Knapp, Merrill   368
Korte, Andrew R.   368
Křetínský, Jan   337
Kugler, Hillel   289

Lamp, Josephine   188
Liu, Fei   302
Lorton, Christopher W.   308
Losová, Barbora   360

Mandon, Hugues   3
Margetiny, Filip   120
Martinelli, Julien   352
Morton, Christine A.   368

Nenzi, Laura   188
Niehren, Joachim   266
Nishita, Denise   368
Nouvian, Morgane   383
Nowicka, Melania   96

Öcal, Kaan   347
Oshurko, Eugenia   322

Pang, Jun   3, 364
Papoušek, Jan   356

Pardo, Jérémie   20
Parvin, Lida   368
Pastva, Samuel   356
Paul, Soumya   364
Paulevé, Loïc   3
Pejznoch, Aleš   356
Pérez-Verona, Isabel Cristina   248
Petrov, Tatjana   155, 383
Phillips, Andrew   224
Piho, Paul   120
Poggio, Andrew   368
Přibylová, Lenka   360
Proctor, Joshua L.   308

Radulescu, Ovidiu   60
Ribchester, Richard R.   120
Roh, Min K.   308

Saccomani, M. P.   329
Šafránek, David   356, 383
Sahab, Ziad J.   368
Sanguinetti, Guido   342, 347
Sevinsky, Christopher J.   368
Sezgin, Ali   155
Siebert, Heike   96

Silvetti, Simone   188
Singh, Abhyudai   140
Soliman, Sylvain   352
Spaccasassi, Carlo   224
Stehr, Mark-Oliver   368
Su, Cui   3, 364

Talcott, Carolyn L.   368
Tribastone, Mirco   248
Troják, Matej   356

Valibouze, Annick   373
Vandin, Andrea   248
Vejpustek, Tomáš   356
Versari, Cristian   266
Vertes, Akos   368

Welkhoff, Philip A.   308
Wolf, Verena   42, 379

Yordanov, Boyan   224

Zavodszky, Maria I.   368

Printed in the United States
By Bookmasters